WITHDRAWN
UTSA LIBRARIES

Springer Series on Environmental Management

Series Editors
Bruce N. Anderson
Planreal Australasia, Keilor, Victoria, Australia

Robert W. Howarth
Cornell University, Ithaca, NY, USA

Lawrence R. Walker
University of Nevada, Las Vegas, NV, 89154

For further volumes:
http://www.springer.com/series/412

Hypoxia in the Northern Gulf of Mexico

Virginia H. Dale
Catherine L. Kling
Judith L. Meyer
James Sanders
Holly Stallworth
Thomas Armitage
David Wangsness
Thomas Bianchi
Alan Blumberg
Walter Boynton
Daniel J. Conley
William Crumpton

Mark David
Denis Gilbert
Robert W. Howarth
Richard Lowrance
Kyle Mankin
James Opaluch
Hans Paerl
Kenneth Reckhow
Andrew N. Sharpley
Thomas W. Simpson
Clifford S. Snyder
Donelson Wright

Virginia H. Dale
Oak Ridge National Laboratory
Division of Environmental Sciences
1 Bethel Valley Road
Oak Ridge TN 37831-6036
USA

Catherine L. Kling
Department of Econmics
Iowa State University
578F Heady Hall
Ames IA 50011-1070
USA

Judith L. Meyer
Department of Plant Biology
University of Georgia
Institute of Ecology
Athens GA 30602
USA

James Sanders
Skidaway Institute of Oceanography
10 Ocean Science Circle
Savannah GA 31411
USA

Holly Stallworth
US Environmental Protection Agency
401 M Street SW.,
Washington DC 20460
USA

Thomas Armitage
US Environmental Protection Agency
401 M Street SW.,
Washington DC 20460
USA

David Wangsness
US Geological Survey
430 National Center
Reston VA 20192-0001
USA

Thomas Bianchi
Department of Oceanography
Texas A & M University
College Station TX
77843-3146
USA

Alan Blumberg
Department of Chemical, Biomedical &
  Materials Engineering
Stevens Institute of Technology
Hobroken NJ 07030
USA

Walter Boynton
University of Maryland
Center for Environmental Science
Chesapeake Biological Lab.
P.O. Box 38
Solomons MD 20688
USA

Daniel J. Conley
Department of Geology
Lund University
GeoBiosphere Science Center
Sölvegatan 12
SE-223 62 Lund
Sweden

William Crumpton
Department of Ecology, Evolution, &
  Organismal Biology (EEOB)
Iowa State University
353 Bessey Hall
Ames IA 50011
USA

Mark David
Department of Natural Resources &
  Environmental Sciences
University of Illinois
Urbana-Champaign
1201 W. Gregory Dr.
Urbana IL 61801
USA

Denis Gilbert
Fisheries and Oceans Canada
Maurice Lamontagne Institute
850 Route de la Mer
Mont-Joli QC G5H 3Z4
Canada

Robert W. Howarth
Department of Ecology & Evolutionary Biology
Cornell University
Corson Hall
Ithaca NY 14853-2701
USA

Richard Lowrance
U.S. Department of Agriculture
Agricultural Research Service
Southeast Watershed
Research Laboratory
P.O.Box 748
Tifton GA 31793
USA

Kyle Mankin
Department of Biological &
 Agricultural Engineering
Kansas State University
147 Seaton Hall
Manhattan KS 66506
USA

James Opaluch
Department of Environmental &
 Natural Resource Economics
University of Rhode Island
Kingston RI 02881
USA

Hans Paerl
University of North Carolina
Institute of Marine Sciences
Moorehead City NC 28557
USA

Kenneth Reckhow
Duke University
Nicholas School of the
 Environment & Earth Science
P.O. Box 90340
Durham NC 27708-0340
USA

Andrew N. Sharpley
Department of Crop, Soil, &
 Environmental Sciences
University of Arkansas
Fayetteville AR 72701
115 Plant Science Building
USA

Thomas W. Simpson
University of Maryland
College of Agriculture & Natural Resources
Symons Hall
College Park MD 20742
USA

Clifford S. Snyder
International Plant Nutrition Institute
P.O. Box 2440
Conway AR 72033
USA

Donelson Wright
Virginia Institute for Marine Science
College of William & Mary
School of Marine Science
1208 Greate Road
Gloucester Point VA 23062
USA

ISSN 0172-6161
ISBN 978-0-387-89685-4   e-ISBN 978-0-387-89686-1
DOI 10.1007/978-0-387-89686-1
Springer New York Dordrecht Heidelberg London

Library of Congress Control Number: 2009941055

© Springer Science+Business Media, LLC 2010
All rights reserved. This work may not be translated or copied in whole or in part without the written permission of the publisher (Springer Science+Business Media, LLC, 233 Spring Street, New York, NY 10013, USA), except for brief excerpts in connection with reviews or scholarly analysis. Use in connection with any form of information storage and retrieval, electronic adaptation, computer software, or by similar or dissimilar methodology now known or hereafter developed is forbidden.
The use in this publication of trade names, trademarks, service marks, and similar terms, even if they are not identified as such, is not to be taken as an expression of opinion as to whether or not they are subject to proprietary rights.

Printed on acid-free paper

Springer is part of Springer Science+Business Media (www.springer.com)

Acknowledgments

This book is based on a series of meetings and report developed by the US Environmental Protection Agency (EPA) Science Advisory Board's (SAB's) Hypoxia Advisory Panel. The book presents the view of the authors and does not represent SAB or EPA policy.

The efforts of many people who contributed to the meetings and reviewed earlier drafts of the manuscript are appreciated. We gratefully acknowledge the many individuals who provided their scientific perspectives for the Panel's consideration in the development of this book.

Invited Speakers:

- Rich Alexander, U.S. Geological Survey, *SPARROW Model*
- Jim Ammerman, Rutgers State University, *Effects of nutrients*
- Jeff Arnold, U.S. Department of Agriculture, *SWAT Model*
- James Baker and Dean Lemke, UMRSHNC, *Upper Mississippi Symposia Summary*
- Robert Dean, University of Florida, *Drawing Louisiana's New Map*
- Steven DiMarco, Texas A&M University, *Physical Oceanography in the Gulf*
- Katie Flahive, U.S. Environmental Protection Agency, *Status of the Management Actions Reassessment Team (MART) Report*
- Rick Greene (EPA) and Alan Lewitus (National Oceanic and Atmospheric Administration), *Gulf Science Symposia Summary*
- Dan Jaynes, U.S. Department of Agriculture, *Agricultural N & P Management Approaches*
- Bob Kellogg, U.S. Department of Agriculture, *Status of the Conservation Effectiveness Assessment Program (CEAP)*
- Tim Miller, U.S. Geological Survey, *Monitoring Activities in the Mississippi River basin*
- Marc Ribaudo, U.S. Department of Agriculture, *Costs and Benefits of Methods to Reduce Nutrient Loads*
- Don Scavia, University of Michigan, (1) *Science and Policy Context* and (2) *Hypoxia Forecast Models*
- Janice Ward, U.S. Geological Survey, *Fate and Transport Symposia Summary*

Invited Technical Reviewers:

- Mark Alley, Virginia Tech
- Walter Dodds, Kansas State University
- Madhu Khanna, University of Illinois
- William Wiseman, Jr., National Science Foundation

Public Commenters:

- James Baker, Iowa Department of Agriculture and Land Stewardship
- Victor Bierman, Donald Boesch, John Day, Robert Diaz, Dubravko Justic, Dennis Keeney, William Mitsch, Nancy Rabalais, Gyles Randall, Donald Scavia, and Eugene Turner, Contributors to the *Integrated Assessment*
- Donald Boesch, University of Maryland Center for Environmental Science
- Darrell Brown, EPA Office of Water
- Daniel Coleman, O'Brien & Gere
- Richard Cruse, Iowa State University
- Doug Daigle, Lower Mississippi River Sub-basin Committee on Hypoxia
- Bob Diaz, Virginia Institute of Marine Sciences
- Michael Duffy, Iowa State University
- Nancy Erickson, Illinois Farm Bureau
- Jason Flickner, Kentucky Waterways Alliance
- Norman Fousey, U.S. Department of Agriculture
- James Fouss, U.S. Department of Agriculture
- Doug Gronau, Iowa Farm Bureau Federation
- Ben Grumbles, Assistant Administrator for EPA's Office of Water
- Stephen Harper, O'Brien & Gere
- Chuck Hartke, Illinois Department of Agriculture
- Susan Heathcote, Iowa Environmental Council
- Matthew Helmers, Iowa State University
- Ed Hopkins, Sierra Club
- Chris Hornback, National Association of Clean Water Agencies
- Illinois Department of Agriculture
- Thomas Isenhart, Iowa State University
- Dan Jaynes, U.S. Department of Agriculture
- Doug Karlen, U.S. Department of Agriculture
- Dennis Keeney, Institute for Agriculture and Trade Policy
- Louis Kollias, Metropolitan Water Reclamation District of Greater Chicago
- Dean Lemke, Iowa Department of Agriculture and Land Stewardship
- Alan Lewitus and David Kidwell, National Oceanic and Atmospheric Administration
- Antonio Mallarino, Iowa State University
- Mark Maslyn, American Farm Bureau Federation
- Dennis McKenna, Illinois Department of Agriculture
- Mississippi River Water Quality Cooperative (MSWQC)

Acknowledgments

- Bill Northey, Iowa Secretary of Agriculture
- Don Parrish, American Farm Bureau
- Paul Patterson, City of Memphis
- Jean Payne, Illinois Fertilizer and Chemical Association
- Michelle Perez, Environmental Working Group
- Bob and Kristen Perry, Missouri Clean Water Commission
- Nancy Rabalais, Louisiana Universities Marine Consortium
- Russell Rasmussen, Wisconsin Department of Natural Resources
- Jack Riessen, Iowa Department of Natural Resources
- Rick Robinson, Iowa Farm Bureau
- Matt Rota, Gulf Restoration Network
- John Sawyer, Iowa State University
- Al Schafbuch, Affiliation not identified
- Tim Strickland, U.S. Department of Agriculture
- Richard Swenson, U.S. Department of Agriculture
- Michael Tate, Kansas Department of Health and Environment
- Steve Taylor, Environmental Resource Coalition
- Mark Tomer, U.S. Department of Agriculture
- Eugene Turner, Louisiana State University
- Ford B. West, The Fertilizer Institute
- Wendy Wintersteen, Iowa State University

The book never could have come to fruition without the efforts of Frederick O'Hara in editing the book. His careful attention to the details and to effective communication is appreciated.

I appreciate the support of the Environmental Sciences Division at Oak Ridge National Laboratory and, especially, of my children.

Oak Ridge, Tennessee
December 2007

Virginia H. Dale

Contents

1	**Introduction**		1
	1.1	Hypoxia and the Northern Gulf of Mexico – A Brief Overview	1
	1.2	Science and Management Goals for Reducing Hypoxia	3
	1.3	Hypoxia Study Group	4
	1.4	The Study Group's Approach	7
2	**Characterization of Hypoxia**		9
	2.1	Historical Patterns and Evidence for Hypoxia on the Shelf	9
	2.2	The Physical Context	12
		2.2.1 Oxygen Budget: General Considerations	12
		2.2.2 Vertical Mixing as a Function of Stratification and Vertical Shear	13
		2.2.3 Changes in Mississippi River Hydrology and Their Effects on Vertical Mixing	15
		2.2.4 Zones of Hypoxia Controls	18
		2.2.5 Shelf Circulation: Local Versus Regional	20
	2.3	Role of N and P in Controlling Primary Production	23
		2.3.1 Nitrogen and Phosphorus Fluxes to the NGOM Background	23
		2.3.2 N and P Limitation in Different Shelf Zones and Linkages Between High Primary Production Inshore and the Hypoxic Regions Farther Offshore . . .	24
	2.4	Other Limiting Factors and the Role of Si	29
	2.5	Sources of Organic Matter to the Hypoxic Zone	31
		2.5.1 Sources of Organic Matter to NGOM: Post 2000 Integrated Assessment	33
		2.5.2 Advances in Organic Matter Understanding: Characterization and Processes	34
		2.5.3 Synthesis Efforts Regarding Organic Matter Sources . .	37
	2.6	Denitrification, P Burial, and Nutrient Recycling	38
	2.7	Possible Regime Shift in the Gulf of Mexico	41
	2.8	Single Versus Dual Nutrient Removal Strategies	44
	2.9	Current State of Forecasting	46

3 Nutrient Fate, Transport, and Sources ... 51
- 3.1 Temporal Characteristics of Streamflow and Nutrient Flux 51
  - 3.1.1 MARB Annual and Seasonal Fluxes 56
  - 3.1.2 Subbasin Annual and Seasonal Flux 65
- 3.2 Mass Balance of Nutrients 76
  - 3.2.1 Cropping Patterns 76
  - 3.2.2 Nonpoint Sources 77
  - 3.2.3 Point Sources 84
- 3.3 Nutrient Transport Processes 87
  - 3.3.1 Aquatic Processes 87
  - 3.3.2 Freshwater Wetlands 93
  - 3.3.3 Nutrient Sources and Sinks in Coastal Wetlands 94
- 3.4 Ability to Route and Predict Nutrient Delivery to the Gulf 96
  - 3.4.1 SPARROW Model 97
  - 3.4.2 SWAT Model 103
  - 3.4.3 IBIS/THMB Model 104
  - 3.4.4 Discussion and Comparison of Models 106
  - 3.4.5 Targeting 106
  - 3.4.6 Model Uncertainty 107

4 Scientific Basis for Goals and Management Options 111
- 4.1 Adaptive Management 111
- 4.2 Setting Targets for Nitrogen and Phosphorus Reduction 115
- 4.3 Protecting Water Quality and Social Welfare in the Basin 120
  - 4.3.1 Assessment and Review of the Cost Estimates from the CENR Integrated Assessment 121
  - 4.3.2 Other Large-Scale Integrated Economic and Biophysical Models for Agricultural Nonpoint Sources . 125
  - 4.3.3 Research Assessing the Basin-Wide Co-benefits 128
  - 4.3.4 Principles of Landscape Design 129
- 4.4 Cost-Effective Approaches for Nonpoint Source Control 133
  - 4.4.1 Voluntary Programs – Without Economic Incentives .. 134
  - 4.4.2 Existing Agricultural Conservation Programs 135
  - 4.4.3 Emissions and Water Quality Trading Programs 137
  - 4.4.4 Agricultural Subsidies and Conservation Compliance Provisions 138
  - 4.4.5 Taxes 140
  - 4.4.6 Eco-labeling and Consumer Driven Demand 141
- 4.5 Options for Managing Nutrients, Co-benefits, and Consequences 143
  - 4.5.1 Agricultural Drainage 143
  - 4.5.2 Freshwater Wetlands 146
  - 4.5.3 Conservation Buffers 151
  - 4.5.4 Cropping Systems 155
  - 4.5.5 Animal Production Systems 158

		4.5.6	In-Field Nutrient Management	164
		4.5.7	Effective Actions for Other Nonpoint Sources	183
		4.5.8	Most Effective Actions for Industrial and Municipal Sources	186
		4.5.9	Ethanol and Water Quality in the MARB	190
		4.5.10	Integrating Conservation Options	195

5 Summary of Findings and Recommendations 205
  5.1 Characterization of Hypoxia 205
  5.2 Nutrient Fate, Transport, and Sources 207
  5.3 Goals and Management Options 209
  5.4 Conclusion 211

Appendices 215

References 239

Subject Index 277

List of Figures

1.1 Map of the frequency of hypoxia in the northern Gulf of Mexico, 1985–2005. Taken from N.N. Rabalais, LUMCON, 2006 2
1.2 Map showing the extent of the Mississippi–Atchafalaya River basin 2
2.1 Plots of the PEB index (%PEB) in sediment cores from the Louisiana shelf. Higher values of the PEB index indicate lower dissolved oxygen contents in bottom waters. Taken from Osterman et al. (2005) 11
2.2 Change in the relative importance of the Atchafalaya flow to the combined flows from the Mississippi and Atchafalaya Rivers over the 20th century. Reprinted from Bratkovich et al. (1994) 15
2.3 Modeled surface salinity showing the freshwater plumes from the Atchafalaya and Mississippi Rivers during upwelling-favorable winds (top panel) and during downwelling favorable winds 8 days later (bottom panel). Adapted from Hetland and DiMarco (2007) 17
2.4 Proposed diversions of Mississippi effluents for coastal protection. From Coastal Protection and Restoration Authority (CPRA) of Louisiana, 2007 Integrated Ecosystem Restoration and Hurricane Protection: Louisiana's Comprehensive Master Plan for a Sustainable Coast. CPRA, Office of the Governor (LA) 117 pp 18
2.5 An illustration depicting different zones (Zones 1–4, numbered above) in the NGOM during the period when hypoxia can occur. These zones are controlled by differing physical, chemical, and biological processes, are variable in size, and move temporally and spatially. Diagram created by D. Gilbert 19
2.6 Response of natural phytoplankton assemblages from coastal NGOM stations to nutrient additions, March through September. All experiments, except those done in September,

	indicate a strong response to P additions. Taken from Sylvan et al., 2006	25
2.7	NASA-SeaWiFS image of the Northern Gulf of Mexico recorded in April, 2000. This image shows the distributions and relative concentrations of chlorophyll *a*, an indicator of phytoplankton biomass in this region. Note the very high concentrations (orange to red) present in the inshore regions of the mouths of the Mississippi and Atchafalaya Rivers	26
3.1	Estimated extent of agricultural drainage based on the distribution of row crops, largely corn and soybean, and poorly drained soils (per D. Jaynes, National Soil Tilth Lab, Ames, IA)	52
3.2	Land cover based on Landsat data (adapted from Crumpton et al., 2006)	53
3.3	Flow-weighted average nitrate concentrations estimated from STORET data selected to exclude point source influences (adapted from Crumpton et al., 2006)	54
3.4	Flow-weighted average nitrate and reduced N versus percent cropland (adapted from Crumpton et al., 2006)	54
3.5	MARB nitrate-N fluxes for 1955 through 2005 water years comparing estimates from various methods for 1979–2005. Based on USGS data from Battaglin (2006) and Aulenbach et al. (2007)	55
3.6	Comparison (percent and absolute basis) of MARB nitrate-N fluxes to LOADEST 5-year method for 1979 through 2005 water years. Based on USGS data from Battaglin (2006) and Aulenbach et al. (2007)	57
3.7	Schematic showing locations of MARB monitoring sites (Aulenbach et al., 2007)	58
3.8	Flow and available nitrogen monitoring data for the MARB for 1955 through 2005 water years (LOWESS, locally weighted scatterplot smooth, curves shown as a *solid line*). LOWESS describes the relationship between Y and X without assuming linearity or normality of residuals and is a robust description of the data pattern (Helsel and Hirsch, 2002)	59
3.9	Flow, available phosphorus, and available silicate monitoring data for the MARB for 1955 through 2005 water years (LOWESS curves shown as a *solid line*). Based on USGS data from Battaglin (2006) and Aulenbach et al. (2007)	60
3.10	Ratio of total N to total P and dissolved silicate to dissolved inorganic N for MARB for the 1980 through 2005 water years. Based on USGS data from Battaglin (2006) and Aulenbach et al. (2007)	61

List of Figures xvii

3.11 Flow and nitrogen flux for the MARB during spring (April, May, and June) for the period 1979–2005 (LOWESS curve shown as a *solid line*). Based on USGS data from Battaglin (2006) and Aulenbach et al. (2007) 62

3.12 Flow, phosphorus, and silicate flux for the MARB during spring (April, May, and June) for the period 1979–2006 (LOWESS curve shown as a *solid line*). Based on USGS data from Battaglin (2006) and Aulenbach et al. (2007) 63

3.13 Sum of April, May and June fluxes as a percent of annual (water year basis) for combined Mississippi mainstem and Atchafalaya River. *Box* plots show median (*line in center* of box), 25th and 75th percentiles (*bottom and top* of box, respectively), 10th and 90th percentiles (*bottom and top error bars*, respectively), and values <10th percentile and >90th percentile (*solid circles* below and above *error bars*, respectively). Based on USGS data from Battaglin (2006) and Aulenbach et al. (2007) 64

3.14 Ratio of total N to total P and silicate to dissolved inorganic N for the MARB during spring (April, May, and June) for the period 1980–2006. Based on USGS data from Battaglin (2006) and Aulenbach et al. (2007) 65

3.15 Location of nine large subbasins comprising the MARB that are used for estimating nutrient fluxes (from Aulenbach et al., 2007) 66

3.16 Net N inputs and annual nitrate-N fluxes and yields for the Ohio River subbasin. (LOWESS curves for riverine nitrate-N shown with *solid lines*.) Based on USGS data from Battaglin (2006) and Aulenbach et al. (2007) 70

3.17 Net N inputs and annual nitrate-N fluxes and yields for the upper Mississippi River subbasin. (LOWESS curves for riverine nitrate-N shown with *solid lines*.) Shown in triangles is a recalculated net N input for the upper Mississippi River basin, increasing soybean N_2 fixation from 50 to 70% of above ground N, and a soil net N mineralization rate from 0 to 10 kg N/ha-year. Based on USGS data from Battaglin (2006) and Aulenbach et al. (2007) 71

3.18 Total P and particulate/organic P fluxes for the Ohio River near Grand Chain, Illinois (LOWESS curves shown in *solid* and *dashed lines*). Based on USGS data from Battaglin (2006) and Aulenbach et al. (2007) 72

3.19 Spring water flux and nitrate-N flux for the Mississippi River at Grafton and the Ohio River at Grand Chain, IL, for water years 1975–2005 (LOWESS curves shown with *solid lines*.) Based on USGS data from Battaglin (2006) and Aulenbach et al. (2007) 73

3.20	Spring nitrate-N flux (sum of April, May, and June) for the Mississippi River at Grafton plus Ohio River at Grand Chain subbasins compared to the combined Mississippi and Atchafalaya River for 1979 through 2005. Based on USGS data from Battaglin (2006) and Aulenbach et al. (2007)	74
3.21	Area of major crops planted in the MARB from 1941 through 2007. Adapted from McIsaac (2006)	76
3.22	Nitrogen mass balance components and net N inputs for the MARB, as calculated by McIsaac et al. (2002) and updated through 2005 by McIsaac (2006)	78
3.23	Net N inputs for the four major regions of the MARB through 2005. Adapted from McIsaac (2006)	79
3.24	Nitrogen mass balance components and net N inputs for the upper Mississippi River basin, as calculated by McIsaac et al. (2002) and updated through 2005 by McIsaac (2006)	81
3.25	Phosphorus mass balance components and net P inputs for the MARB. Adapted from McIsaac (2006)	82
3.26	Net P inputs for the four major subbasins of the MARB through 2005. Adaptive from McIsaac (2006)	83
3.27	Phosphorus mass balance components and net N inputs for the upper Mississippi River basin. Adapted from McIsaac (2006)	85
3.28	Total phosphorus point source fluxes as a percent of total flux for the MARB for 2004 by hydrologic region	86
3.29	Percentage of nutrient inputs to streams that are removed by in-stream and reservoir processes as predicted by the SPARROW model (Alexander et al., 2008)	88
3.30	N removed in aquatic ecosystems (as a % of inputs) as a function of ecosystem depth/water travel time (modified from David et al., 2006). Values shown are for 23 years in an Illinois reservoir (David et al., 2006), French reservoirs (Garnier et al., 1999), Illinois streams (an average from Royer et al., 2004), agricultural streams (Opdyke et al., 2006), and rivers (Seitzinger et al., 2002). The curve from Seitzinger et al. (2002) is not as steep as the curve that includes information from reservoirs in an agricultural region	90
4.1	A conceptual framework for hypoxia in the northern Gulf of Mexico	112
4.2	Percent mass nitrate removal in wetlands as a function of hydraulic loading rate. Best fit for percent mass loss = 103 * (hydraulic loading rate)$^{-0.33}$ ($R^2 = 0.69$). Adapted from Crumpton et al. (2006, 2008)	147
4.3	Observed NO_3 mass removal (*blue points*) versus predicted NO_3 mass removal (*blue surface*) based on the function [mass NO_3 removed = 10.3*(HLR)$^{0.67}$ * FWA] for	

	which $R^2 = 0.94$. *Blue lines* are isopleths of predicted mass removal at intervals of 250 kg/ha-year. The *dashed, red line* represents the isopleth for mass removal rate of 290 kg/ha-year suggested by Mitsch et al. (2005a). The *green plane* intersecting function surface represents organic N export. Adapted from Crumpton et al. (2006, 2008)	148
4.4	Recoverable manure N, assuming no export of manure from the farm, using 1997 census data. Adapted from USDA (2003) with the author's permission	158
4.5	Recoverable manure P, assuming no export of manure from the farm, using 1997 census data. Adapted from USDA (2003) with the author's permission	159
4.6	Fertilizer N consumption as anhydrous ammonia in leading corn-producing states for years ending June 30	165
4.7	Changes in the consumption of principal fertilizer N sources used in the six leading corn-producing states (IA, IL, IN, MN, NE, and OH) for years ending June 30	165
4.8	Percentage of N-fertilized corn acreage that received some amount of N in the fall	166
4.9	USDA ARMS data for the three states with highest fall N application, showing total amount of fall-applied N for that crop. Also shown are Illinois sales data for the same period	167
4.10	Fraction of annual fertilizer N tonnage in Illinois sold in the fall	168
4.11	Average corn yields in six leading corn-producing states (IA, IL, IN, MN, NE, and OH), 1990–2006 (Source: USDA National Agricultural Statistics Service)	171
4.12	Variability in soil test P levels in typical farmer fields in Minnesota (2007 personal communication with Dr. Gary Malzer, University of Minnesota)	178
4.13	Effect of variable-rate versus uniform-rate application of liquid swine manure on changes in soil test phosphorus in Iowa fields [2007 personal communication with Dr. Antonio Mallarino, Iowa State University and Wittry and Mallarino (2002)]	179
4.14	Effect of variable-rate versus uniform-rate application of fertilizer P on soil test P in multiple Iowa fields across multiple years	180
B.1	Nitrogen cycle flow diagram. Taken from Encyclopedia of Earth (2007) at http://www.eoearth.org/global_material_cycles . . .	223
B.2	Phosphorus cycle flow diagram. Taken from Encyclopedia of Earth (2007) at http://www.eoearth.org/global_material_cycles . . .	224
B.3	Silicon cycle flow diagram. Taken from Encyclopedia of Earth (2007) at http://www.eoearth.org/globa_material_cycles . . .	225

List of Tables

2.1 A partial summary of papers published following the *Integrated Assessment* related to sources of organic matter to the Gulf of Mexico 34

3.1 Average annual nutrient fluxes in 1000 metric tons for the five large subbasins in the MARB for the 2001–2005 water years. (Percent of total basin flux shown in parentheses) 67

3.2 Average annual nutrient fluxes for 10 subbasins in the MARB for the 2001–2005 water years. Some subbasin fluxes are calculated as the difference between the upstream and the downstream monitoring station. (Percent of total basin flux shown in parentheses) 68

3.3 Average annual nutrient yields in kg/ha-year for the five large subbasins in the MARB for water years 2001–2005 68

3.4 Average annual nutrient yields for nine subbasins in the MARB for the 2001–2005 water years. Some subbasin yields are calculated as the difference between the upstream and the downstream monitoring stations 69

3.5 Acres of wetlands created, restored, or enhanced in major subbasins of the Mississippi River from 2000 to 2006 under the Wetland Reserve Program (WRP), Conservation Reserve Program (CRP), Conservation Reserve Enhancement Program (CREP), Environmental Quality Incentive Program (EQIP), and Conservation Technical Assistance (CTA). (Personal communication, Mike Sullivan, USDA) 93

3.6 Attributes of models used to estimate sources, transport, and/or delivery of nutrients to the Gulf of Mexico 98

4.1 Annual and spring (sum of April, May, June) average flow and N and P fluxes for the MARB for the 1980–1996 reference period compared to the most recent 5-year period (2001–2005). Load reductions in mass of N or P also shown 116

4.2 Summary of study features of basin-wide integrated economic-biophysical models 123

4.3	Summary of policies and findings from integrated economic-biophysical models	124
4.4	Areas (ha) of conservation buffers installed in the six subbasins of the MARB for FY 2000–FY2006	154
4.5	Status of implementation of permits under the 2003 CAFO rule for states within the MARB. Data provided by USEPA Office of Wastewater Management, 2007	160
4.6	Estimates of manure production and N and P loss to water and air from Animal feeding operations within the Mississippi River basin. Total manure in millions of milligrams; other materials in millions of kilograms. Based on information from the 2002 US Census of Agriculture (adapted from Aillery et al., 2005)	162
4.7	Partial N balance for 4-year rate study by Jaynes et al. (2001). The last two columns were added here and were not part of original table	174
4.8	Estimated changes in N losses from cropping changes predicted by FAPRI from 2007 to 2013	192
4.9	Potential total nitrogen (TN) and phosphorus (TP) efficiencies (percent change) produced by *nutrient-use* conservation practices on surface runoff, subsurface flow, and tile drainage. Estimates are average values for a multiple-year basis, and some of the numbers in this table are based on a very small amount of field information. Shading highlights the methods producing the greatest reduction efficiencies within the three types of N and P loss (surface runoff, subsurface, and tile drainage)	197
4.10	Potential total nitrogen (TN) and phosphorus (TP) efficiencies (percent change) produced by *in-field* conservation practices on surface runoff, subsurface flow, and tile drainage. Estimates are average values for a multiple-year basis, and some of the numbers in this table are based on a very small amount of field information. Shading highlights the methods producing the greatest reduction efficiencies within the three types of N and P loss (surface runoff, subsurface, and tile drainage)	198
4.11	Potential total nitrogen (TN) and phosphorus (TP) efficiencies (percent change) produced by *off-site* conservation practices on surface runoff, subsurface flow, and tile drainage. Estimates are average values for a multiple-year basis, and some of the numbers in this table are based on a very small amount of field information. Shading highlights the methods producing the greatest reduction efficiencies within the three types of N and P loss (surface runoff, subsurface, and tile drainage)	199
4.12	Anticipated benefits associated with different agricultural management options	200

4.13	Anticipated benefits associated with other management options	201
C.1	Farming system and nutrient budget; amounts given in kg ha^{-1} $year^{-1}$	228
C.2	Number of animals and amount of manure produced and N and P excreted within the MARB states based on information from the 1997 US Census of Agriculture (data obtained from USDA-ERS, http://ers.usda.gov/data/MANURE/)	228
D.1	Comparison of MART estimated sewage treatment plant annual effluent loads of total N and P and values from measurements at each plant for 2004	232

Contributors

Dr. Thomas Armitage Environmental Protection Agency Science Advisory Board Staff Office, Washington, DC, USA

Dr. Thomas Bianchi Professor, Oceanography, Geosciences, Texas A&M University, College Station, TX, USA

Dr. Alan Blumberg Professor, Civil, Environmental and Ocean Engineering, Stevens Institute of Technology, Hoboken, NJ, USA

Dr. Walter Boynton Professor, Chesapeake Biological Laboratory, Center for Environmental Science, University of Maryland, Solomons, MD, USA

Dr. Daniel Joseph Conley Professor, Marie Curie Chair, GeoBiosphere Centre, Department of Geology, Lund University, Lund, Sweden

Dr. William Crumpton Associate Professor & Coordinator of Environmental Programs, Ecology, Evolution, and Organismal Biology, Iowa State University, Ames, IA, USA

Dr. Virginia Dale Corporate Fellow, Environmental Sciences Division, Oak Ridge National Laboratory, Oak Ridge, TN, USA

Dr. Mark David Professor, Natural Resources & Environmental Sciences, University of Illinois, Urbana, IL, USA

Dr. Denis Gilbert Research Scientist, Ocean and Environment Science Branch, Maurice-Lamontagne Institute, Department of Fisheries and Oceans Canada, Mont-Joli, Quebec, Canada

Dr. Robert W. Howarth David R. Atkinson Professor, Department of Ecology and Evolutionary Biology, Cornell University, Ithaca, NY, USA

Dr. Catherine Kling Professor, Department of Economics, Iowa State University, Ames, IA, USA

Dr. Richard Lowrance Research Ecologist, Southeast Watershed, Agricultural Research Service, USDA, Tifton, GA, USA

Dr. Kyle Mankin Associate Professor, Biological and Agricultural Engineering, Kansas State University, Manhattan, KS, USA

Dr. Judith L. Meyer Distinguished Research Professor Emeritus, Institute of Ecology, University of Georgia, Athens, GA, USA

Dr. James Opaluch Professor, Department of Environmental and Natural Resource Economics, College of the Environment and Life Sciences, University of Rhode Island, Kingston, RI, USA

Dr. Hans Paerl Professor of Marine and Environmental Sciences, Institute of Marine Sciences, University of North Carolina, Chapel Hill, Morehead City, NC, USA

Dr. Kenneth Reckhow Professor and Chair, Environmental Science & Policy, Nicholas School, Duke University, Durham, NC, USA

Dr. James Sanders Director, Skidaway Institute of Oceanography, Savannah, GA, USA

Dr. Andrew N. Sharpley Research Soil Scientist, Department of Crop, Soil and Environmental Sciences, University of Arkansas, Fayetteville, AR, USA

Dr. Thomas W. Simpson Professor and Coordinator, Chesapeake Bay Programs, College of Agriculture and Natural Resources, University of Maryland, College Park, MD, USA

Dr. Clifford Snyder Nitrogen Program Director, International Plant Nutrition Institute, Conway, AR, USA

Dr. Holly Stallworth Environmental Protection Agency Science Advisory Board Staff Office, Washington, DC, USA

Mr. David Wangsness U.S. Geological Survey, Atlanta, GA, USA

Dr. Donelson Wright Chancellor Professor Emeritus, School of Marine Science, Virginia Institute of Marine Science, College of William and Mary, Gloucester Point, VA, USA

Contributors

Authors of *Hypoxia in the Northern Gulf of Mexico*

To the HAP Panelists

Serving on the Hypoxia Advisory Panel
gave you each the unique opportunity to channel
your experience, knowledge, perspective, wisdom, and thought
into a set of key recommendations of what ought
to be done to sample, learn about, manage, and protect
resource use in the Mississippi Basin that affects
low-oxygen conditions in the Gulf of Mexico
and other co-benefits, such as clean air, the flow
of water, recreation, and rural amenities.

The natural system will benefit from your expertise
if the many suggestions and key recommendations
(for which are provided lengthy, detailed explanations)
will be used to improve those river and Gulf conditions
that allow the Mississippi Basin to transition
to a healthy and sustainable ecosystem
that supports life and our economy with vigor and vim.

Virginia H. Dale, June 2007

Glossary

Algae A group of chiefly aquatic plants (e.g., seaweed, pond scum, stonewort, phytoplankton) that contain chlorophyll and may passively drift, weakly swim, grow on a substrate, or establish root-like anchors (steadfasts) in a water body.

Anaerobic digestion Decomposition of biological wastes by micro-organisms, usually under wet conditions, in the absence of air (oxygen), to produce a gas comprising mostly methane and carbon dioxide.

Animal feeding operation (AFO) An agricultural enterprises where animals are kept and raised in confined situations. AFOs congregate animals, feed, manure, urine, dead animals, and production operations on a small land area. Feed is brought to the animals rather than the animals grazing or otherwise seeking feed in pastures, in fields, or on rangeland. Winter feeding of animals on pasture or rangeland is not normally considered an AFO.

Anoxia The absence of dissolved oxygen.

Bacterioplankton The bacterial component of the plankton that drifts in the water column.

Benthic organisms Organisms living in association with the bottom of aquatic environments (e.g., polychaetes, clams, snails).

Best Management Practices (*BMPs*) Effective, practical, structural, or nonstructural methods that are designed to prevent or reduce the movement of sediment, nutrients, pesticides, and other chemical contaminants from the land to surface or ground water, or which otherwise protect water quality from potential adverse effects of agricultural activities. These practices are developed to achieve a cost-effective balance between the water quality protection and the agricultural production (e.g., crop, forage, animal, forest).

Bioenergy Useful, renewable energy produced from organic matter – the conversion of the complex carbohydrates in organic matter to energy. Organic matter may either be used directly as a fuel, processed into liquids and gasses, or be a residual of processing and conversion.

Biogas A combustible gas derived from decomposing biological waste under anaerobic conditions. Biogas normally consists of 50–60% methane. See also landfill gas.

Biomass Any organic matter that is available on a renewable or recurring basis, including agricultural crops and trees, wood and wood residues, plants (including aquatic plants), grasses, animal residues, municipal residues, and other residue materials. Biomass is generally produced in a sustainable manner from water and carbon dioxide by photosynthesis. There are three main categories of biomass – primary, secondary, and tertiary.

Bioreactor A container in which a biological reaction takes place. As used in this book, a bioreactor is a container or a trench filled with a biodegradeable carbon source used to enhance biological denitrification for removal of nitrate from drainage water.

Biosolids Nutrient-rich soil-like materials resulting from the treatment of domestic sewage in a treatment facility. During treatment, bacteria and other tiny organisms break sewage down into organic matter, sometimes used as fertilizer.

Cellulosic ethanol Ethanol that is produced from cellulose material; a long chain of simple sugar molecules and the principal chemical constituent of cell walls of plants.

Chlorophyll Pigment found in plant cells that are active in harnessing energy during photosynthesis.

Conservation Reserve Program (CRP) CRP provides farm owners or operators with an annual per-acre rental payment and half the cost of establishing a permanent land cover, in exchange for retiring environmentally sensitive cropland from production for 10–15 years. In 1996, Congress reauthorized CRP for an additional round of contracts, limiting enrollment to 36.4 million acres at any time. The 2002 Farm Act increased the enrollment limit to 39 million acres. Producers can offer land for competitive bidding based on an Environmental Benefits Index (EBI) during periodic signups or can automatically enroll more limited acreages in practices such as riparian buffers, field windbreaks, and grass strips on a continuous basis. CRP is funded through the Commodity Credit Corporation (CCC).

Conservation practices (CPs) Any action taken to produce environmental improvements, particularly with respect to agricultural nonpoint source emissions. The term is used broadly to refer to structural practices, such as buffers, as well as nonstructural preactices, such as in-field nutrient management planning and application. Conservation practice standards have been developed by NRCS and are available at http://www.nrcs.usda.gov/Technical/Standards/nhcp.html

Corn stover Corn stocks that remain after the corn is harvested. Such stocks are low in water content and very bulky.

Glossary

Cyanobacteria A phylum (or "division") of bacteria that obtain their energy through photosynthesis. They are often referred to as blue-green algae, although they are in fact prokaryotes, not algae. The description is primarily used to reflect their appearance and ecological role rather than their evolutionary lineage. The name "cyanobacteria" comes from the color of the bacteria, cyan.

Demersal organisms Organisms that are, at times, associated with the bottom of aquatic environments, but capable of moving away from it (e.g., blue crabs, shrimp, red drum).

Denitrification Nitrogen transformations in water and soil that make nitrogen effectively unavailable for plant uptake, usually returning it to the atmosphere as nitrogen gas.

Diatom A major phytoplankton group characterized by cells enclosed in silicon frustules, or shells.

Dinoflagellates Mostly single-celled photosynthetic algae that bear flagella (long cell extensions that function in swimming) and live in fresh or marine waters.

Edge-of-field nitrogen loss A term that refers to the nitrogen that is lost or exported from fields in agricultural production.

Effluent The liquid or gas discharged from a process or chemical reactor, usually containing residues from that process.

Emissions Waste substances released into the air or water. See also effluent.

Eutrophic Waters, soils, or habitats that are high in nutrients; in aquatic systems, associated with wide swings in dissolved oxygen concentrations and frequent algal blooms.

Eutrophication An increase in the rate of supply of organic matter to an ecosystem.

Greenhouse gases Gases that trap the heat of the sun in the Earth's atmosphere, producing the greenhouse effect. The two major greenhouse gases are water vapor and carbon dioxide. Other greenhouse gases include methane, ozone, chlorofluorocarbons, and nitrous oxide.

Hydrogen sulfide A chemical, toxic to oxygen-dependent organisms, that diffuses into the water as the oxygen levels above the seabed sediments become zero.

Hypoxia Very low dissolved oxygen concentrations, generally less than 2 mg/L.

Lignocellulose A combination of lignin and cellulose that strengthens woody plant cells.

Nitrate An inorganic form of nitrogen; chemically NO_3.

Nitrogen fixation The transformation of atmospheric nitrogen into nitrogen compounds that can be used by growing plants.

Nonpoint source A diffuse source of chemical and/or nutrient inputs not attributable to any single discharge (e.g., agricultural runoff, urban runoff, atmospheric deposition).

Nutrients Inorganic chemicals (particularly nitrogen, phosphorus, and silicon) required for the growth of plants, including crops and phytoplankton.

Phytoplankton Plant life (e.g., algae), usually containing chlorophyll, that passively drifts in a water body.

Plankton Organisms living suspended in the water column, incapable of moving against currents.

Point source Readily identifiable inputs where treated wastes are discharged from municipal, industrial, and agricultural facilities to the receiving waters through a pipe or drain.

Pre-sidedress-nitrate test (PSNT) A soil nitrate-N test determined in surface soil samples (usually 0–30 cm or 0–12 in deep), collected between corn rows when the corn is about 15 cm (6 in) tall. Adjustments in the rate of sidedressed N can be made if the soil test indicates elevated nitrate-N levels, based upon calibrations that vary among growing regions. When successfully calibrated, the test results can be used as an index of the amount of N that may be released during the course of the growing season by organic sources, such as soil organic matter, manure, and crop residues.

Productivity The conversion of light energy and carbon dioxide into living organic material.

Pycnocline The region of the water column characterized by the strongest vertical gradient in density, attributable to temperature, salinity, or both.

Recoverable manure The portion of manure as excreted that could be collected from buildings and lots where livestock are held, and thus would be available for land application.

Recoverable manure nutrients The amounts of nitrogen and phosphorus in manure that would be expected to be available for land application. They are estimated by adjusting the quantity of recoverable manure for nutrient loss during collection, transfer, storage, and treatment, but are not adjusted for losses of nutrients at the time of land application.

Respiration The consumption of oxygen during energy utilization by cells and organisms.

Riparian floodplain Area adjacent to a river or other body of water subject to frequent flooding.

Soil tilth The physical condition of the soil as related to its ease of tillage, fitness as a seedbed, and impedance to seedling emergence and root penetration. A soil with good "tilth" has large pore spaces for adequate air infiltration and water movement and holds a reasonable supply of water and nutrients. Soil tilth is a factor of soil texture, soil structure, and the interplay with organic content and the living organisms that help make up the soil ecosystem.

Stratification A multilayered water column, delineated by pycnoclines.

Sustainable An ecosystem condition in which biodiversity, renewability, and resource productivity are maintained over time.

Urease and nitrification inhibitors Urease is a ubiquitous soil microbial enzyme that facilitates the hydrolysis of urine and urea to form ammonia. In the soil, ammonia readily hydrolyzes to ammonium. Soil ammonium also is formed by the mineralization of soil organic matter and manures. Ammonium is then oxidized or "nitrified" first to nitrite (NO_2) and then to nitrate (NO_3), which is highly soluble and subject to movement in the soil with the moisture front, or leaching under certain conditions. Under anaerobic conditions, NO_3 can be "denitrified" to the gases nitrous oxide (N_2O) and nitrogen (N_2) and released to the atmosphere. Urease inhibitors are chemicals applied to fertilizers or manures to reduce urease activity. Under certain environmental conditions urease inhibitors can temporarily inhibit or reduce ammonia loss (volatilization) to the atmosphere from urea-containing fertilizers or manures. Nitrification inhibitors are chemicals which can temporarily inhibit or reduce nitrification of anhydrous ammonia, ammonium-containing, or urea-containing fertilizers applied to the soil, which may indirectly help to reduce denitrification losses of N. Under certain environmental conditions, urease and nitrification inhibitors help to improve soil retention and crop recovery of applied N, which may reduce potential environmental N losses.

Voluntary programs Voluntary conservation programs that have no significant financial incentive (positive or negative) to encourage the adoption of conservation practices.

Watershed The drainage basin contributing water, organic matter, dissolved nutrients, and sediments to a stream or lake.

Zooplankton Animal life that drifts or weakly swims in a water body, often feeding on phytoplankton.

List of Acronyms and Symbols

ADCP	acoustic Doppler current profiler
AFO	animal feeding operation
AMLE	adjusted maximum likelihood estimate
ANNAMOX	anaerobic ammonia oxidation
A/P ratio	agglutinated to porcelaneous ratio (based on the relative abundance of three low-oxygen tolerant species of benthic foraminifers; *Pseudononin altlanticum, Epistominella vitrea, and Buliminella morgani*)
ARS	Agricultural Research Service (USDA)
AU	animal unit
BBL	benthic boundary layer
BMP	best management practice
BNR	biological nutrient removal
BOD	biochemical oxygen demand
Bu/A	Bushels per acre
C	carbon
CAFO	concentrated animal feeding operation
CASTnet	clean air status and trends network
CC or Ccc	continuous corn
CCC	Commodity Credit Corporation
CCOA	corn–corn–oat–alfalfa (crop rotation)
CDOM	colored dissolved organic matter
CEAP	Conservation Effectiveness Assessment Program
CENR	Committee on Environment and Natural Resources
Cm	corn–meadow (crop rotation)
CMAQ	community multiscale air quality model
COAA	corn–oat–alfalfa–alfalfa (crop rotation)
CO_2	carbon dioxide
cph	cycles per hour
CPRA	Coastal Protection and Restoration Authority
CREP	Conservation Reserve Enhancement Program

CRN	controlled and slow release N fertilizers
CRP	Conservation Reserve Program
CRPA	Coastal Protection and Restoration Authority
CS or CSb	corn–soybean rotation
CSP	Conservation Security Program
CTA	conservation technical assistance
CTD	conductivity, temperature, and depth instrumentation
CV	coefficients of variation
DDG	dried distillers grain
DIN:DIP	dissolved inorganic nitrogen:dissolved inorganic phosphorus
DO	dissolved oxygen
DOC	dissolved organic carbon
DOE	Department of Energy
DOM	dissolved organic matter
DON	dissolved organic nitrogen
DRP	dissolved reactive phosphorus
EBI	Environmental Benefits Index
ECa	electrical conductivity
ENR	enhanced nutrient removal
EPC_0	equilibrium P concentration
EPIC	environment productivity impact calculator model
EQIP	Environmental Quality Incentives Program
ERS	Economic Research Service (USDA)
Fe^{+2}	ferrous iron
FR	Federal Register
FWA	flow-weighted average
GAO	General Accounting Office
GCOOS	Gulf of Mexico Coastal Ocean Observing System
GCTM	Global Chemistry Transport Model
GHG	greenhouse gases
GIS	Geographic Information System
GLWQA	Great Lakes Water Quality Agreement
GOM	Gulf of Mexico
GPS	Global Positioning System
GWW	grass waterways
HAB	harmful algal bloom
HEL	highly erodable land
HLR	hydraulic loading rate
HRU	hydraulic response unit
HUC	hydrologic unit code
HYDRA	hydrological routing algorithm
IATP	Institute of Agricultural and Trade Policy
IBIS	integrated biosphere simulator model
IJC	International Joint Commission

IPCC	Intergovernmental Panel on Climate Change
ISNT	Illinois Soil Nitrogen Test
LOADEST	load estimator model
LOWESS	locally weighted scatterplot smooth curves
LSNT	late spring nitrate test
LUMCON	Louisiana Universities Marine Consortium
M	million
MGD	million gallons per day
MARB	Mississippi–Atchafalaya River basin
MART	management action reassessment team
Mn^{+2}	manganese (oxidation state common in aquatic-biological systems)
MRB	Mississippi River basin
MR/GMWNTF	Mississippi River/Gulf of Mexico Watershed Nutrient Task Force
MSEA	management system evaluation area
N	nitrogen
N_2	nitrogen gas (colorless, odorless, and tasteless gas that makes up 78.09% of air)
N_2O	nitrous oxide
NADP	National Air Deposition Program
NANI	net anthropogenic nitrogen inputs
NAS	National Academy of Sciences
NASA	National Aeronautics and Space Administration
NASA-SeaWiFS	NASA Sea-viewing Wide Field-of-view Sensor (project providing qualitative data on global ocean bio-optical properties)
NASQAN	National Stream Quality Accounting Network (USGS water-quality monitoring program)
NECOP	nutrient-enhanced coastal ocean productivity
NGOM	Northern Gulf of Mexico
NH_3	ammonia
NH_4^+	ammonium
NHx	the total atmospheric concentration of ammonia (NH_3) and ammonium (NH_4^+)
NOAA	National Oceanic and Atmospheric Administration
NO_2	nitrite nitrogen (NO_2^-) if in water and nitrogen dioxide (NO_2) if in air
NO_3	nitrate nitrogen
NOx	mono-nitrogen oxides, or the total concentration of nitric oxide (NO) plus nitrogen dioxide (NO_2)
NOy	reactive odd nitrogen or the sum of NOx plus compounds produced from the oxidation of NOx, which includes nitric acid, peroxyacetyl nitrate, and other compounds
NPDES	National Pollutant Discharge Elimination System

NPS	nonpoint source
NRC	National Research Council
NRCS	Natural Resource Conservation Service
NRI	National Resources Inventory
NSTC	National Science and Technology Council
O_2	diatomic oxygen (makes up 20.95% of air)
OM	organic matter
P	phosphorus
PEB index	an index based on the relative abundance of three low-oxygen tolerant species of benthic foraminifers; *Pseudononin altlanticum, Epistominella vitrea,* and *Buliminella morgani*
POC	particulate organic carbon
ppmv	parts per million by volume
ppt	parts per thousand
PS	point source
PSNT	Pre-Sidedress Nitrate Test
RivR-N	a regression model that predicts the proportion of N removed from streams and reservoirs as an inverse function of the water displacement time of the water body (ratio of water body depth to water time of travel)
SAB	Science Advisory Board
SCOPE	Science Committee on Problems of the Environment
SD	standard deviation
Si	silicon
SOC	soil organic carbon
SOM	soil organic matter
SON	soil organic nitrogen
SPARROW	spatially referenced regression on watershed attributes model
SRP or DRP or ortho P	soluble reactive phosphorus, dissolved reactive phosphorus, orthophosphate
STATSGO	State Soil Geographic database
STORET	STOrage and RETrieval data system (USEPA's largest computerized environmental data system)
STP	sewage treatment plant
SWAT	soil and water assessment tool model
THMB	terrestrial hydrology model with biogeochemistry
TKN	total Kjeldahl nitrogen
TM3	Tracer Model version 3 (a global atmospheric chemistry/transport model)
TN	total nitrogen
TP	total phosphorus
TPC	typical pollutant concentration
TSS	total suspended solids
UAN	urea–ammonium nitrate

UMR	Upper Mississippi River
UMRB	Upper Mississippi River basin
UMRSHNC	Upper Mississippi River Sub-basin Hypoxia Nutrient Committee
USMP	US Agriculture Sector Mathematical Programming model
USACE	United States Army Corps of Engineers
USDA	United States Department of Agriculture
USEPA or EPA	United States Environmental Protection Agency
USGS	United States Geological Survey
WRP	Wetlands Reserve Program

Conversion Factors and Abbreviations

Multiply	By	To obtain
centimeter (cm)	0.3937	inch (in)
millimeter (mm)	0.0394	inch (in)
meter (m)	3.281	foot (ft)
kilometer (km)	0.6214	mile (mi)
square kilometer (km^2)	0.3861	square mile (mi^2)
hectare (ha)	2.471	acre (ac)
hectare (ha)	0.01	square kilometer (km^2)
liter (L)	1.057	quart (qt)
liter (L)	0.0284	bushel (bu) US, dry
gram (g)	0.0353	ounce (oz)
gram per cubic meter (g/m^3)	0.00169	pound per cubic yard (lb/yd^3)
kilogram (kg)	2.205	pound (lb), avoirdupois
metric tonne (ton)	2,205.0	pound (lb), avoirdupois
metric tonne (ton)	1.1023	U.S. short ton (ton)
cubic meter per second (m3/s)	35.31	cubic foot per second (cfs)
kilogram per hectare (kg/ha)	0.893	pound per acre (lb/ac)
Concentration unit milligram per liter (mg/L)	Approximately equals part per million (ppm)	

The following equation was used to compute flux of chemicals:

$$\text{concentration (mg/L)} \times \text{flow (m}^3\text{/s)} \times 8.64 \times 10^{-2} = \text{metric tonne per day (ton/d)}$$

Executive Summary

Since 1985, scientists have been documenting a hypoxic zone in the Gulf of Mexico each year. The hypoxic zone, an area of low dissolved oxygen that cannot support marine life, generally manifests itself in the spring. Since marine species either die or flee the hypoxic zone, the spread of hypoxia reduces the available habitat for marine species, which are important for the ecosystem as well as commercial and recreational fishing in the Gulf. Since 2001, the hypoxic zone has averaged 16,500 km^2 during its peak summer months[1], an area slightly larger than the state of Connecticut, and ranged from a low of 8,500 km^2 to a high of 22,000 km^2. To address the hypoxia problem, the Mississippi River/Gulf of Mexico Watershed Nutrient Task Force (or Task Force) was formed to bring together representatives from federal agencies, states, and tribes to consider options for responding to hypoxia. The Task Force asked the White House Office of Science and Technology Policy to conduct a scientific assessment of the causes and consequences of Gulf hypoxia through its Committee on Environment and Natural Resources (CENR). In 2000 the CENR completed *An Integrated Assessment: Hypoxia in the Northern Gulf of Mexico* (*or Integrated Assessment*), which formed the scientific basis for the Task Force's *Action Plan for Reducing, Mitigating, and Controlling Hypoxia in the Northern Gulf of Mexico* (*Action Plan*, 2001). In its *Action Plan*, the Task Force pledged to implement ten management actions and to assess progress every 5 years. This reassessment would address the nutrient load reductions achieved, the responses of the hypoxic zone and associated water quality and habitat conditions, and economic and social effects. The Task Force began its reassessment in 2005.

In 2006 as part of the reassessment, USEPA's Office of Water, on behalf of the Task Force, requested that the U.S. Environmental Protection Agency (USEPA) Science Advisory Board (SAB) convene an independent panel to evaluate the

---

[1] The areal extent of the full hypoxic region has not been mapped with sufficient frequency to completely understand its temporal variability. The limited number of observations that have been taken more than once per year suggests that the hypoxic region reaches its maximum extent in late summer. There are physical and biological reasons to expect such a pattern of temporal variation but available data provide a conservative estimate of the maximum extent of hypoxia. The actual areal extent may be larger than estimated.

state-of-the-science regarding hypoxia in the Northern Gulf of Mexico and potential nutrient mitigation and control options in the Mississippi–Atchafalaya River basin (MARB). The Task Force was particularly interested in scientific advances since the *Integrated Assessment* and posed questions in three areas: characterization of hypoxia; nutrient fate, transport and sources; and the scientific basis for goals and management options. The Hypoxia Study Group began its deliberations in September of 2006 and completed its report in August of 2007 while operating under the "sunshine" requirements of the Federal Advisory Committee Act, which include providing public access to advisory meetings and opportunities for public comment. This Executive Summary summarizes the Hypoxia Study Group's major findings and recommendations.

Findings

Since publication of the *Integrated Assessment*, scientific understanding of the causes of hypoxia has grown while actions to control hypoxia have lagged. Recent science has affirmed the basic conclusion that contemporary changes in the hypoxic area in the northern Gulf of Mexico (NGOM) are primarily related to nutrient fluxes from the MARB. Moreover, new research provides early warnings about the deleterious long-term effects of hypoxia on living resources in the Gulf.

The Study Group was asked to comment on the *Action Plan*'s goal to reduce the hypoxic zone to a 5-year running average of 5,000 km^2 by 2015. The 5,000 km^2 target remains a reasonable end point for continued use in an adaptive management context; however, it may no longer be possible to achieve this goal by 2015 for two reasons. There is limited current movement toward the goal in either implementing policies or changing technologies. In addition, there are time lags in response of the ecological system. In August of 2007, the hypoxic zone was measured to be 20,500 km^2 (LUMCON, 2007), the third largest hypoxic zone since measurements began in 1985. Accordingly, it is even more important to proceed in a directionally correct fashion to manage factors affecting hypoxia than to wait for greater precision in setting the goal for the size of the zone. Much can be learned by implementing management plans, documenting practices, and measuring their effects with appropriate monitoring programs.

To reduce the size of the hypoxic zone and improve water quality in the MARB, the Study Group recommends a dual nutrient strategy targeting at least a 45% reduction in riverine total nitrogen flux (to approximately 870,000 metric tons/year or 960,000 tons/year) and at least a 45% reduction in riverine total phosphorus flux (to approximately 75,000 metric tonne/yr or 83,000 tons/year). Both of these reductions refer to changes measured against average flux over the 1980–1996 time period. For both nutrients, incremental annual reductions will be needed to achieve the 45% reduction goals over the long run. For nitrogen, the greatest emphasis should be placed on reducing spring flux, the time period most correlated with the size of the hypoxic zone. While the state of predictive and process models of NGOM hypoxia has continued to develop since 2000, models similar to those in place at that time are still the best tools for producing *dose–response* estimates for nitrogen (N) reductions, with most recent model runs showing a 45–55% required reduction for N in order to reduce the size of the hypoxic zone. A number of studies have suggested that climate change will create conditions for which larger nutrient reductions, e.g., 50–60% for nitrogen, would be required to reduce the size of the hypoxic zone.

New information has emerged that more precisely demonstrates the role of phosphorus (P) in determining the size of the hypoxic zone. Contrary to conventional wisdom that N typically limits phytoplankton production in near-coastal waters, the NGOM exhibits an unusual phenomenon whereby P is an important limiting constituent during the spring and summer in the lower salinity, near-shore regions. Phosphorus limitation is now occurring because over the past 50 years excessive N loadings have dramatically altered nitrogen to phosphorus ratios. Taken together, N and P both contribute to excess phytoplankton production and the hypoxia associated with such production, and they will need to be reduced concurrently to make progress in reducing the size of the hypoxic zone. The Study Group's best professional judgment is that phosphorus reductions will need to be comparable (in percentage terms) to nitrogen reductions to reduce the size of the hypoxic zone.

Scientific advances have improved our understanding of the physical factors that contribute to hypoxia. One physical factor that has changed substantially over the past century is river hydrology. The hydrologic regime of the Mississippi and Atchafalaya Rivers and the timing of freshwater inputs to the continental shelf are critical to mixing and hypoxia development. The most important hydrological change over the past century has been the diversion of a large amount of freshwater from the Mississippi River through the Atchafalaya River to the Atchafalaya Bay and maintenance of this diversion by the US Army Corps of Engineers. The major injection of freshwater into Atchafalaya Bay, some 200 km to the west of the Mississippi River Delta, has profoundly modified the spatial distribution of freshwater inputs, nutrient loadings, and stratification on the Louisiana–Texas continental shelf.

Methods used by the US Geological Survey (USGS) to calculate nutrient fluxes in the MARB have changed since the *Integrated Assessment*. The latest USGS estimates show that total N flux averaged 1.24 million metric tons/year (1.37 million ton/year) from 2001 to 2005 (65% of the flux is nitrate), and the total P flux averaged 154,000 metric tons/year (170,000 tons/year). This change represents a 21% decline in total N flux and a 12% increase in total P flux when compared to the averages from the 1980 to 1996 time period. The spring (April–June) flux of nutrients appears to be an important determinant of hypoxia, for that is when the river is disproportionately enriched with both N (especially nitrate) and P. Spring total N flux has declined since the 1980s; whereas total P flux shows a 9.5% increase (when average total P flux for 2001–2005 is compared to the 1980–1996 average). USGS data also show that during the past 5 years, the upper Mississippi and Ohio–Tennessee River subbasins contributed about 82% of nitrate-N flux, 69% of the TKN flux, and 58% of total P flux, although these subbasins represent only 31% of the entire MARB drainage area.

The Study Group's estimates of point source discharge show that point sources represented 22% of total annual average N flux and 34% of total annual average P flux discharged to the NGOM during the past 5 years. New methods also have been used to calculate nutrient mass balances (net anthropogenic N inputs, NANI). NANI for the MARB has declined in the past decade because of increased crop yields, reduced or redistributed livestock populations, and little change in N fertilizer inputs. From 1999 to 2005, NANI calculations show 54% of nonpoint N inputs in the MARB were from fertilizer, 37% from nitrogen fixation, and 9% from atmospheric deposition.

The Study Group finds that the Gulf of Mexico ecosystem appears to have gone through a regime shift with hypoxia such that today the system is more sensitive to inputs of nutrients than in the past, with nutrient inputs inducing a larger response in hypoxia as shown for other coastal marine ecosystems such as the Chesapeake Bay and Danish coastal waters. Changes in benthic and fish communities with the change in frequency of hypoxia are cause for concern. The recovery of hypoxic ecosystems may occur only after long time periods or with further reductions in nutrient inputs. If actions to control hypoxia are not taken, further ecosystem impacts could occur within the Gulf, as has been observed in other ecosystems.

Certain aspects of the nation's current agricultural and energy policies are at odds with the goals of hypoxia reduction and improving water quality. Since the *Integrated Assessment*, an emerging national strategy on renewable fuels has granted economic incentives to corn-based ethanol production. The projected increase in corn production from this strategy has profound implications for water quality in the MARB, as well as hypoxia in the NGOM. Recent energy policies, combined with pre-existing crop subsidies, tax policies, global market conditions, and trade barriers all provide economic incentives for conversion of retired and other cropland to corn production for use in ethanol production. Such conversions are projected to lead to corn production on an additional 6.5 million ha (16 million acres) in coming years with the majority of this increase occurring in the MARB. Without some change to the current structure of economic incentives favoring corn-based ethanol, N loadings to the MARB from increased corn production could increase dramatically in coming years, rather than decreasing, as needed for the NGOM.

Recommendations for Monitoring and Research

Most of the research and monitoring needs identified in the *Integrated Assessment* have not been met, and fewer rivers and streams are monitored today than in 2000. The majority of monitoring recommendations in the *Integrated Assessment* remain relevant and should be heeded. The Study Group affirms and reiterates the CENR's call to improve and expand monitoring of the temporal and spatial extent of hypoxia and the processes controlling its formation; the flux of nutrients, carbon, and other constituents from nonpoint sources throughout the MARB and to the NGOM; and measured (rather than estimated) nitrogen and phosphorus fluxes from municipal and industrial point sources.

The Study Group affirms the need for research in the following areas identified in the *Integrated Assessment*: ecological effects of hypoxia; watershed nutrient dynamics; effects of different agricultural practices on nutrient losses from land, particularly at the small watershed scale; nutrient cycling and carbon dynamics; long-term changes in hydrology and climate; and economic and social impacts of hypoxia.

A suite of models is needed to simulate the processes and linkages that regulate the onset, duration, and extent of hypoxia. Emerging coastal ocean observation and prediction systems should be encouraged to monitor dissolved oxygen and other physical and biogeochemical parameters needed to continue improving hypoxia models.

To advance the science characterizing hypoxia and its causes, the Study Group finds that research is also needed to

- collect and analyze additional sediment core data needed to develop a better understanding of spatial and temporal trends in hypoxia;
- investigate freshwater plume dispersal, vertical mixing processes, and stratification over the Louisiana–Texas continental shelf and Mississippi Sound, and use three-dimensional hydrodynamic models to study the consequences of past and future flow diversions to NGOM distributaries;
- advance the understanding of biogeochemical and transport processes affecting the load of biologically available nutrients and organic matter to the Gulf of Mexico, and develop a suite of models that integrate physics and biogeochemistry;
- elucidate the role of P relative to N in regulating phytoplankton production in various zones and seasons, and investigate the linkages between inshore primary production, offshore production, and the fate of carbon produced in each zone;
- improve models that characterize the onset, volume, extent, and duration of the hypoxic zone, and develop modeling capability to capture the importance of P, N, and P–N interactions in hypoxia formation;

To advance the science on sources, fate, and transport of nutrients, the Study Group recommends research to:

- develop models to simulate fluvial processes and estimate N and P transfer to stream channels under different management scenarios;
- improve the understanding of temporal and seasonal nutrient fluxes and develop nutrient, sediment, and organic matter budgets within the MARB;

To enhance the scientific basis for implementation of management options, the Study Group finds that research is needed to

- examine the efficacy of dual nutrient control practices;
- determine the extent, pattern, and intensity of agricultural drainage as well as opportunities to reduce nutrient discharge by improving drainage management;

- integrate monitoring, modeling, experimental results, and ongoing management into an improved conceptual understanding of how the forces at key management scales influence the formation of the hypoxia zone;
- develop integrated economic and watershed models to support adaptive management at multiple scales; and
- evaluate co-benefits of reducing nutrient loading to the Gulf.

Developments in the biofuels industry have created new questions for researchers to address. More research is needed on biofuel life cycles in order to identify system efficiency with respect to environmental effects, economics, and resource availability of biofuel alternatives. That is, research needs to evaluate the environmental effects of different biofuel production processes on soil, water quality, and climate under realistic strategies of deploying production facilities and moving the biofuels to the market. Current incentives favor corn-based ethanol production, although research has thus far shown fewer environmental consequences with other feedstocks, e.g., cellulosic feedstocks, such as switchgrass. Yet the technology for conversion of cellulosic feedstocks to biofuel is not yet commercially viable. Policies of all kinds (taxes, subsidies, trade) could be used to support research and technological developments for those biofuels that balance high energy yields with the lowest environmental impacts.

Recommendations for Adaptive Management

Adaptive management provides a framework for ongoing management in the face of uncertainty. It requires that conceptual models be developed to guide management and that management actions be treated like well-monitored experiments that answer questions for improving decisions with each successive cycle of learning. The most urgent need is to decrease nutrient discharge. In fact, nutrients should be decreased as soon as possible before the system requires even larger nutrient reductions to reduce the area of hypoxia. Already many taxa are lost during the peak of hypoxia, and there has been a shift in the relative abundance of fish species. Increases in certain pelagic species can disrupt food web structure, and the new system may respond in a quite different way to changes in nutrient level. The Study Group thus agrees with the CENR's emphasis on decreasing nutrient discharge in the context of adaptive management.

These adaptive management actions must be interpreted in view of both field measures and models of their effects. Conceptual models are needed for nutrient management at several spatial resolutions from small catchments, to large watersheds, to the entire MARB in order to guide research and ongoing adaptive management at each of the relevant scales. To the greatest extent possible, feedbacks should be incorporated into the models so that management is accompanied by learning about the full systems of linkages between human activities and hypoxia as well as the full range of co-benefits of N and P reductions.

Management Options

Large N and P reductions, on the order of 45% or more, are needed to reduce the size of the hypoxic zone. The most significant opportunities for N and P reductions occur in five areas:

- promotion, via research and economic incentives, of environmentally sustainable approaches to biofuel production and associated cropping systems (e.g., perennials);
- improved management of nutrients by emphasizing infield nutrient management efficiency and effectiveness to reduce losses;
- construction and restoration of wetlands, as well as criteria for targeting those wetlands that may have a higher priority for reducing nutrient losses;
- introduction of tighter N and P limits on municipal point sources; and
- improved targeting of conservation buffers, including riparian buffers, filter strips, and grassed waterways to control surface-borne nutrients.

Importantly, not all approaches will be cost-effective in all locations; the optimal combination and location of these practices will vary across and within watersheds.

In terms of cropping systems, research comparing nutrient discharge between alternative cropping systems (including row crops and nonrow crops, such as perennials) and a corn–soybean rotation shows that significant nutrient loss reductions could be achieved by converting current corn–soybean rotations to alternative crops or alternative rotations. Moreover, since corn crops require more nitrogen input, cellulosic sources (e.g., perennial grasses, fast-growing woody species) could, by comparison, provide alternative energy while protecting water quality. However, the technology for converting cellulosic sources to biofuel is not yet commercially viable. Significant reductions in nutrient runoff could also be achieved if nutrients are managed more efficiently on farms, for example, by moving to spring fertilization rather than fall. More wetlands are needed, especially in those areas that promise the greatest N and P reductions. Since the greatest N and P runoff is coming from upper Mississippi and Ohio–Tennessee River subbasins, where the highest proportion of tile drainage occurs, measures to improve drainage water management are urgently needed. In fact, improved targeting of almost all agricultural conservation practices in the region (e.g., conservation buffers, wetlands, land set aside in the Conservation Reserve Program [CRP], drainage water management) could achieve greater local water quality benefits and simultaneously contribute to hypoxia reduction. Nearly all of these opportunities were recognized in the *Integrated Assessment*.

The CENR did not emphasize tighter limits on municipal point sources; however, new calculations from the Study Group indicate that 22% of annual average total N flux and 34% of annual average total P flux to the Gulf comes from permitted point-source dischargers. The Study Group's calculations further demonstrate that tighter limits on N and P in effluent (3 mg N/L and 0.3 mg P/L) from sewage treatment plants could realize an estimated 11% reduction in annual average total N flux and a

21% reduction in total annual average P flux to the Gulf. Although the exact N and P limit could be debated, clearly there are regulatory opportunities to significantly reduce N and P fluxes to the Gulf. The cost associated with such regulations could be reduced if trading programs for point and nonpoint sources are properly developed and implemented concurrently with regulations.

Protecting and Enhancing Social Welfare in the Basin

Implementing the management options needed to reduce nutrients will clearly affect the social welfare of many who live in the basin. On the positive side, N and P reductions will improve environmental quality within the basin and, as the *Integrated Assessment* documented, these co-benefits can be highly valuable. Second, if the costs of implementing these management options are borne largely by residents in the region, then preserving/enhancing social welfare will require implementing policies that target the most cost-effective sources and locations for nutrient reductions.

Subsidies, not regulation, have been the government's primary tool for managing agricultural production and income support in the United States, as well as conservation in agriculture. Hence restructuring subsidies and conservation programs represents an important tool for reducing nutrient runoff from agricultural production. The *Integrated Assessment* recognized numerous agricultural management practices that improve water quality but did not discuss the efficiency of the tools for their implementation. A large body of economics literature exists regarding the relative merits and cost-effectiveness of taxes, regulations, voluntary approaches, permit trading, subsidies, and other instruments that could apply to reducing nutrient losses. This research indicates that if significant behavioral changes are to be realized, incentives are needed across a wide range of sectors. Such incentives can be positive (e.g., subsidies) or negative (e.g., taxes or direction regulation with enforcement actions), but they must be strong enough to change behavior. A thorough and quantitative comparison of all possible incentives for all sectors was beyond the Study Group's scope; however, research indicates that the following approaches are cost-effective.

First, the establishment (and continuation where appropriate) of targeting and competitive bidding mechanisms results in lands enrolled in conservation programs (e.g., the Conservation Reserve Program, the Environmental Quality Incentives Program, and the Conservation Security Program) that achieve maximum environmental benefits. Moreover, conservation compliance requirements extended to nutrient management, if adequately monitored and enforced, could be cost-effective. Targeting conservation practices to the locations within a watershed where they produce the most N and P reductions (and co-benefits) and targeting entire watersheds that have relatively high N and/or high P contributions are both cost-effective targeting approaches.

Second, economic incentives are needed for the full range of conservation options. Incentives for development of technologies to convert cellulosic perennials to biofuels would be needed to greatly reduce N and P losses from agricultural systems. Restructuring eligibility requirements for existing subsidies to reward conservation in all its forms (in-field nutrient management, cover crops, conservation buffers, wetlands, alternative drainage, manure management) could help mitigate the unintended consequences of agricultural production.

Conclusion

In sum, environmental decisions and improvements require a balance between research, monitoring, and action. In the Gulf of Mexico, the action component lags behind the growing body of science. Moreover, certain aspects of current agricultural and energy policies conflict with measures needed for hypoxia reduction. Although uncertainty remains, there is an abundance of information on how to reduce hypoxia in the Gulf of Mexico and to improve water quality in the MARB, much of it highlighted in the *Integrated Assessment*. To utilize that information, it may be necessary to confront the conflicts between certain aspects of current agricultural and energy policies on the one hand and the goals of hypoxia reduction and improving water quality on the other hand. This dilemma is particularly relevant with respect to those policies that create economic incentives. The Study Group's recommendation to address the structure of economic incentives stems from sound science.

Basing management decisions on sound science means taking action at several different scales, addressing conflicts between policies, and acting in the face of uncertainties. Lessons learned from current actions can inform and improve future decisions. While actions must come first, they must also be coupled with monitoring and modeling of management activities within a conceptual framework to improve understanding of the system. Done well, this process of adaptive management means that, over time, society will benefit from cost-effective environmental decisions that reduce hypoxia in the Gulf and improve water quality in the MARB.

Chapter 1
Introduction

1.1 Hypoxia and the Northern Gulf of Mexico – A Brief Overview

Nutrient over-enrichment from anthropogenic sources is a major stressor of aquatic, estuarine, and marine ecosystems. Nutrients enter ecosystems through off-target migration of fertilizer from agricultural fields, golf courses, and lawns; disposal of animal manure; atmospheric deposition of nitrogen; erosion of soil containing nutrients; sewage treatment plant discharges; and other industrial discharges. Excessive nutrients promote nuisance blooms (excessive growth) of opportunistic bacteria, cyanobacteria, and algae. When the available nutrients in the water column have been sequestered in plant biomass, the nuisance blooms die, decompose, and deplete dissolved oxygen in the water column and at the sediment water interface. This oxygen depletion, known as *hypoxia,* occurs when normal dissolved oxygen concentrations in shallow coastal and estuarine systems decrease below the level required to support many estuarine and marine organisms (≤ 2 mg/L).

Hypoxia can occur naturally in deep basins, fjords, and oxygen-minimal coastal zones associated with upwelling. However, nutrient-induced hypoxia in shallow coastal and estuarine systems is increasing worldwide. A large hypoxic area, averaging about 16,500 km^2 (10,250 mi^2) and ranging from 8,500 to 22,000 km^2 (3,100–7,700 mi^2) forms annually between May and September in the northern Gulf of Mexico. Shown in Fig. 1.1, the northern Gulf hypoxic zone is the largest in the United States and the second largest worldwide. Hypoxic conditions result from complex interactions between climate, weather, basin morphology, circulation patterns, water retention times, freshwater inflows, stratification, mixing, and nutrient loadings. Nutrient fluxes from the Mississippi–Atchafalaya River basin (MARB), coupled with temperature and density-induced stratification, have been implicated as the primary cause of hypoxia in the northern Gulf of Mexico (NGOM) (CENR, 2000).

The MARB is one of the largest river systems in the world (Fig. 1.2), draining approximately 40% of the contiguous United States, and is the largest contributor of freshwater and nutrients to the NGOM. About two-thirds of the total Mississippi River flow enters the northern Gulf via the Mississippi River delta. The remaining

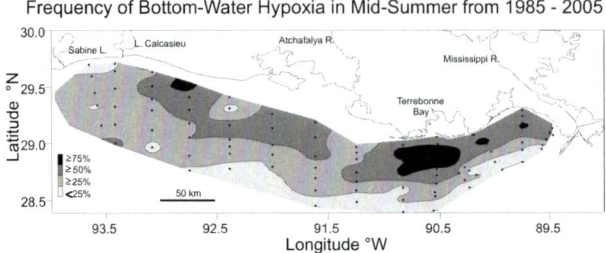

Fig. 1.1 Map of the frequency of hypoxia in the northern Gulf of Mexico, 1985–2005. Taken from N.N. Rabalais, LUMCON, 2006

Fig. 1.2 Map showing the extent of the Mississippi–Atchafalaya River basin

third is diverted to the Atchafalaya River and eventually enters the northern Gulf about 200 km west of the main Mississippi River delta. Prevailing east-to-west currents in the Gulf move much of the freshwater, suspended sediments, and dissolved and particulate nutrients onto the Louisiana–Texas continental shelf.

Land-use activities in the MARB influence water quality in the entire watershed as well as in the NGOM. Low-oxygen events on the Louisiana–Texas continental shelf have been reconstructed over the past 180 years using the relative abundance of low-oxygen tolerant benthic foraminifera in sediment cores (Osterman et al., 2005). These data show that the prevalence of low-oxygen events has increased over the past 50 years. Several hypoxic events from 1870 and 1910 (prior to widespread fertilizer use) were attributed to natural variation in river flow that enhanced freshwater and nutrient transport. The increased prevalence over the past several decades is clearly related to increased nutrient loads. However, there is substantial variation in year-to-year inputs of both freshwater and nutrients from the MARB. Since these are correlated, it is not possible to tease apart the relative importance of increased eutrophication versus increased stratification in any given year over the recent past (Gooday et al., 2009, Middelburg and Levin, 2009). Clearly, land-use practices in the MARB affect watershed dynamics and water quality within the Basin as well as the northern Gulf. Land-use practices in the Basin are also influenced by various, and conflicting, national environmental, conservation, and agricultural policies.

1.2 Science and Management Goals for Reducing Hypoxia

In 1997, the US Environmental Protection Agency (USEPA) established the Mississippi River/Gulf of Mexico Watershed Nutrient Task Force (or Task Force). The Task Force brought together federal agencies, states, and tribes to consider options for reducing, mitigating, and controlling hypoxia in the NGOM. The Task Force requested that the White House National Science and Technology Council (NSTC) conduct a scientific assessment of the causes and consequences of Gulf hypoxia. The NSTC Committee on Environment and Natural Resources (CENR) formed a federal intra-agency Hypoxia Working Group to plan and conduct the assessment. The need for the assessment was given additional impetus by passage of the Harmful Algal Bloom and Hypoxia Research and Control Act of 1998. The Act specifically called for an integrated scientific assessment of causes and consequences of hypoxia in the Gulf of Mexico and a plan of action to reduce, mitigate, and control hypoxia.

The scientific assessment was led by the National Oceanic and Atmospheric Administration (NOAA) with oversight among several federal agencies. As a first step, six reports (available at http://www.nos.noaa.gov/products/pub_hypox.html) covering key topics were developed. These include characterization of hypoxia (Rabalais et al., 1999a); ecological and economic consequences of hypoxia (Diaz and Solow, 1999); flux and sources of nutrients in the Mississippi–Atchafalaya River basin (Goolsby et al., 1999); effects of reducing nutrient loads to surface waters within the Mississippi River basin and Gulf of Mexico (Brezonik et al., 1999); reducing nutrient fluxes, especially nitrate–nitrogen, to surface water, ground water, and the Gulf of Mexico (Mitsch et al., 1999); and evaluation of the economic costs and benefits of the methods for reducing nutrient fluxes to the Gulf of Mexico (Doering et al., 1999).

The six NOAA reports provided the scientific foundation for the Integrated Assessment of Hypoxia in the Northern Gulf of Mexico (CENR, 2000) (or *Integrated Assessment*, available at http://oceanservice.noaa.gov/products/pubs_hypox.html). The *Integrated Assessment* concluded that hypoxia in the northern Gulf was caused by excess nitrogen from the MARB, in combination with stratification of Gulf waters. Informed by the *Integrated Assessment*, in 2001 the Task Force completed its *Action Plan for Reducing, Mitigating, and*

Controlling Hypoxia in the Northern Gulf of Mexico (MR/GMWNTF, 2001) (or *Action Plan*, available at http://www.epa.gov/msbasin/taskforce/actionplan.htm). The *Action Plan* described three primary hypoxia management goals.

1. Coastal goal: By the year 2015, subject to the availability of additional resources, reduce the 5-year running average of the areal extent of the Gulf of Mexico hypoxic zone to less than 5,000 km^2 (1,930 mi^2) through implementation of specific, practical, and cost-effective voluntary actions by all states, tribes, and all categories of sources and removals within the Mississippi–Atchafalaya River basin to reduce the annual discharge of nitrogen into the Gulf.
2. Within-basin goal: To restore and protect the waters of the 31 states and tribes within the MARB through implementation of nutrient- and sediment-reduction actions to protect public health and aquatic life as well as reduce negative impacts of water pollution on the Gulf of Mexico.
3. Quality of life goal: To improve the communities and economic conditions across the Mississippi–Atchafalaya River basin, in particular the agriculture, fisheries, and recreation sectors, through improved public and private land management and a cooperative incentive-based approach.

In 2005, the Task Force recognized a need to update the *Integrated Assessment* and *Action Plan* with more recent science. Accordingly, the Task Force sponsored four symposia on the upper Mississippi River basin; Gulf Hypoxia; the lower Mississippi River basin; and Nutrient Sources, Fate, and Transport. Each of the symposia focused on scientific developments since 1999. In conjunction with the symposia, the Task Force also developed a bibliography of recent literature on hypoxia causes, effects, and control options since the year 2000 (available at http://www.epa.gov/msbasin/taskforce/reassess2005.htm). In addition to science activities, the Task Force also compiled information necessary for nutrient management and control in the MARB in two reports. The *Management Action Review Team Report* (MART, 2006a) summarized federal programs that encouraged watershed planning and land-use practices to reduce nutrient loadings. The *Reassessment of Point Source Nutrient Mass Loadings to the Mississippi River Basin* report (MART, 2006b) updated annual mass loading estimates for total nitrogen (TN), total phosphorus (TP), and biochemical oxygen demand (BOD). (Task Force documents are available at http://www.epa.gov/msbasin/taskforce/reassess2005.htm.) The Task Force is also working with the US Department of Agriculture's (USDA) Conservation Effects Assessment Program (CEAP) to encourage the quantification and documentation of environmental effects and benefits of conservation practices on agricultural lands to control nutrients in the MARB. CEAP documents are available at http://www.nrcs.usda.gov/Technical/nri/ceap/.

1.3 Hypoxia Study Group

On behalf of the Task Force, USEPA's Office of Water requested that the Science Advisory Board (SAB) evaluate the state-of-the-science regarding hypoxia in

1.3 Hypoxia Study Group

the Gulf of Mexico and potential nutrient mitigation and control options in the Mississippi–Atchafalaya River basin. In response to this request, the SAB asked the Hypoxia Study Group to evaluate the following issues and questions. The specific charges to the Study Group appear below in italic type.

1. **Characterization of hypoxia**: *The development, persistence, and areal extent of hypoxia are thought to result from interactions in physical, chemical, and biological oceanographic processes along the northern Gulf continental shelf and changes in the Mississippi River basin that affect nutrient loads and fresh water flow.*

   A. *Address the state of the science and the importance of various processes in the formation of hypoxia in the Gulf of Mexico. These issues include the following:*

      i. *increased volume or funneling of fresh water discharges from the Mississippi River;*
      ii. *changes in hydrologic or geomorphic processes in the Gulf of Mexico and the Mississippi River basin;*
      iii. *increased nutrient loads due to coastal wetlands losses, upwelling, or increased loadings from the Mississippi River basin;*
      iv. *increased stratification and seasonal changes in magnitude and spatial distribution of stratification and nutrient concentrations in the Gulf;*
      v. *temporal and spatial changes in nutrient limitation or co-limitation, for nitrogen or phosphorus, as significant factors in the development of the hypoxic zone; and*
      vi. *the implications of reduction of phosphorus or nitrogen without concomitant reduction of the other.*

   B. *Comment on the state of the science for characterizing the onset, volume, extent, and duration of the hypoxic zone.*

2. **Characterization of Nutrient Fate, Transport, and Sources:** *Nutrient loads, concentrations, speciation, seasonality, and biogeochemical recycling processes have been suggested as important causal factors in the development and persistence of hypoxia in the Gulf. The Integrated Assessment (CENR 2000) presented information on the geographic locations of nutrient loads to the Gulf and the human and natural activities that contribute nutrient loadings.*

   A. *Given the available literature and information (especially since 2000), data, and models on the loads, fate, and transport and effects of nutrients evaluate the importance of various processes in nutrient delivery and effects. These may include the following:*

      i. *the pertinent temporal (annual and seasonal) characteristics of nutrient loads/fluxes throughout the Mississippi River basin and, ultimately, to the Gulf of Mexico;*

ii. *the ability to determine an accurate mass balance of the nutrient loads throughout the basin; and*
iii. *nutrient transport processes (fate/transport, sources/sinks, transformations, etc.) through the basin, the deltaic zone, and into the Gulf.*

B. *Given the available literature and information (especially since 2000) on nutrient sources and delivery within and from the basin, evaluate capabilities to*

i. *predict nutrient delivery to the Gulf, using currently available scientific tools and models; and*
ii. *route nutrients from their various sources and account for the transport processes throughout the basin and deltaic zone, using currently available scientific tools and models.*

3. **Scientific Basis for Goals and Management Options:** *The Task Force has stated goals of reducing the 5-year running average areal extent of the Gulf of Mexico hypoxic zone to less than 5,000 km^2 by the year 2015, improving water quality within the basin and protecting the communities and economic conditions within the basin. Additionally, nutrient loads from various sources in the Mississippi River basin have been suggested as the major driver for the formation, extent, and duration of the Gulf hypoxic zone.*

    A. *Are these goals supported by present scientific knowledge and understanding of the hypoxic zone, nutrient loads, fate and transport, sources, and control options?*

    i. *Based on the current state of the science, should the reduction goal for the size of the hypoxia zone be revised?*
    ii. *Based on the current state of the science, can the areal extent of Gulf hypoxia be reduced while also protecting water quality and social welfare in the basin?*

    B. *Based on the current state of the science, what level of reduction in causal agents (nutrients/discharge) will be needed to achieve the current reduction goal for the size of the hypoxic zone?*
    C. *Given the available literature and information (especially since 2000) on technologies and practices to reduce nutrient loss from agriculture, runoff from other nonpoint sources and point source discharges, discuss options (and combinations of options) for reducing nutrient flux in terms of cost, feasibility, and any other social welfare considerations. These options may include*

    i. *the most effective agricultural practices, considering maintenance of soil sustainability and avoiding unintended negative environmental consequences;*
    ii. *the most effective actions for other nonpoint sources; and*
    iii. *the most effective technologies for industrial and municipal point sources.*

In all three areas, please address research and information gaps (expanded monitoring, documentation of sources and management practices, effects of practices, further model development and validation, etc.) that should be addressed prior to the next 5-year review.

1.4 The Study Group's Approach

The NOAA, CENR, and Task Force documents (see Section 1.2 above) provide a comprehensive scientific review of hypoxia causes, and potential mitigation and control actions through about 1999–2000. Further, more recent science and management information on the Gulf and MARB has been captured in the Task Force sponsored symposia, literature search, MART reports, and CEAP activities. Accordingly, the Study Group initiated its deliberations by reviewing these documents. The Study Group invited the chairs of the four symposia to present summaries of key findings, and also invited selected researchers (see acknowledgements) currently working on hypoxia issues to present their recent work. The Study Group also relied on the individual and collective experience and expertise of its members to provide additional relevant publications and information to assist its deliberations. The Study Group convened 4 public face-to-face meetings and 15 public teleconferences to deliberate and develop this state-of-the-science report (background and other materials for the meetings may be found at: http://www.epa.gov/sab/panels/hypoxia_adv_panel.htm).

The Study Group recognized the inherent complexity and connectivity between the Mississippi–Atchafalaya River basin and Gulf of Mexico and agreed that a systems perspective within an adaptive management framework was needed. The systems approach allowed understanding of feedback loops so that perturbations in one part of a system affect the interrelationships and stability of the system as a whole. Adaptive management seeks to maximize flexibility in management so that learning and adjustments can occur. Adaptive management employs six basic operating principles: (1) resources of concern are clearly defined; (2) conceptual models are developed during planning and assessment; (3) management questions are formulated as testable hypotheses to guide inquiry; (4) management actions are treated like experiments that test hypotheses to answer questions and provide future management guidance; (5) ongoing monitoring and evaluation is necessary to improve accuracy and completeness of knowledge; and (6) management actions are revised with new cycles of learning.

This book considers models as essential for understanding the inherent complexities of the MARB and the NGOM. Additionally, the collection of critical data at appropriate spatial and temporal scales is absolutely necessary to optimize future research and management actions. Data collection should be based on a well-defined conceptual model of the overall system. Monitoring programs will often provide data for existing models and assist with broader interpretations of data and information. In summary, a systems perspective combined with an adaptive management approach will greatly enhance scientific understanding and management of hypoxia in the MARB and the NGOM.

This book deals largely with the review of research and findings since the *Integrated Assessment*. Background material and findings prior to 2000 are used when appropriate or when instrumental to understanding the relative importance of more recent work. However, those interested in the details of the *Integrated Assessment* and the six topical reports that provided the scientific basis for the assessment are referred directly to those documents.

Chapter 2
Characterization of Hypoxia

The hypoxic region along the northern Gulf of Mexico (NGOM) extends up to 125 km offshore and to 60 m water depth, has substantial variability with an average midsummer areal extent of 16,500 km^2 (2001–2007), and extends in some years from the Mississippi River mouth westward to Texas coastal waters (Rabalais et al., 2007a, 2007b). This hypoxic region (Fig. 1.1) occurs along a relatively shallow, open coastline with complex circulation and water column structure typical of many coastal regions and includes massive inputs of freshwater, weak tidal energies, seasonally varying stratification strength, generally high water temperature, wind effects from both frontal weather systems and hurricanes, and mixing of river plumes from the Atchafalaya and Mississippi Rivers and other smaller sources (DiMarco et al., 2006; Hetland and DiMarco, 2007). The plumes of the Mississippi and Atchafalya Rivers can be observed as areas of highly turbid low-salinity surface water. The limits of these plumes have been defined in different ways, but in satellite imagery their boundaries can be clearly observed as sharp color discontinuities. Since the release of the *Integrated Assessment* and the *Action Plan* in 2001, the measured areal extent of the hypoxic region has averaged 16,500 km^2, with a range of 8,500–22,000 km^2. Many reports from both the *Integrated Assessment* and the post-*Integrated Assessment* periods concluded that physical and morphological characteristics such as these make the NGOM prone to hypoxic conditions.

2.1 Historical Patterns and Evidence for Hypoxia on the Shelf

An important question regarding hypoxia on the Mississippi River shelf is how far back in time has hypoxia been observed? Is it a recent phenomenon or has hypoxia been a regular natural feature of a productive shelf region? Unfortunately the monitoring data are not entirely sufficient to address this question, for only a limited number of measurements are available prior to the time when widespread hypoxia was first observed on the Louisiana shelf in the mid-1980s (Rabalais et al., 1999a). However, a limited number of additional paleoecological studies have been carried out on the Mississippi River shelf since the *Integrated Assessment* (e.g., Swarzenski et al., 2008). All studies from dated sediment cores show recent increases in low oxygen concentrations with time, although the precise timing and response varies depending upon the proxy studied and the dating of cores. The accumulated body of evidence shows that the pattern of change is concomitant with recent (since the 1960s) increases in nutrient loading from the Mississippi River causing increasingly severe hypoxia on the shelf. The spatial distribution of reliably dated sediment cores, with most cores taken on the southeastern Louisiana shelf just west of the Mississippi River delta, is not sufficient to determine the increases in the spatial extent of hypoxia with time.

A limiting factor in all paleoecological studies is the availability of undisturbed sediment cores to provide an accurate picture of changes through time. This is a particular challenge in a hydrologically dynamic, relatively shallow environment as found on the Mississippi River shelf with resuspension processes, movement of fluid muds, mixing by benthic organisms, and more recently sediment disturbance of upper sediment layers through bottom trawling. Despite these challenges, a number of reasonably dated sediment cores, primarily within the Louisiana bight, have provided a coherent picture of changes in hypoxia with time.

Bacterial pigments measured in sediments at one location on the Louisiana shelf were characteristic of anoxygenic phototrophic sulfur bacteria and have their highest concentrations between 1960 and the present (Chen et al., 2001). These bacterial pigments were not present prior to 1900. Further evidence of increased hypoxia is provided by Chen et al. (2001) using algal pigments, which show increases in the 1960s. The increase in these pigments reflects enhanced preservation with hypoxia as well as nutrient-driven increases in production. Rabalais et al. (2004, 2007a) also report increases in algal pigment concentrations over time from a number of sediment cores, with gradual changes from 1955 to 1970, followed by a steady increase to the late 1990s. However, the patterns observed by Rabalais et al. (2004, 2007a) are confounded by the rapid degradation of carbon and algal pigments in upper surface sediments with most studies of sediment pigments correcting for diagenesis by normalizing pigments with organic carbon (Leavitt and Hodson, 2001). In addition, there is some evidence for spatial increases in hypoxic extent through time: increases in pigment concentrations from one sediment core from west of the Atchafalaya River outflow suggests that nutrient-driven increases in production occurred later at this location than in the Mississippi River Bight (Rabalais et al., 2004). There has been an increased accumulation of total organic carbon and biogenic silica in recent sediments near the mouth of the Mississippi River (Turner et al., 2004; Turner and Rabalais, 1994), although the spatial and temporal variations observed between dated sediment cores are large.

Several studies have examined changes in the benthic foraminiferal community in dated sediment cores (Osterman et al., 2005; Platon et al., 2005; Platon and Sen Gupta, 2001). Different species of bottom-living benthic foraminifera are particularly sensitive to changes in bottom water oxygen concentrations, and the abundance of these species is a widely used indicator of hypoxia. Significant changes in the composition of the benthic foraminiferal community have occurred in the past century. Several indicators, e.g., the PEB index (the relative abundance of three low-oxygen tolerant species of benthic foraminifers: *Pseudononin*

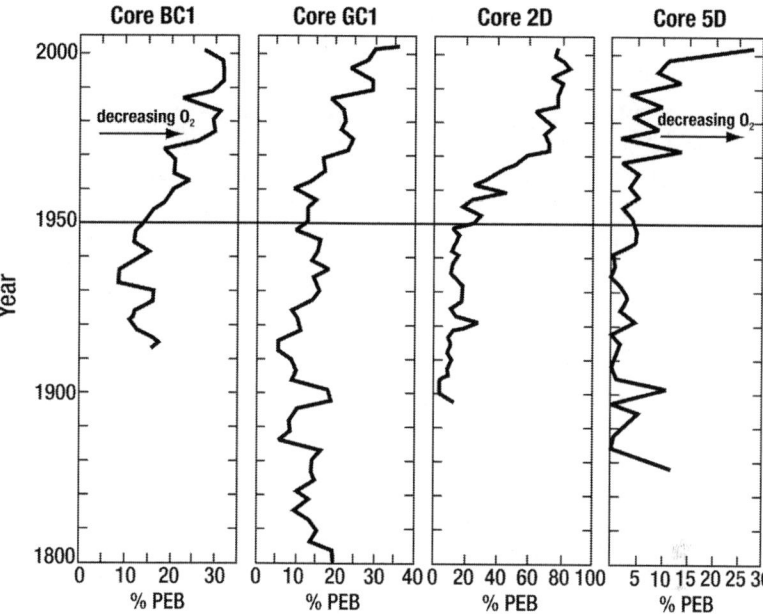

Fig. 2.1 Plots of the PEB index (%PEB) in sediment cores from the Louisiana shelf. Higher values of the PEB index indicate lower dissolved oxygen contents in bottom waters. Taken from Osterman et al. (2005)

altlanticum, Epistominella vitrea, and *Buliminella morgani*) (Osterman et al., 2005) and the A/P ratio (agglutinated to porcelaneous orders) (Platon et al., 2005) indicate that increases in the occurrence of low oxygen events have occurred over the past 50 years (Fig. 2.1). In addition, the porcelaneous genus *Quinqueloculina*, an organism that occurs where dissolved oxygen concentrations are higher than 2 mg/l, was present but has disappeared from the foraminiferal community since 1900, indicating that prior to this time there was sufficient oxygen at the sediment–water interface to enable survival of such species (Rabalais et al., 2007a). Osterman et al. (2005) have shown that several probable low oxygen events that occurred in the past 180 years are associated with high Mississippi River discharge rates, although the recent changes in foraminiferal communities are more extreme than any that occurred in the past. The data support the interpretation that hypoxia is a recent phenomenon and has been amplified from an otherwise naturally occurring process.

Key Findings and Recommendations

The Study Group finds that the paleoecological data are consistent with increased prevalence of hypoxic conditions in recent decades. However, the spatial distribution of sediment cores is not sufficient to determine the increases in the spatial extent of hypoxia with time. Although given the complex nature of disturbance, there may be limited opportunities to determine temporal changes in the extent of hypoxia. To advance the understanding of spatial and temporal trends in hypoxia in the NGOM, the Study Group offers the following recommendations.

- In future research on the Mississippi River shelf, more attention should be focused on establishing reliable chronologies in additional sediment cores.
- In order to establish spatial changes in hypoxia over time, where possible, additional sediment cores should be collected over a broader area of the Mississippi River shelf.

2.2 The Physical Context

2.2.1 Oxygen Budget: General Considerations

The oxygen budget on the NGOM shelf is influenced by several sink and source terms. Oxygen (O_2) concentration in the bottom layer will decrease and possibly become hypoxic or even anoxic when the export and consumption of oxygen by respiration exceed the import or production of "new" oxygenated water by photosynthesis. Mathematically, this relationship can be expressed in its simplest form by the following oxygen balance equation:

$$\frac{\partial O_2}{\partial t} = -u\frac{\partial O_2}{\partial x} - v\frac{\partial O_2}{\partial y} - w\frac{\partial O_2}{\partial z} + K_z\frac{\partial^2 O_2}{\partial z^2} + K_H\left(\frac{\partial^2 O_2}{\partial x^2} + \frac{\partial^2 O_2}{\partial y^2}\right)$$
$$\text{Change} \quad (1) \quad\quad (2) \quad\quad (3) \quad\quad (4) \quad\quad\quad (5)$$
$$+ \vec{F}_{as} - \text{resp.} + \text{photosynthesis}$$
$$(6) \quad\quad (7) \quad\quad (8)$$

(2.1)

in which the left-hand term represents the change of oxygen concentration with time; term (1) on the right represents the horizontal advection by across-shelf currents, u; term (2) represents the horizontal advection by along-shelf currents, v; term (3) represents vertical transport by upwelling or downwelling; term (4) represents vertical mixing, and K_z (x,y,z) is the vertical eddy diffusivity; term (5) represents horizontal diffusion, and K_H (x,y,z) is the horizontal eddy diffusivity; term (6) is oxygen flux across the air–sea interface; term (7) is the nonconservative sink (i.e., oxygen consumption); and term (8) refers to in situ production of oxygen by photosynthesis. The horizontal advection terms may reflect contributions from tides, wind stress, buoyancy, and momentum input from rivers, large-scale and mesoscale eddies, or topographically trapped shelf waves. Three-dimensional hydrodynamic models are required to adequately account for these contributions (Hetland and DiMarco, 2007; Morey et al., 2003a, 2003b). The respiration term (7) relates directly to organic matter mineralization and must be understood in the context of water column and sediment biogeochemical processes described in later sections. As depicted in Eq. 2.1, the change in oxygen concentration with time at any point in the water column is affected by sources and sinks of oxygen at and below the surface. Term (6) (oxygen flux across the air–sea interface) represents a surface source and sink, while term (8) (photosynthesis) is a source of oxygen in waters beneath the air–sea interface. Although Eq. 2.1 above suggests that alongshore and cross-shore dispersion coefficients are of equal magnitude, the Study Group notes that this has not been demonstrated. The effects of cross-shore dispersion processes must be parameterized and additional research on lateral mixing processes must be completed before such parameterization can be performed with confidence.

2.2.2 Vertical Mixing as a Function of Stratification and Vertical Shear

Over the Louisiana–Texas shelf, the vertical mixing term (4) plays a key role in the local oxygen balance. Its magnitude depends on the value of vertical eddy diffusivity K_z, which is highly variable in both space and time and depends on the gradient Richardson number Ri (MacKinnon and Gregg, 2005), defined by

$$Ri = \frac{N^2}{\left(\frac{\partial u}{\partial z}\right)^2 + \left(\frac{\partial v}{\partial z}\right)^2} = \frac{\left(\frac{-g}{\rho}\frac{\partial \rho}{\partial z}\right)}{\left(\frac{\partial V}{\partial z}\right)^2} \qquad (2.2)$$

where N is an index of stratification strength known as the buoyancy frequency, ρ is the water density, g is the gravitational acceleration (9.8 m/s^2), and $\partial V/\partial z$ is the vertical shear of horizontal current. The gradient Richardson number, Ri, expresses the ratio of turbulence suppression by stratification (numerator) relative to vertical shear production of turbulence (denominator). When $Ri > 1/4$, turbulence is suppressed, and vertical transport of oxygen from surface to bottom layers by turbulent mixing is unlikely to occur. Thus, strong vertical density gradients (for example, when freshwater sits on top of salty water) and/or weak current shears can suppress vertical mixing and be favorable to hypoxia. Key physical factors that produce stronger vertical density gradients ($\partial \rho/\partial z$) and thus reduce vertical mixing include freshwater inputs from rivers or precipitation, warmer surface temperatures from absorption of solar radiation or sensible heat input, and near-bed suspended sediment (which causes benthic stratification). Factors responsible for producing enhanced vertical shear ($\partial V/\partial z$) and enhanced vertical mixing include tidal and wind-driven currents, inertial waves, internal tides, surface waves, and Langmuir cells (Kantha and Clayson, 2000). Although no field studies of vertical mixing by microstructure measurements of the turbulent dissipation rates of velocity, salinity, and temperature fluctuations have been reported for the NGOM, many of the physical mechanisms described on the New England shelf (MacKinnon and Gregg, 2005) and in Monterrey Bay (Carter et al., 2005) are at play on the NGOM as well.

While the *tributaries* within the Mississippi River basin are the sources of nutrient loading to the river trunk, the *distributaries* within the Mississippi Delta are critical to the final dispersal of nutrients, buoyancy, and sediment into the Gulf of Mexico. The multiple distributary mouths of the Mississippi and Atchafalaya Rivers are, for the most part, highly stratified "salt wedge" estuaries, and their combined effluent debouches onto the shelf as a discrete layer of fresh water that is spread into the surface layer. Exceptions occur where smaller distributaries enter shallow bays where salinity is nearly uniform from top to bottom. Total buoyancy fluxes are, of course, proportional to river discharge and cause the turbulence suppressing stratification of the upper water column that is strongly implicated in hypoxia. In most inner shelf environments, tidal currents are the major source of mixing, and the position of temperature fronts (sharp horizontal temperature gradients) can often be accurately predicted from the h/U_t^3 criterion of Simpson and Hunter (1974), where h is the local depth and U_t represents the depth-averaged tidal velocity. Unfortunately, the Simpson–Hunter criterion of tidal mixing has not yet been mapped for the northern Gulf of Mexico. Nevertheless, it is generally agreed that tidal mixing over the Louisiana–Texas shelf is very weak because the tidal range is only about 40 cm and tidal currents typically do not exceed 10 cm/s (Kantha, 2005). So the contribution of tidal mixing to the vertical exchange of oxygen is minimal over the shelf, particularly off the mouths of the larger distributaries, such as Southwest and South Passes, which debouch into deep water. Wind-driven currents are stronger than tidal currents but occur episodically (Ohlmann and Niiler, 2005). Winds also cause breaking and white capping waves as well as vertical circulation (Langmuir) cells (Thorpe, 2004) that contribute to mixing in the upper water column.

The hydrologic regime of the Mississippi River and the spatial distribution and timing of freshwater inputs to the shelf relative to the occurrence of energetic currents and waves are critical to vertical mixing intensity, stratification, and hypoxia. These influences were recognized in the CENR report (Rabalais et al., 1999a). Using oxygen measurements within 2 m of the bottom and vertical profiles of temperature and salinity collected during the 1992–1994 LaTex experiment on the Louisiana–Texas shelf and during the 1996–1998 NECOP (Northeastern Gulf of Mexico Chemical Oceanography Program) in the region east of the Mississippi delta and north of Tampa Bay, Belabbassi (2006) performed an evaluation of the empirical relationships between the maximum value of the buoyancy frequency N_{max} in the water column, bottom silicate concentration as a proxy of phytoplankton remineralization, and the occurrence of hypoxic waters (< 2 mg/L) or low-oxygen waters (< 3.4 mg/L). She found that low-oxygen and hypoxic bottom waters only occurred when N_{max} evaluated at a vertical resolution of 0.5 m was greater than 40 cycles per hour (cph), which corresponds to a buoyancy period shorter than 1.5 min. This result confirms that strong density stratification is a prerequisite for hypoxia occurrence on the northern Gulf of Mexico shelf. She also found that low-salinity water from the Mississippi and Atchalafaya rivers was generally the main contributor to stratification in spring and summer, although temperature was more important than salinity in determining stratification during summer at all depths west of Galveston Bay and at depths greater than 20 m between Galveston Bay and Terrebonne Bay. Interestingly, stations with strong stratification (N_{max} greater than 40 cph) but low bottom silicate concentrations (less than 18 mmol m^{-3}) did not have low-oxygen or hypoxic bottom waters. The analyses of Belabbassi (2006) thus indicate that strong stratification (N_{max} greater than 40 cph) is a necessary but not sufficient condition for bottom layer hypoxia; a second necessary condition for hypoxia occurrence is high bottom water remineralization as indicated by the proxy of high concentrations of bottom water silicates (greater than 18 mmol m^{-3}). Simply put, there cannot be hypoxia without both density stratification and degradation of labile organic matter.

Stow et al. (2005) attempted to disentangle the relative contributions of eutrophication and stratification as drivers of hypoxia in the NGOM. Their analysis indicates that the probability of observing bottom hypoxia increases rapidly when the top to bottom salinity difference reaches a threshold of 4.1. Stow et al. (2005) also showed that this salinity threshold decreased from 1982 to 2002. Concurrently, they highlighted that surface temperature had increased, while surface dissolved oxygen decreased, suggesting that changes in surface mixed layer properties may be partly responsible for oxygen decrease in the bottom layer.

2.2.3 Changes in Mississippi River Hydrology and Their Effects on Vertical Mixing

By far the most important change in local hydrology has been the increased flow of the Atchafalaya River during the 20th century. Available data show that in the early 1900s the discharge from the Atchafalaya River accounted for less than 15% of the combined Atchafalaya–Mississippi River discharge (Fig. 2.2). This proportion progressively increased to reach about 30% in 1960, peaked at 35% in 1975, and since then was reduced to 30% by means of regulatory measures (Bratkovich et al., 1994). To understand the significance of this change on circulation patterns and on the strength of stratification on the Louisiana–Texas shelf, it must be kept in mind that the Mississippi River plume enters the shelf near the shelf edge and typically

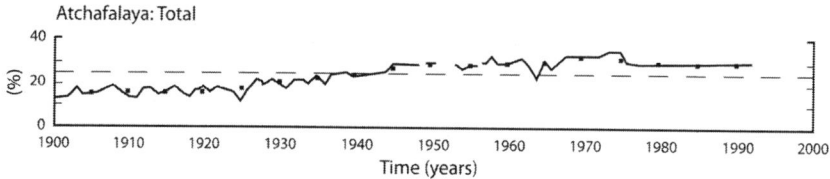

Fig. 2.2 Change in the relative importance of the Atchafalaya flow to the combined flows from the Mississippi and Atchafalaya Rivers over the 20th century. Reprinted from Bratkovich et al. (1994)

does not extend to the bottom, even near the river mouth. On the other hand, the Atchafalaya River plume enters a broader shelf, is more diffuse, and extends to the bottom over a larger distance from the river mouth.

The short distances (10–30 km) separating Mississippi River delta passes from the shelf break facilitate the export of plume waters offshore and to the east by sporadic wind events or by eddies present on the upper continental slope, some of which may have been spun off by the Loop Current (Oey et al., 2005a, 2005b; Ohlmann and Niiler, 2005). The modeling study of Morey et al. (2003a) shows that a prime export pathway for river freshwater during the summer months is to the east and offshore of the Mississippi River delta. During nonsummer months, the main freshwater export pathway consists of a coastal jet flowing westward to Texas and then southward. Etter et al. (2004) estimate that $43 \pm 10\%$ of the Mississippi River discharge is carried westward to the Louisiana–Texas continental shelf, the remainder being carried offshore and/or eastward. While this proportion is slightly lower than the earlier estimate of $53 \pm 10\%$ from Dinnel and Wiseman (1986), both studies indicate that roughly half of the freshwater from the Mississippi River goes westward, toward the Louisiana–Texas continental shelf.

In contrast, 100% of the Atchafalaya River discharge of freshwater, nutrients, and sediments is delivered to the Louisiana–Texas continental shelf. Moreover, the very broad shelf near Atchafalaya Bay implies longer residence times of this freshwater source on the shelf compared with freshwater from the Mississippi River delta. A "back-of-the-envelope" calculation helps capture the full significance of the increased Atchafalaya River flow. In the early 1900s, for every 100 m^3 of water discharged, 85 m^3 took the Mississippi River delta route. Of these, roughly 42.5 m^3 went westward and 42.5 m^3 went offshore or eastward. The 42.5 m^3 that went westward were added to the 15 m^3 that took the Atchafalaya River route to give a grand total of 57.5 m^3 of freshwater on the Louisiana–Texas continental shelf. By contrast, in the post-1970s, for every 100 m^3 of combined Atchafalaya and Mississippi River outflows, 70 m^3 took the Mississippi River route. Of these, roughly 35 m^3 went westward and 35 m^3 went offshore or eastward. The 35 m^3 that went westward were added to the 30 m^3 that took the Atchafalaya River route to give a grand total of 65 m^3 of freshwater on the Louisiana–Texas continental shelf. This simple calculation reveals two things. First, it suggests that even in the absence of a temporal trend in combined Atchafalaya–Mississippi River freshwater discharge, the amount of freshwater delivered to the Louisiana–Texas continental shelf would have increased by 13% ($65/57.5 = 1.13$). Second and more importantly, it reveals that in the 1920s, the Atchafalaya River contributed about one-quarter ($15/57.5 = 0.26$) of the freshwater discharge to the Louisiana–Texas continental shelf. Between 1920 and about 1960, the Atchafalaya River's contribution markedly increased to about one-half ($30/65 = 0.46$) of the freshwater discharge to the Louisiana–Texas continental shelf. While this probably made the Louisiana–Texas continental shelf more prone to hypoxia, the timing of this change occurred 15–20 years earlier than the onset of regular summer hypoxia (Section 2.1.1).

Future physical modeling studies are needed to investigate the effects of past and proposed future changes in the distribution of freshwater flows, including inputs to

Atchafalaya Bay some 200 km to the west of the Mississippi River delta, on changes in the spatial distribution of surface salinity, temperature, and stratification on the Louisiana–Texas continental shelf and on the Mississippi Sound to the east of the "bird's foot" delta. Physical oceanographic models that can adequately answer such questions about the impacts of flow diversions already exist but have only been run using the post-1970s flow conditions (30% Atchalafaya River, 70% Mississippi River). One such modeling study by Hetland and DiMarco (2007) suggests that the freshwater plumes from the Atchafalaya and Mississippi Rivers are often distinct from one another (Fig. 2.3) and that both contribute significantly to the development of hypoxia (Fig. 1.1) on the shelf through their influence on stratification and nutrient delivery (Rabalais et al., 2002). In addition, maps of observed surface salinity and satellite images of chlorophyll (e.g., Fig. 2.7) show the same result. It thus appears likely that increases in freshwater discharge from the Atchafalaya River and resulting increased stratification from the early 1900s to the mid-1970s have increased the area of the Louisiana–Texas continental shelf that is prone to bottom layer hypoxia.

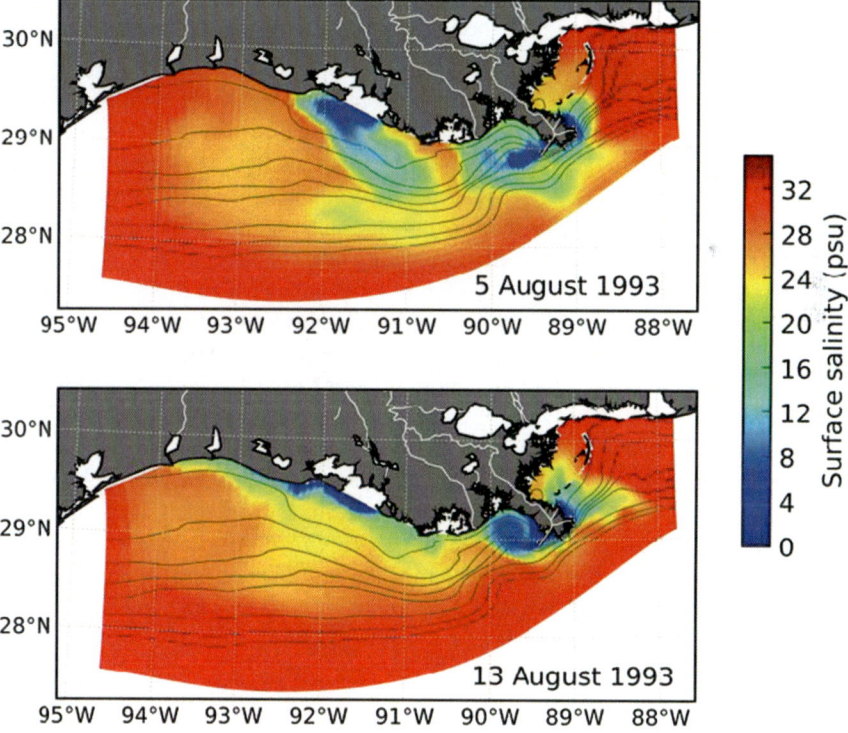

Fig. 2.3 Modeled surface salinity showing the freshwater plumes from the Atchafalaya and Mississippi Rivers during upwelling-favorable winds (top panel) and during downwelling favorable winds 8 days later (bottom panel). Adapted from Hetland and DiMarco (2007)

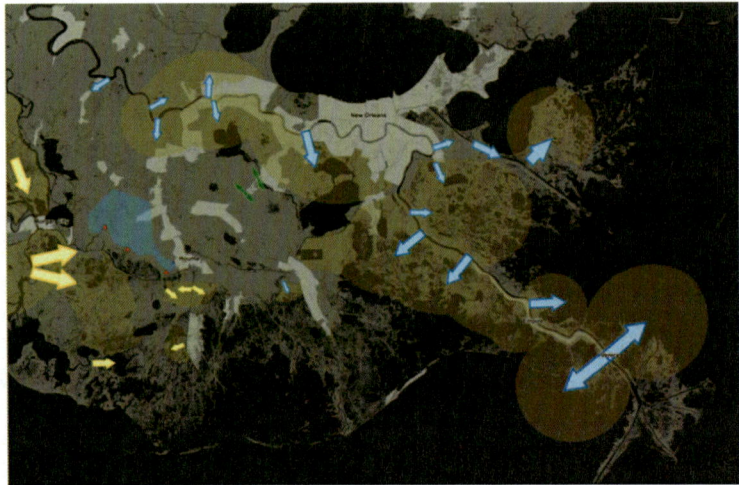

Fig. 2.4 Proposed diversions of Mississippi effluents for coastal protection. From Coastal Protection and Restoration Authority (CPRA) of Louisiana, 2007 Integrated Ecosystem Restoration and Hurricane Protection: Louisiana's Comprehensive Master Plan for a Sustainable Coast. CPRA, Office of the Governor (LA) 117 pp

Recently evolved plans for protecting coastal Louisiana (CPRA, 2007) propose significant diversions of the water, nutrients, and sediment outflow from the Mississippi River into the Gulf. Figure 2.4 illustrates a diversion scenario that involves redirecting a large part of the outflow into shallow bays upstream of the present day "bird's foot" delta. This scenario could alter the shelf hydrodynamics, particularly if more of the buoyancy is directed into shallow water instead of the deep water off the active river mouths, which are near the shelf edge. It is important that three-dimensional numerical circulation models be applied to these scenarios. Future management strategies may be able to utilize engineered modulations of the timing of freshwater releases to coincide more closely with more energetic waves and current conditions, thereby reducing the strength of stratification (i.e., Ri). This approach will, of course, rely on engineering innovations and effective diversion management. The opportunity exists for USEPA and other federal and management agencies to urge flow diversion strategies that also consider the goal of reducing the volume and bottom area of hypoxic waters on the NGOM shelf without endangering other estuarine and coastal waters. The CPRA/US Army Corp of Engineers proposals also highlight the need for interagency coordination and for an integrated approach to management strategies for jointly addressing multiple issues including hypoxia, coastal protection, and coastal inundation.

2.2.4 Zones of Hypoxia Controls

The resulting stratified region influenced by the Mississippi and Atchafalaya River plumes exerts strong control on the extent and spatial distribution of hypoxia and is

2.2 The Physical Context

an important factor in determining where hypoxia may occur (Rabalais and Turner, 2006). The buoyancy fluxes from the rivers also contribute to regional circulation in the form of baroclinic flows (Morey et al., 2003a, 2003b). Following a similar line of reasoning used in earlier work by Rhoads et al. (1985) off the mouth of the Changjiang (Yangtze) River, Rowe and Chapman (2002) defined three zones of hypoxia control in the NGOM. The boundaries between these three zones are admittedly fuzzy and change through time; however, Fig. 2.5 illustrates the Study Group's view of these concepts as represented by four zones. In zone 1, which is most proximal to river mouth sources, strongly stratified and light as well as nutrient limited, respiration of organic carbon coming both directly from the river efflux and from nutrient-dominated eutrophication dominates. The relative importance of these organic carbon sources as the cause of hypoxia remains somewhat uncertain, although the model of Green et al. (2006b) indicates a major dominance by in situ phytoplankton production even in the immediate plume of the Mississippi River. In the intermediate zone 2, stratification is also strong; light limitation is less than in zone 1; very high rates of phytoplankton production occur; and water column respiration fuels bottom layer hypoxia. Farther along the coast from the river mouths but within the low-salinity coastal plume (zone 3), local phytoplankton production is less, but labile organic matter may have been imported from zone 2 and deposited on the bottom. In zone 3, stratification remains strong, and oxygen consumption in the

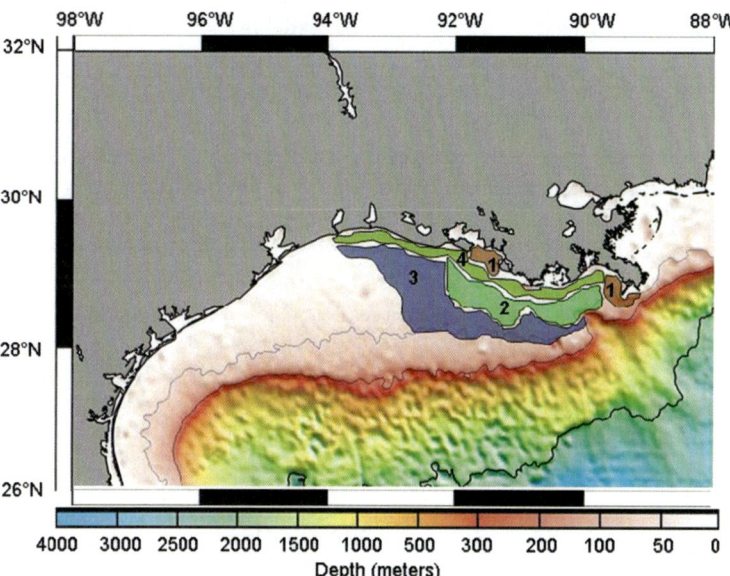

Fig. 2.5 An illustration depicting different zones (Zones 1–4, numbered above) in the NGOM during the period when hypoxia can occur. These zones are controlled by differing physical, chemical, and biological processes, are variable in size, and move temporally and spatially. Diagram created by D. Gilbert

sediment is more important than water column respiration in driving hypoxia. Zone 4 depicts the highly productive, coastal current, as suggested by Boesch (2003).

Boesch (2003) strongly criticized the physical, biological, and chemical reasoning behind the delineation of the Louisiana–Texas continental shelf into these three distinct zones of hypoxia control. He also argued that these zones did not capture well the physics and biology of the Louisiana coastal current, which is characterized by low salinities and high nutrient and chlorophyll levels (Wiseman et al., 2004). Nevertheless, Rowe and Chapman (2002) stimulated new research into the role that stratification plays in the reduction of vertical mixing rates and the flux of oxygen through the pycnocline in the regions of the Louisiana–Texas continental shelf under the influence of the Mississippi and Atchafalaya River plumes. Using realistic three-dimensional physics (Eq. 2.1) with simple representations of water column and benthic respiration for the zones A, B, and C of Rowe and Chapman (2002), Hetland and DiMarco (2007) were able to represent the bottom area, thickness, and volume of hypoxic waters over the NGOM fairly well.

So far as we are aware, time series measurements of physical oceanographic parameters are inadequate to support or refute hypotheses regarding changes in shelf circulation, stratification, and vertical mixing during the 20th century. Initial planning for a Gulf of Mexico Coastal Ocean Observing System (GCOOS) has begun (for additional information see http://www.gcoos.org). As these GCOOS plans continue to evolve and implementation begins over the next few years, it is important that physical parameters relevant to oxygen dynamics be included among the measurements. Empirical parameterizations of vertical eddy diffusivity K_z as a function of vertical shear and density stratification are available for shallow continental shelf environments (MacKinnon and Gregg, 2005). These parameterizations enable quantification of vertical mixing [term (4) in Eq. 2.1] with vertical shear measurements from moored Acoustic Doppler Current Profilers (ADCPs) and vertically profiling conductivity, temperature, and depth instrumentation (CTDs) tethered on a cable. Ship-based microstructure measurements of the turbulent rates of dissipation of velocity, salinity, and temperature fluctuations (Gregg, 1999) should also be conducted occasionally to complement the moored ADCP and profiling CTD measurements. Physics-based models of ocean mixing and turbulence exist today and are part of three-dimensional circulation models (Mellor and Yamada, 1982). These models need to be rigorously tested using ADCP, CTD, and microstructure data because vertical mixing is the most important physical process to model correctly when hypoxia is under consideration.

2.2.5 Shelf Circulation: Local Versus Regional

Circulation in the NGOM can be considered on two scales: Gulf-wide deep-sea circulation and shelf circulation near the coast. Among the most prominent features of the large-scale Gulf-wide circulation are the Loop Current and the Loop Current Eddy System (Oey et al., 2005a, 2005b). Although these features impinge on and affect the outer shelf, Rabalais et al. (1999a) conclude that local wind forcing and buoyancy are more important to shelf circulation inshore of the 50 m isobath. Direct shipboard observations by Jarosz and Murray (2005) during five separate cruises led those authors to conclude that the momentum balance on the inner and mid-shelf to the west of the active "bird's foot" delta is indeed dominated by wind stress. During summer, alongshore sea-surface slope caused by buoyancy forcing was also important in forcing currents. On the 20 m isobath off Terrebonne Bay, ADCP measurements (Wiseman et al., 2004) show periods of several days with negligible vertical shear followed by other periods of a few days with much more elevated vertical shear and reduced density gradients, suggestive of more intense vertical mixing.

Several physical oceanographic models taking into account the crucial baroclinic effects that typify the Louisiana–Texas continental shelf are now available (e.g., Morey et al., 2003a, 2003b; Zavala-Hidalgo et al., 2003). The model results of Hetland and DiMarco (2007) show that the plume from the Mississippi River, which enters the shelf near the shelf edge, forms a recirculating gyre in Louisiana Bight and does not interact with the seabed, whereas the Atchafalaya River plume interacts with the shallow coastal topography (Hetland and DiMarco, 2007). Both plumes respond directly to local winds and are advected seaward during upwelling-favorable winds (Fig. 2.3). The distinct plumes from the Mississippi and Atchafalaya Rivers influence the spatial pattern of bottom hypoxia on the Louisiana–Texas continental shelf. This influence is clearly seen on the 1985–2005 map of hypoxia frequency of occurrence (Fig. 1.1) and is even more obvious in certain years (e.g., 1986, Rabalais and Turner, 2006). Given this interaction, planned diversions of Mississippi River and Atchafalaya River flow may alter shelf circulation and the spatial pattern of bottom hypoxia. Applications of three-dimensional baroclinic models to future scenarios such as that portrayed in Fig. 2.4 are thus important to planning for future strategies for coastal restoration (CPRA, 2007).

In their analysis of low-frequency (occurring over a timescale greater than 24 h) currents over the shelf, Nowlin et al. (2005) distinguished between currents that respond within the "weather band" of 2–10 days and those within the mesoscale band of 10–100 days corresponding to large-scale eddies off the shelf. Inshore of the 50 m isobath, the local winds within the weather band dominated and drove currents from east to west during nonsummer months influenced by the passage of frontal systems. Current fluctuations seaward of the 50 m isobath were primarily within the mesoscale band and predominantly oriented from west to east but with high variability. Along-shelf and across-shelf currents in the upper layer over the inner shelf, as reported by Nowlin et al. (2005), averaged about 10 and 1 cm/s, respectively. Over the outer shelf and near the seabed, flows were weaker.

Key Findings and Recommendations

The Study Group finds that 20th century changes in the hydrologic regime of the Mississippi and Atchalafaya Rivers and the timing of freshwater inputs to the Louisiana–Texas continental shelf have likely increased the shelf area with potential for hypoxia, although these changes occurred mostly from the 1920s to the 1960s, before the measured onset of hypoxia in the mid-1970s. Additional work is needed to advance the understanding of the relative importance of physical factors in the formation of hypoxia in the NGOM. The Study Group therefore provides the following recommendations.

- The development of a new suite of models that integrate physics and biogeochemistry should be encouraged and supported. This suite should include multiple types of models [i.e., relatively simple models such as those developed by Scavia et al. (2003) as well as more complex three-dimensional types, such as Hetland and DiMarco (2007)].
- A comparative impact study of past, present, and future river flow diversions and scenarios of altered nutrient supply to the river mouths should be encouraged and supported. Three-dimensional hydrodynamic modeling studies are needed to compare the spatial distribution of salinity and stratification with 15% (early 1900s) and 30% (post-1970s) Atchafalaya River contributions to the combined Atchafalaya–Mississippi River outflow. Coupling of this three-dimensional hydrodynamic model with a biogeochemical model would allow quantification of the impacts of past river flow diversions on the spatiotemporal extent of hypoxia. In addition, to anticipate the possible effects of proposed future effluent diversion plans via rerouted deltaic distributaries (CPRA, 2007), these three-dimensional biogeochemical and baroclinic shelf circulation models need to be applied to scenarios such as that shown in Fig. 2.4 while also considering the effects of nutrient-rich Mississippi River waters discharged into local bays and estuaries.
- Emerging coastal ocean observing and predicting systems in the Gulf of Mexico (http://www.gcoos.org) should be encouraged to measure and disseminate information needed by hypoxia modelers and those charged with adaptive management. Direct measurements of physical and biogeochemical parameters as well as direct time series measurement of dissolved oxygen in the bottom boundary layer should be routinely provided by the next generation of shelf moorings.
- Studies of turbulent mixing processes involving the effects of stratification over the Louisiana–Texas shelf with instruments and techniques capable of quantifying turbulent dissipation rates of velocity, salinity, and temperature fluctuations should also be encouraged. Studies of the importance of lateral mixing processes should be encouraged.

2.3 Role of N and P in Controlling Primary Production

2.3.1 Nitrogen and Phosphorus Fluxes to the NGOM Background

Excessive nutrient loading, dominated by discharge from the MARB, enhances planktonic primary production in the shallow near-shore receiving waters of the NGOM (Lohrenz et al., 1990, 1992; Rabalais et al., 1999a; Turner and Rabalais, 1994). The nutrients of concern are nitrogen (N), phosphorus (P), and silicon (Si) in the form of silicate. Both primary productivity and phytoplankton biomass are stimulated by these nutrient sources (Ammerman and Sylvan, 2004; Lohrenz et al., 1990, 1992; Sylvan et al., 2006). The spatial and temporal extent and magnitudes of this stimulation vary significantly, and their patterns and size appear to be related to (1) amounts of freshwater discharge and their nutrient loads; (2) the nature and frequencies of discharge (i.e., acute, storm- and flood-based versus more gradual, chronic, seasonal discharge); and (3) the direction and spatial patterns of discharge plumes as they enter and disperse in the NGOM (Justić et al., 1993; Lohrenz et al., 1994; Rabalais et al., 1999b). The *Integrated Assessment* concluded that N loading from the MARB was the primary driver for hypoxia in the NGOM. Since the *Integrated Assessment*, however, considerable knowledge has been gained concerning the processes that influence primary production and the relative importance of elements other than N as is discussed below.

A proportion of the freshwater discharge transits via freshwater and coastal wetlands and coastal groundwater aquifers, which modify the concentrations and total loads of nutrients entering the NGOM (Day et al., 2003; Turner, 2005). The extent to which wetlands alter nutrient loads and the effects wetland losses have had on changes in nutrient processing and loading are subjects of considerable debate (Day et al., 2003; Mitsch et al., 2001; Turner, 2005). Nutrients can also enter this region from deeper offshore sources, by advective transport over the shelf, a modified form of "upwelling" (Cai and Lohrenz, 2005; Chen et al., 2000), although this input is estimated to be only 7% of the nitrogen coming down the Mississippi River (Howarth, 1998). Lastly, nutrients can be derived from atmospheric deposition directly onto nutrient-sensitive NGOM waters (deposition onto the MARB and subsequent downstream export to the Gulf is considered in later sections). For nitrogen, this direct deposition is estimated to be 13% of the amount of nitrogen that flows down the river (Howarth, 1998).

Historic analyses indicate a great deal of variability in seasonal, interannual, and decadal-scale patterns and amounts of freshwater and nutrient discharge to the NGOM (Rabalais et al., 2002; Turner and Rabalais, 1991). As a result, primary productivity and phytoplankton biomass response can vary dramatically on similar timescales, which poses a significant challenge to interpreting trends in nutrient-driven eutrophication in the NGOM as in other systems (Boynton and Kemp, 2000; Harding, 1994; Paerl et al., 2006b). Furthermore, in the turbid and highly colored waters (containing colored dissolved organic matter or CDOM) of the river plumes entering the NGOM, nutrient and light availability strongly interact as controls of primary production and biomass. These interactive controls modulate the relationships between nutrient inputs and phytoplankton growth responses in this region (Justić et al., 2003a, 2003b; Lohrenz et al., 1994). Ultimately these interactions affect the formation and fate of autochthonously produced organic carbon that provides an important source of the "fuel" for bottom water hypoxia in this region.

2.3.2 N and P Limitation in Different Shelf Zones and Linkages Between High Primary Production Inshore and the Hypoxic Regions Farther Offshore

Physically, chemically, and biologically, the NGOM region is highly complex, and nutrient limitation reflects this complexity. Along the freshwater to full-salinity hydrologic continuum representing the coastal NGOM influenced by river discharge, ratios of nutrient concentrations vary significantly, both in time and in space. For example, depending on the season, specific hydrologic events, and conditions (storms, floods, droughts), molar ratios of total N to P (N:P) supplied to these waters can vary from over 300 to less than 5 (Ammerman and Sylvan, 2004; Sylvan et al., 2006; Turner et al., 1999; Turner et al., 2007a). Furthermore, additional environmental factors, such as flushing rate (residence time), turbidity and water color (light limitation), internal nutrient recycling, and vertical mixing, strongly interact to determine which nutrient(s) may be controlling primary production (Lohrenz et al., 1999b). Compounding this complexity is the frequent spatial separation among high nutrient loads, the zones of maximum productivity, and hypoxia (e.g., Fig. 2.5). Conceivably, primary production and algal biomass accumulation limited by a specific nutrient in the river plume region near-shore may constitute the "fuel" for hypoxia further offshore in the next zone, where productivity in the overlying water column may be limited by another nutrient. Limitation by different nutrients in different areas appears to be the case during the spring to summer transitional period, when primary production in the river plume region near-shore is P limited (Ammerman and Sylvan, 2004; Lohrenz et al., 1992, 1997; Sylvan et al., 2006), but offshore productivity is largely N limited (Dortch and Whitledge, 1992; Lohrenz 1992, 1997). The relevant questions concerning causes of hypoxia are what are the relative amounts of inshore river plume (largely P-limited) versus offshore (largely N-limited) productivity and what roles do these different sources of productivity play in "fueling" hypoxia?

Early work on NGOM nutrient limitation tended to focus on the waters overlying the hypoxic zone; typically, these waters are over the shelf but farther offshore than the river plume waters. Stoichiometric N:P ratios indicated that, during summer months when hypoxia was most pronounced, N should be the most limiting nutrient (Justić et al., 1995; Rabalais et al., 2002). This work has been the basis for the general conclusion that N is most limiting and that reductions in N loading would be most effective in reducing "new" carbon (C) fixation and resultant phytoplankton biomass supporting hypoxia (Rabalais et al., 2002, 2004). This conclusion, coupled with the nutrient loading trend data over the past 40–50 years, which showed N loading increasing more rapidly than P loading, has formed the basis for arguing that N input reductions would be most effective in reducing the eutrophication potential and hence formation of "new" C supporting hypoxic conditions. The 2000 report from the National Academy of Sciences' Committee on Causes and Management of Coastal Eutrophication (National Research Council, 2000) concluded that nitrogen is the primary cause of eutrophication in most coastal marine systems in the United States at salinities greater than 5–10 parts per thousand (ppt), including the NGOM.

While it is likely that N limitation characterizes coastal shelf and offshore waters, more recent nutrient addition bioassays (Ammerman and Sylvan, 2004; Sylvan et al., 2006) and examinations of nutrient stoichiometric ratios have shown that river plume-influenced inshore productivity appears to be more P limited, especially during periods of highest productivity and phytoplankton biomass formation (February–May) (Fig. 2.6) when freshwater discharge and total nutrient loading are also highest (Lohrenz et al., 1999a, 1999b; Sylvan et al., 2006).

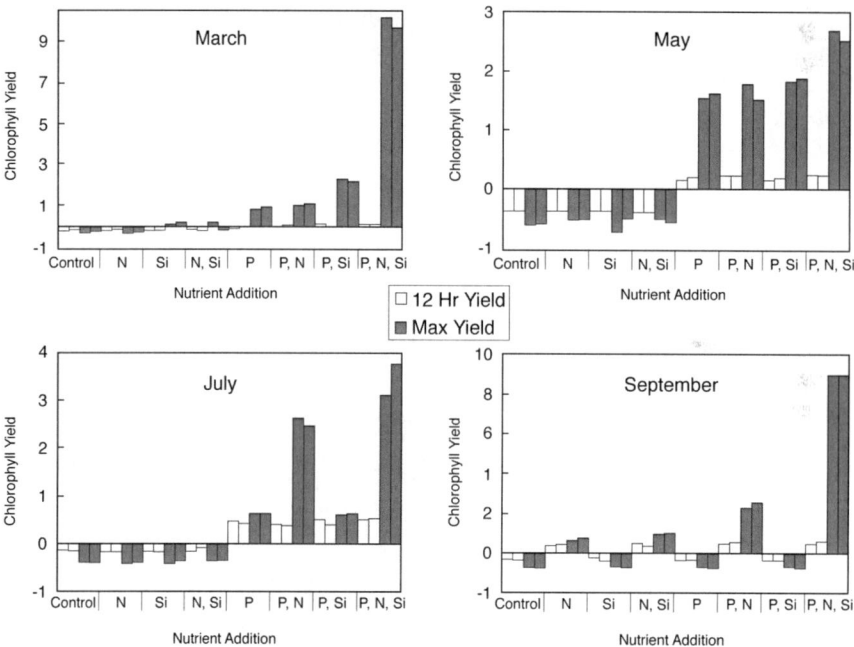

Fig. 2.6 Response of natural phytoplankton assemblages from coastal NGOM stations to nutrient additions, March through September. All experiments, except those done in September, indicate a strong response to P additions. Taken from Sylvan et al., 2006

The strong P limitation during this period appears to be a result of the very high rates of N loading that have increased more rapidly than P loading over recent history (the past 50 years) (Turner et al., 1999; Turner and Rabalais, 1991). This situation is exacerbated during periods of high freshwater runoff, which typically contain very high N:P ratios. Primary productivity in the river plume region nearshore tends to shift into a more N-limited mode once freshwater discharge decreases during the drier summer–fall period (June–October). However, total primary production and phytoplankton biomass accumulation are far lower during this more N-limited period than during the earlier P-limited period. Overall, maximum "new" organic C formation in recent years tends to coincide with periods of highest N:P, which are P limited (Ammerman and Sylvan, 2004; Lohrenz et al., 1992, 1997, 1999a; Sylvan et al., 2006).

Field data and remote-sensing imagery indicate that in situ phytoplankton biomass (as chlorophyll *a*) concentrations can be quite high in river plume-influenced inshore waters that have been shown to be P limited. This pattern is evident in Fig. 2.7, an image provided by the National Oceanic and Atmospheric Administration Sea-viewing Wide Field-of-view Sensor Project (NASA-SeaWiFS, 2007). Therefore, the following question emerges. What is the spatiotemporal linkage of this P-limited high primary production and phytoplankton biomass accumulation to hypoxic bottom waters located further offshore? Furthermore, what are the relationships between N-limited production later in the summer and hypoxic conditions, which typically are most extensive during this period? These potential "relationships" are complicated by the fact that there are strong, co-occurring physical drivers of hypoxia, including vertical density stratification and respiration rates, which tend to be maximal during periods of maximum development of hypoxia (c.f. Hetland and DiMarco, 2008; Rowe and Chapman, 2002; Wiseman et al., 2004).

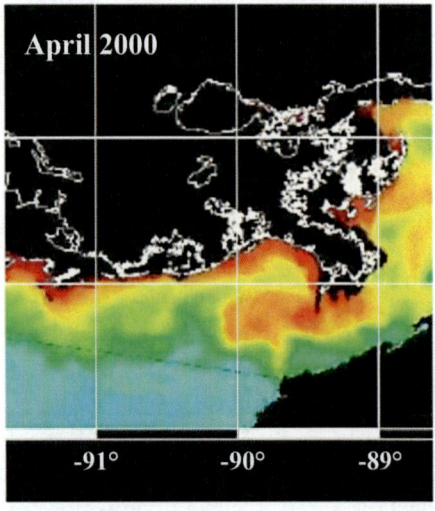

Fig. 2.7 NASA-SeaWiFS image of the Northern Gulf of Mexico recorded in April, 2000. This image shows the distributions and relative concentrations of chlorophyll *a*, an indicator of phytoplankton biomass in this region. Note the very high concentrations (orange to red) present in the inshore regions of the mouths of the Mississippi and Atchafalaya Rivers

2.3 Role of N and P in Controlling Primary Production

There are likely to be periods when both P and N are supplied at very low levels and co-limit phytoplankton production. These periods occur during the transition from spring to summer. A similar condition is observed in large estuarine systems with a history of eutrophication, such as Chesapeake Bay (Fisher et al., 1992). Spatially, the upstream, freshwater segments of Chesapeake Bay tend to be most P limited, especially during spring runoff conditions, while the more saline down-estuarine waters tend to be most N limited. In Chesapeake Bay, the more turbid upstream freshwater component tends to exhibit interactive light and P limitation or N+P co-limitation (Fisher et al., 1992; Harding et al., 2002). Farther downstream, light limitation plays a less important role. This scenario could prove similar to the riverine-coastal continuum in the NGOM, where the most turbid upstream river plume waters are likely to exhibit the highest probability for light-nutrient interactive limitation of primary production (Lohrenz et al., 1999a, b).

While bioassay data tend to indicate P limitation during springtime in the lower salinity portions of this continuum and N and P co-limitation and N limitation in the more saline offshore waters during summer months, the bioassays do not account for sediment–water column exchange because sediments are excluded during the course of incubation. It is possible, although unlikely because of short incubation times, that sediment–water column P cycling in the shallow NGOM water column may minimize P limitation in situ. In order for this scenario to be operative, parallel N recycling would have to be far less efficient than P cycling, which numerous studies suggest is the case (Bode and Dortch, 1996; Cai and Lohrenz, 2005; Gardner et al., 1994; Jochem et al., 2004; Pakulski et al., 2000; Wawrik et al., 2004). Bioassay-based N limitation results might also be influenced by the elimination of "internal" sediment–water column N recycling, although this situation seems unlikely as well, especially if denitrification is operative (Childs et al., 2002). Sediment-based denitrification would lead to N "losses" from the system, thereby exacerbating N limitation. This influence would not be captured in bioassays, which isolate the sediments from the water column during incubation. The relatively short incubation times of bioassays probably preclude these potential artifacts. They offer a "snapshot" of nutrient limitation to complement longer-term, ecosystem-scale assessments.

The degree of N and P limitation can be calculated from bioassays, and the data can be used to create ratios of N and P limitation (Dodds et al., 2004). Interestingly, N and P limitation inferred from stoichiometric ratios of soluble (and hence biologically available) inorganic or total N or P concentrations and inputs (loads) tends to confirm bioassay-based conclusions concerning specific nutrient limitations. For example, inshore, river-influenced waters exhibit quite high molar N:P ratios, often exceeding 50 [Nutrient Enhanced Coastal Ocean Productivity (NECOP) Reports, NOAA, 2007]. Nutrient addition bioassays initially conducted in these waters by Lohrenz et al. (1999a) and more recently by Sylvan et al. (2006), consistently revealed P limitation, especially during spring periods of maximum primary production and phytoplankton biomass accumulation. These same studies also indicated a tendency toward N and P co-limitation and exclusive N limitation during later summer months, when soluble and total N:P values dipped below 15. It should also be noted, however, that rates of primary production and phytoplankton biomass during this more N-limited period are at least 5-fold lower than spring values, according the Gulf of Mexico NECOP data (Lohrenz et al., 1999a, b). Sylvan et al. (2006) point out that P-limited spring production of "new" C may play a proportionately greater role than N-limited summer production as a source of "fuel" supporting hypoxia in the NGOM. The degree and extent to which C from this nutrient-enhanced elevated spring production is transported and accounts for summer hypoxia need to be quantified. Developing an understanding of processes that link zones and periods of high primary production and phytoplankton biomass to zones exhibiting bottom water hypoxia is a fundamentally important and challenging area of research. Such research is necessary to improve understanding of the linkage between nutrient-enhanced production and bottom water hypoxia in the NGOM. Extrapolation of C production to hypoxia data along the entire riverine-coastal shelf continuum, where zones and periods of maximum productivity and bottom water hypoxia do not necessarily coincide or overlap, depends on knowing C transport and storage (including burial), internal nutrient, and C cycling and C consumption (heterotrophic metabolism and respiration) processes along this continuum (Cai and Lohrenz, 2005; Redalje et al., 1992). Quantifying the links between locations and periods of specific nutrient limitation (or stimulation) of production and the fate of this production relative to hypoxia will contribute to long-term, effective nutrient management strategies for this region.

Key Findings and Recommendations

The Study Group finds that there is compelling evidence that the near-shore Mississippi/Atchafalaya River plume-influenced waters are P limited and P–N co-limited during the spring periods of highest primary production. Nitrogen limitation of primary production prevails during summer periods. Recent research results indicate that the spring period of maximum primary production is P limited in at least the plumes of the rivers, largely due to excessive N input. As a result of this man-made imbalance in nutrient loading during this crucial period, P availability plays an important role in contributing to the production of "new" organic carbon in the spring time and quite likely contributing in a major way to the "fueling" of summer hypoxia in the NGOM. However, as stressed elsewhere in this book, there is great uncertainty over the coupling in space or time of phytoplankton production and its decomposition leading to hypoxia. Therefore, a better understanding of the spatial extent and temporal patterns of these nutrient limitations is needed. The Study Group recommends that the following work be undertaken to advance knowledge of the importance of nutrient limitation and co-limitation as factors in the formation of Gulf hypoxia.

- Research should be conducted to develop a more complete understanding of the spatial and temporal linkages between river plume-influenced inshore P (in spring) and/or N-limited (in summer) primary production, and offshore coastal shelf, more N-limited production, as well the fate of C produced in each zone throughout the year.
- Research should be conducted to link in time and space near-shore river-plume-influenced production to O_2 depletion farther offshore. Green et al. (2006b) suggest that the small region that the central Mississippi River plume could supply is responsible for about 25% of the C necessary to fuel hypoxia. The role of the Atchafalaya plume and other riverine influenced inshore high productivity regions in offshore hypoxia needs to be clarified.
- Research should be conducted to address the following questions. How closely linked are the periods of high productivity and hypoxic events throughout the regions in which they occur? What is the lag between C production and its ultimate degradation?

2.4 Other Limiting Factors and the Role of Si

While excessive N and P loading are implicated in eutrophication of the NGOM, these nutrients also play a role in the balance, availability, and ecological manifestations of other potentially limiting nutrients, most notably Si. In the Mississippi River plume region, N is supplied in excess of the stoichiometric nutrient ratios needed to support phytoplankton and higher plant growth (i.e., Redfield ratio, Redfield, 1958). If N over-enrichment persists for days to weeks, other nutrient limitations may, at times, result and seasonally dominate; the most obvious and important is P limitation, which has recently been demonstrated in bioassays (Ammerman and Sylvan, 2004; Sylvan et al., 2006). In addition to P limitation, N and P co-limitation and Si limitation (of diatom growth) have been observed in the fresh and brackish water components of riverine plumes that can extend more than 100 km into the receiving waters (Dortch et al., 2001; Dortch and Whitledge, 1992; Lohrenz et al., 1999a). A similar scenario is evident in the Chesapeake Bay, where elevated N loading accompanying the spring maximal freshwater runoff period increases the potential for P limitation (Fisher and Gustafson, 2004). The biogeochemical and trophic ramifications of such shifts are discussed below.

With regard to nutrient primary production interactions, it is important to know who the dominant primary producers are, where they reside, what their contributions to new production are, and what their fate is. In NGOM waters downstream of the rivers, wetlands, and intertidal regions, microalgae are by far the dominant primary producers (Lohrenz et al., 1992, 1997; Rabalais et al., 1999a; Redalje et al., 1992). The microalgal communities are dominated by phytoplankton (Chen et al.,

2000; Redalje et al., 1994a, 1994b), although benthic microalgal communities can also be important sites of primary production and nutrient cycling, especially in near-shore regions (Jochem et al., 2004). As nutrient loads and limitations change over time and space, the proportions of planktonic versus benthic microalgae may also change; i.e., as nutrient inputs are reduced and planktonic primary production is reduced, the microalgal community may shift to a more benthic-dominated one. This process could yield significant implications for biogeochemical (nutrients, carbon, and oxygen) cycling and trophodynamics (Darrow et al., 2003; Rizzo et al., 1992).

Historic and contemporary evidence supports the contention that anthropogenically and climatically induced changes in N and P loading have increased NGOM primary productivity and phytoplankton biomass and altered phytoplankton community composition. There are several reasons why phytoplankton community composition may have been altered by changes in nutrient loading: (1) competitive interactions among phytoplankton taxa based on varying nutrient supply rates and differing affinities for nutrient uptake and assimilation (i.e., varying nutrient uptake affinities and kinetics); (2) competitive interactions based on the relationships between nutrient supply rates and photosynthetically available light (i.e., low versus high light adapted taxa); (3) competitive interactions based on changes in N versus P supply rates (e.g., differential N versus P uptake capabilities and selection for nitrogen fixing cyanobacteria); (4) competition based on the ratios of N and P versus Si (silicious versus nonsilicious taxa and heavily versus lightly silicified diatoms); (5) differential grazing on phytoplankton taxa (top-down controls); and (6) nutrient-salinity controls (interactive effects of changes in freshwater discharge on NGOM salinity and nutrient regimes due to climatic and watershed hydrologic control changes). Each set of controls can influence the amounts and composition of primary producers. These controls can also interact in time and space, greatly compounding and confounding the interpretation of their combined effects.

One important aspect of differential nutrient loading is the well-documented increase in N and P relative to Si loading. While N and P loads tend to reflect human activities in and alterations of the watershed, Si loads tend to reflect the mineral (bedrock and soil) composition of the watershed; a geochemical aspect that is less influenced by human watershed perturbations. Agricultural, urban, and industrial development and hydrologic alterations in the MARB have led to dramatic increases in N and P relative to Si loading. In addition, the construction of reservoirs on tributaries of these river systems has further exacerbated this situation by trapping Si relative to N and P. This anthropogenic biogeochemical change has been shown to alter phytoplankton community structure (i.e., away from diatom dominance), with subsequent impacts on nutrient and carbon cycling and food web dynamics (Humborg et al., 2000; Ragueneau et al., 2006a, 2006b). The overall result has been an increase in N:Si and P:Si ratios that can influence both the amounts and the composition of phytoplankton; including potential shifts from diatoms to flagellates and dinoflagellates (Justić et al., 1995; Rabalais and Turner, 2001; Turner et al., 1998). Diatoms are a highly desired food item for a variety of planktonic and benthic grazers, including key zooplankton species serving an intermediate role in the NGOM food web (Dagg, 1995). The dinoflagellates, cyanobacteria, and even a few diatom species, while serving important roles in the food web, also contain species that may be toxic and/or inedible (Anderson and Garrison, 1997; Paerl and Fulton, 2006). Some of these species can rapidly proliferate or "bloom" under nutrient sufficient and enriched conditions, and thus constitute harmful algal bloom (HAB) species. Toxicity may directly and negatively impact consumers of phytoplankton as well as higher-ranked consumers, including finfish, shellfish, and mammals (including humans). If nontoxic but inedible (due to size, shape, coloniality) phytoplankton taxa increase in dominance, trophic transfer may be impaired. Planktonic invertebrates, shellfish, and finfish consumers (whose diets are highly dependent on the composition and abundance of specific phytoplankton food species and groups) may then be affected (Turner et al., 1998). This could have consequences for C flux, with a relatively higher fraction of C being processed through microbial pathways (i.e., the "microbial loop") or sedimented to the bottom. In either case, a greater fraction of the primary production would remain in the system, as opposed to being exported out of the system by transfer to higher trophic level and fisheries. The net result would be more C metabolized within the system, leading to enhanced oxygen consumption and increased hypoxia potentials.

Key Findings and Recommendations

Research has shown the potential importance of silicate in structuring phytoplankton communities. Based on this finding, the Study Group offers the following recommendation.

- The potential for silicate limitation and its effects on phytoplankton production and composition on the Louisiana–Texas continental shelf should be explored when carrying out experiments on the importance of N and P as limiting factors and when considering nutrient management scenarios.

2.5 Sources of Organic Matter to the Hypoxic Zone

As noted earlier, the physical and geomorphological conditions found along the Louisiana coast make the NGOM prone to hypoxic conditions if there is an organic matter supply sufficient to consume deep water dissolved oxygen (DO) at rates exceeding DO replenishment rates. Ecosystems such as the NGOM shelf have available to them an array of organic matter sources, including those transported from the basin by rivers and those produced in situ. These include particulate and dissolved organic carbon/colored dissolved organic matter (POC and DOC/CDOM) from terrestrial sources in the basin, POC, and DOC from coastal wetland losses, and in situ production by phytoplankton, macrophytes, and benthic microalgae.

The *Integrated Assessment* largely supported the argument that hypoxia in the NGOM was driven by increased N loading to the Gulf of Mexico, which, in turn, stimulated increased in situ phytoplanktonic production of labile (i.e., readily decomposed) organic matter. A portion of this organic matter sinks to deeper, sub-pycnoclinal waters and is used by the heterotrophic community at rates sufficient to deplete DO concentrations to hypoxic levels. Emphasis at that time focused on N, but more recent work has indicated that P also plays a role in regulating organic matter (OM) supply from phytoplankton (see Section 2.1.3). In addition, a number of investigators have noted that changes in the relative supply rates of N, P, and Si lead to changes in species composition of phytoplankton communities, and this would likely modify some aspects of deposition of OM to deep waters. Substantial rates of primary production have been measured along the NGOM shelf, and these rates are comparable to those observed in other eutrophic coastal systems (e.g., Lohrenz et al., 1990, 1997; Nixon, 1992).

In Rabalais et al. (1999a) and the *Integrated Assessment*, organic matter from the major rivers was discounted as a major source because (1) there have not been changes in river OM loads since the beginning of the hypoxic period that account for the current hypoxic zone size and expansion; (2) dissolved organic matter (DOM) sources from rivers, while large, would need to be converted into particulate forms, with attendant losses from this microbial transformation, and hence would be much reduced; (3) much, but not all, of this terrestrially derived material is far less labile than phytoplanktonic debris and hence is not readily respired at timescales associated with shelf hypoxia (weeks to months). Using an estimated annual load of river OM ($\sim 2.6 \times 10^{12}$ g C/year) delivered to an average hypoxic area (15,000 km^2), and assuming that even as much as 30% of this material were labile, suggests a small impact on DO conditions (~ 0.3 g O_2/m^2/day). Additionally, while there is substantial POC and DOC coming down the Mississippi River, there was undoubtedly far more 100–130 years ago when the Mississippi River basin was first cleared for agriculture and before the dams in the basin were built. While this process apparently has not been modeled in the Mississippi River basin, modeling in other basins strongly suggests a huge increase in organic carbon fluxes at the time of land-use conversion to agriculture, followed by decreasing fluxes as agricultural practices improve (Swaney et al., 1996), and globally the flux of carbon in rivers is tied to agricultural land use (Schlesinger and Melack, 1981). This historical land-use change may well have contributed to the paucity of low oxygen conditions seen in the paleoecological record in the late 1800s (Osterman et al., 2005). Given this historical pattern, Mississippi River derived OM is unlikely to be the trigger for the level of hypoxia that developed in the NGOM during the past 35 years. This period does coincide well with the time N loads increased, due mainly to the use of synthetic N fertilizer in the Mississippi River basin. Given experience in many other coastal and estuarine regions (e.g., National Research Council, 2000), there are strong reasons to believe that in situ NGOM primary productivity exploded in response to increased N inputs over this timescale.

2.5 Sources of Organic Matter to the Hypoxic Zone   33

The influence of organic matter losses from coastal wetlands on coastal hypoxia is still debated but seems unlikely to be a primary factor. Whether or not wetlands lose more organic C as they degrade is not well known, but at present this also seems unlikely. While the timing of wetland loss does not coincide with the onset of hypoxia in the 1970s (marsh loss has been occurring since the 1940s), stable isotope and lignin analyses of OM over much of the shelf indicates that terrestrially derived OM is dispersed along and across the shelf (Goni et al., 1998; Gordon et al., 2001). However, marsh particulate organic material is refractory (i.e., resistant to decay) and does not contribute much to hypoxia creation on timescales of weeks to months. Thus, while the conclusion that the main OM source fueling hypoxia is in situ production of marine phytoplankton and that this production increased in response to enhanced nutrient loads from the MARB remains sound, a better understanding of the possible role of other sources would further refine understanding of hypoxia.

2.5.1 Sources of Organic Matter to NGOM: Post 2000 Integrated Assessment

Since the *Integrated Assessment*, there has been substantial research activity in the NGOM regarding organic matter sources, characterization of organic matter, and related issues. Some of this new work has utilized advanced analytical methods and improved field techniques. However, as with the advent of sophisticated imaging devices in medicine, where small and interesting structures in the human body can now be readily observed but not necessarily interpreted in terms of health threats, in marine waters we now have an emerging and more detailed description of the complex mix of organic compounds, which has in the past simply been called organic matter. But it is not yet clear how important some of this material is with respect to hypoxia issues. This elaboration of understanding of OM adds interesting and useful dimensions to this story but does not change the basic theme, which is that enhanced phytoplanktonic production, based on much increased nutrient loading, is the main biological trigger of NGOM hypoxia.

In addition, there have been at least two varieties of what can be called synthesis studies. Studies of the first variety tend to be "review like" wherein the growing time series of observations and new data have been revisited and/or reanalyzed. Several other efforts of this type have also developed revised conceptual models of the role of OM in hypoxia, and these will prove especially useful in time. Studies of the second variety, and these are rarer, involve development of quantitative budgets or models of various sorts. These efforts indicate that the information base regarding many aspects of OM and hypoxia is rich enough to begin these more rigorous examinations. But, in virtually all these efforts, authors conclude that results are preliminary and that more process-based information is critically needed.

2.5.2 Advances in Organic Matter Understanding: Characterization and Processes

A detailed review of these diverse studies is beyond the scope of this effort. However, Table 2.1 summarizes a selection of those works to provide an indication of the diversity of information that is becoming available. Some findings of particular relevance to OM sources are provided below:

Table 2.1 A partial summary of papers published following the *Integrated Assessment* related to sources of organic matter to the Gulf of Mexico

General topics and issues	Comments regarding OM/hypoxia	References
Landside sources		
POC in river sands	Similar in magnitude to suspended POC load in river	Bianchi et al., 2002
Sedimentation of river POC	High deposition of terrestrial POC in plume region	Corbett et al., 2004
Relict peats	Source of old organic matter to plume area	Galler et al., 2003
Seasonal transport of POC	Fluid muds are transported seasonally to GOM	McKee et al., 2004
Sediment storage and transport	Seasonal transport of mobile muds from delta to shelf	Corbett et al., 2006
River OM loads	DOC and DON loads to GOM	Bianchi et al., 2004; Duan et al., 2007
River inputs	Transport of river diatoms to plume area	Duan and Bianchi, 2006; Wysocki et al., 2006
Terrestrial OM	Fate of lignin	Hernes and Benner, 2003
Riverine DON	Photoammonification of DON to DIN	Pakulski et al., 2000
Riverine OM and nutrients	Effects of flow through coastal wetlands	Xu, 2006
Riverine DOM	CDOM analysis	Chen and Gardner, 2004
Marsh/estuary DOC	High DOC concentrations in these systems	Engelhaupt and Bianchi, 2001
OM distribution	Sources and fate of OM from rivers to shelf	Gordon et al., 2001
Water column/sediment processes		
Flocculation and sedimentation	Enhanced process in plume area; high rates	Dagg et al., 2004
Light field	Light absorption/scattering limiting production	D'Sa and Miller, 2003
Plankton characteristics	Satellite-based relations between N loads and chlorophyll	Walker and Rabalais, 2006
Plume budget	CO_2 budget in plume	Cai, 2003
OM source	High rates of plankton production west of plume	Dagg et al., 2007
Deposition	Influence of larvaceans on deposition	Dagg and Brown, 2005

2.5 Sources of Organic Matter to the Hypoxic Zone

Table 2.1 (continued)

General topics and issues	Comments regarding OM/hypoxia	References
DOM characteristics	Lability of DOM in region II	Benner and Opsahl, 2001
Sediment DOC	Release of DOC from shelf sediments	Sutula et al., 2004
Fate of benthic diatoms	Benthic diatom shunted to MR canyon; cleansing effect	Bianchi et al., 2006
Hurricane effects	Storm transport of deposited materials – decadal scale	Corbett et al., 2006
Sediment processes	Ammonium flux from sediments important for plankton	Eldridge and Morse, 2008
Plankton composition	Diatom occurrence in western regions of hypoxic zone	Wawrik et al., 2004
Plankton composition	Microbial processes in shelf waters	Liu et al., 2004
Synthesis/overviews		
OM budget	Carbon budget for plume area	Green et al., 2006b
Conceptual model/synthesis	Planktonic dynamics of region outside plume	Dagg and Breed, 2003
Model analysis	Differences between water and sediment respiration	Hetland and DiMarco, 2007
Statistical model	Relates N load to hypoxia; phytoplankton OM implied	Scavia et al., 2003
Water column synthesis	Plume contributions to hypoxia; gaps in understanding	Dagg et al., 2007
Review/synthesis	New monitoring data strengthens nutrient/hypoxia model	Rabalais et al., 2007a
Nutrient/Organic loads	Confirms *Integrated Assessment*, wetland loss small OM source	Turner et al., 2007
Forecasting hypoxia	Examines models and suggests nutrients major driver	Justić et al., 2007
Primary production-nitrate model	Model indicates buffered response to N-load reductions	Green et al., 2008
Concepts of hypoxic zones	Suggests spatial dimensions/processes in hypoxic zones	Rowe and Chapman, 2002

*Entries are shown for a variety of topics and comments are focused on issues related to organic matter in the GOM. This table is not a complete summary of all papers published on this subject; rather it provides an indication of the great diversity of studies conducted since the *Integrated Assessment*.

- POC associated with sand transport in bottom waters in the lower Mississippi River is similar in magnitude to loading of suspended POC (Bianchi et al., 2007).
- The vertical flux of terrestrially derived particles in the Mississippi River plume is typically very high and mainly deposits locally (Corbett et al., 2004).
- Recent analyses suggested that woody angiosperm material (^{13}C-depleted) preferentially settled within the lower Mississippi River and in the river plume

(Bianchi et al., 2002). Other work has demonstrated that erosion of relict peat in transgressional facies of the lower Mississippi River provide a source of "old" vascular plant detritus to the river plume (Galler et al., 2003).
- High sedimentation rates in the river plume result in the formation of mobile mud, commonly observed in other large river–ocean interfaces (McKee et al., 2004). It is estimated that about 50% of the sediments (and associated OM) delivered to this region are temporarily stored near the delta – with a large fraction transported along/across the shelf in the benthic boundary layer (Corbett et al., 2004, 2006).
- Diatom signals in surface sediments suggested possible inputs of riverine diatom phytodetritus to the inner shelf (Wysocki et al., 2006). Previous work showed higher phytoplankton biomass, mostly as diatoms, than expected in the lower river (Dagg et al., 2007; Duan and Bianchi, 2006) with conversion, via lysis, to DOC. Hence, river nutrients were converted to river phytoplankton biomass and then ultimately to DOC, providing a labile food resource for bacterioplankton.
- An analysis of OM production to the west of the plume found phytoplankton at the outer edge of this region declined due to nutrient limitation, microzooplankton followed trends in phytoplankton, most particle sinking was associated with mesoplankton fecal pellets, phytoplankton-derived DOM reached a peak and was correlated with bacterioplankton, and water column recycling was most intense in this region (Dagg and Breed, 2003).
- Estimates suggested 10–52% of the DOM in the region west of the plume is quite labile (Benner and Opsahl, 2001). More recent data indicated that most riverine DOC was photochemically converted to dissolved inorganic carbon (DIC) over a period of weeks in this region (Dagg et al., 2007). More terrestrially derived components such as lignin had similar fates (Hernes and Benner, 2003).
- Some labile sedimentary organic matter, from in situ diatom production, was rapidly (day to weeks) shunted to the Mississippi River Canyon (Bianchi et al., 2006), essentially bypassing the hypoxic zone to the west. The supply rate of this phytodetritus was sufficient to support macrobenthic polychaete populations that do not exist in near-shore waters off the Louisiana coast. The removal of labile OM by winter season and hurricane events may act as a cleansing mechanism, reducing the potential for hypoxia (Bianchi et al., 2006).
- There are plumes from rivers and local estuaries along the coast containing colored dissolved organic matter (Chen and Gardner, 2004). DOC concentrations are also generally high (Engelhaupt and Bianchi, 2001) but higher still in the Atchafalaya River than the Mississippi River (Bianchi et al., 2004; Chen and Gardner, 2004; Pakulski et al., 2000).

These brief comments hardly do justice to the vast amount of work completed since the *Integrated Assessment*. However, they do provide evidence of improved understanding and elaboration of the role of different forms of OM in the NGOM ecosystem.

2.5.3 Synthesis Efforts Regarding Organic Matter Sources

In most environmental analyses, synthesis of diverse data sets is essential for clarifying cause–effect couplings and sorting out primary from secondary effects. Hypoxia and the role of various OM sources in NGOM hypoxia are no exception. Fortunately, a variety of descriptive and more quantitative syntheses/reviews have been developed since the *Integrated Assessment*.

Several studies, including those of Rabalais et al. (2002), Turner et al. (2007), Justić et al. (2007), and Rabalais et al. (2007a) largely reaffirm the primacy of river nutrients in supporting high rates of in situ primary production as the dominant source of OM supporting intense ecosystem respiration and development of hypoxic conditions. Walker and Rabalais (2006) analyzed SeaWiFS algal biomass data in relationship to river flow, nitrate loads from rivers, and hypoxia. Results confirmed strong relationships between nutrient loading and algal biomass distributions; direct relationships to hypoxic waters remained elusive for a variety of reasons. The importance of this work lies in the fact that the whole hypoxic-prone zone was assessed in a synoptic fashion and data were available for both low and high nutrient load periods. Dagg et al. (2007) also reviewed data to determine Mississippi River plume contributions to hypoxia. Results were largely consistent with those noted above, but Dagg et al. (2007) focused on the important role of the plume in both producing and consuming organic matter and dissolved oxygen and in building a case for the importance of coastal wetlands as an important organic matter source. However, there are problems with the magnitude of wetland OM contributions suggested by these calculations, including conversion of wetland sediment losses to OM mass, no consideration for on-marsh respiration of this material, and no consideration of the refractory nature of the particulate material, a major portion of this OM. Based on present understanding of the issue, it seems unlikely that wetland loss could be a prime source of OM to the hypoxic zone.

Finally, there have been several quantitative assessments of OM for portions of the hypoxic zone, and these are emphasized here because it seems that these types of syntheses are especially useful in understanding hypoxia and could serve as templates for designing future data acquisition programs. Several other studies, including those of Rowe and Chapman (2002) and Dagg and Breed (2003), have proposed broader conceptual models for the plume and the full hypoxic zone, respectively, and these might also be useful in study design and improving our vocabulary when discussing the hypoxic zone and the role of various OM sources. Gordon et al. (2001) used a variety of measurements to evaluate the distribution and accumulation of organic matter on the shelf west of the Atchafalaya River. They reported inputs from rivers and in situ production (in situ production dominated), estimated OM losses due to water column and sediment respiration (OM substrates being marine and riverine, respectively) and long-term burial (<5% of total inputs). Green et al. (2006b) used careful delineation of the Mississippi River turbidity plume coupled to a biological model to investigate OM budgets for this zone. They reported that labile OM was mainly from autochthonous phytoplankton production and that riverine OM inputs to the plume were three times as large but quite refractory. Losses of OM were mainly from microbial respiration, and, importantly, the plume as a whole was net autotrophic, again suggesting the primacy of in situ production. Finally, while the plume is a small fraction of the full hypoxic zone, Green et al. (2006b) estimated that plume-derived OM was equivalent to about 23% of the OM needed to create observed hypoxia on the full shelf.

> **Key Findings and Recommendations**
>
> The Study Group concludes this section with several findings. First, there is general and strong support for the conclusion that riverine nutrients support levels of plankton production capable of creating observed hypoxic conditions. However, some aspects of the relationship between in situ phytoplankton production and hypoxia remain uncertain. There is need for additional study of the hypoxia issue that emphasizes process studies and better coupling of physics to the chemical and biological features of the hypoxic zone. The Study Group therefore provides the following recommendations.
>
> - Continued research should be conducted to further elucidate the role of N and P from the MARB in stimulating phytoplankton production, the primary drivers creating excess OM and thus hypoxia in the Gulf.
> - A series of consistent, well-placed, and well-timed process studies should be conducted in the NGOM. Virtually all the OM review/synthesis papers referenced above state that their analyses suffer from a lack of pertinent process data.
> - DOM and POM delivered to the NGOM by rivers and from coastal wetland losses represent potential OM sources. The weight of evidence currently available suggests that it is unlikely these were triggers for hypoxia development or primary OM sources for hypoxia maintenance. However, the magnitude of river OM sources is large, and hence further characterization of this material is warranted.

2.6 Denitrification, P Burial, and Nutrient Recycling

The availability of N and P in an ecosystem is controlled both by external loadings and internal biogeochemical processes. Ideally information is needed on the load of biologically available nutrients, which is not necessarily well reflected by either the load of dissolved inorganic nutrients or the load of total nutrients. Internal biogeochemical processes are poorly known for the NGOM. Some, but not all, of the dissolved organic nutrients and particle-bound nutrients delivered to coastal waters become biologically available on ecologically meaningful timescales (days to months). In the Mississippi River, the fate of the particle-bound P is of particular

2.6 Denitrification, P Burial, and Nutrient Recycling

interest since it is the most common form of P in the river (Sutula et al., 2004). The bioavailability of this form of P is low within the freshwater portions of the Mississippi River, but, as the particles encounter the increasingly more saline waters of the Gulf of Mexico, the high ion abundances of seawater cause much of the adsorbed inorganic P to desorb, converting it into highly bioavailable dissolved inorganic P (Fox et al., 1985; Froelich, 1988; Howarth et al., 1995; Sutula et al., 2004). In addition, sediment diagenetic processes further increase the biological availability of particle-bound P delivered to the Gulf (Sutula et al., 2004).

For many coastal marine systems, the tendency is for benthic processes to make N limitation more prevalent since the N sink through denitrification is relatively larger than is the loss of P through permanent sediment burial (Blomqvist et al., 2004; Howarth and Marino, 2006; National Research Council, 2000). Phosphorus release from sediments is frequently less than the rate of P remineralization, due to P adsorption and storage in surface sediments (Howarth and Marino, 2006; National Research Council, 2000). Variations in P release are probably due to differences in the amount and forms of iron in the sediments, the extent of sulfate reduction, and mixing by the benthic fauna, particularly as this affects microscale variation in pH (Howarth et al., 1995). The dynamics of P-sediment exchanges in the Louisiana shelf region are sufficiently complex that in a recently published model of sediment diagenesis (Morse and Eldridge, 2007), P processes were deliberately not considered (John Morse, personal communication, 10/27/06). Given the recent evidence of the role of P in controlling phytoplankton production in the plume and near-plume regions, this process needs further examination.

Sulfate reduction is particularly important in affecting the P cycle of coastal marine sediments because it can transform highly adsorptive forms of iron (III) oxides and hydroxides into nonsorptive iron (II) sulfides (Blomqvist et al., 2004; Caraco et al., 1989, 1990; Krom and Berner, 1980). Sulfate reduction may also release P from covalently bound minerals as diagenesis proceeds (Sutula et al., 2004). Sulfate reduction dominates the metabolism of the sediments to the west of the Mississippi River on the Louisiana shelf away from the immediate plume of the river (Rowe et al., 2002), as is true for many coastal marine sediments (Howarth, 1984). Sutula et al. (2004) have demonstrated that the P content of these sediments is only half that of the riverine sediments in the Mississippi from which they are derived due to losses during diagenesis. Sulfate reduction and the concomitant changes in sediment iron chemistry may not be the only factor involved. Sutula et al. (2004) noted that significant sediment P is lost in the immediate plume area of the Mississippi River, a high-energy environment subject to physical mixing and sediment reworking, which may make sulfate reduction unlikely [the "sub-oxic fluidized bed reactor" processes that Aller (1998) described for other riverine plumes].

Studies in the Gulf of Mexico have shown that aerobic respiration in the sediments is low during hypoxic events (Rowe et al., 2002). This result suggests that anaerobic respiration, the accumulation of reduced compounds, and subsequent oxidation of these reduced species in the benthic boundary layer (BBL) and sediments may account for a large percentage of the oxygen draw down in this area (Morse and Rowe, 1999). Other work has found that the balance between the frequency of seabed disturbance, rate of geochemical reactions, and reactant concentrations work together to promote efficient remineralization through redox cycling in highly mobile muds near large rivers (Aller et al., 2004; Chen et al., 2005; Chen and Gardner, 2004; McKee et al., 2004). This frequent cycling of reduced and oxidized compounds is likely to have a profound effect on short-term oxygen consumption in the BBL, which could influence development of bottom hypoxia.

Hypoxia and bottom water oxygen deficiency influence not only the habitat of living resources but also the biogeochemical processes that control nutrient concentrations in the water column. Internal feedbacks on biogeochemical processes occur with oxygen depletion. Increased P flux from sediments into overlying waters with hypoxia is a classic response in freshwater systems (Mortimer, 1941) and has been well-documented in coastal marine ecosystems (Conley et al., 2002a, 2002b; Nixon et al., 1980). However, relatively little work has been done on the Mississippi River shelf on estimating the magnitude of enhanced P release with hypoxia and the impact on the overall P biogeochemical cycle. Higher P levels do accumulate in the bottom waters of the NGOM during hypoxia, but there is no evidence that this mixes into the overlying photic zone where it could be available to phytoplankton. This is critical information as P can be an important limiting nutrient in the plume (Sylvan et al., 2006).

Hypoxia also may influence rates of denitrification. Denitrification is one of the major losses of fixed nitrogen in the oceans (Seitzinger and Giblin, 1996), however, its measurement is difficult (Groffman et al., 2006). Denitrification is the reductive respiration of nitrate or nitrite to N_2 or N_2O and includes the recently discovered anaerobic ammonia oxidation (ANNAMOX) process (Dalsgaard et al., 2003). The rates of denitrification are dependent on a variety of factors, but a major control is the availability of starting products [e.g., nitrate (Kemp et al., 1990) and carbon (Sloth et al., 1995; Smith and Hollibaugh, 1989)]. Note that denitrification is favored by the absence of oxygen, but most coastal marine sediments are anoxic below the top few mm. Given that large-scale increases in nitrate concentrations and in productivity that have occurred on the Mississippi River shelf, it is likely that the rates of denitrification have also increased through time. Very few measurements on this important process are available, however.

An open question is how much hypoxia affects the annual rates of denitrification. Few direct measurements of denitrification exist for the Mississippi River shelf, with most previous estimates using potential denitrification rates. Lower rates of potential denitrification were observed in the Gulf of Mexico zone of hypoxia when low oxygen concentrations were encountered (Childs et al., 2002, 2003), although the observed rates were at the low end of rates reported for other systems (Herbert, 1999). Denitrification can be limited by the availability of nitrate, and hypoxia may reduce the supply rate of nitrate by slowing rates of nitrification (the oxidation of ammonium to nitrate); however, nitrate concentrations in the hypoxic area were high enough in the Childs et al.'s (2002) study not to be limiting. In addition, sulfide, which is commonly found in anoxic environments, acts to inhibit nitrification (the oxidation of ammonium to nitrate) (Joye and Hollibaugh, 1995), thus reducing the availability of nitrate. In Danish coastal waters, rates of denitrification are highest during winter when nitrate concentrations are at their annual maximum (Nielsen et al., 1995), and low rates are observed during the summer. There are no seasonal measurements of denitrification available for the NGOM to estimate the overall effect of hypoxia. In general, the overall rates of denitrification are believed to be lower with hypoxia (Graco et al., 2001; Sørensen et al., 1987) and eutrophication (Smith and Hollibaugh, 1989), although Vahtera et al. (2007) suggest that denitrification has potentially increased with hypoxia. Water column rates of denitrification in the oceans are high in mid-water hypoxia areas (Deutsch et al., 2007). Further investigations of the effects of hypoxia on the rates of denitrification are sorely needed on the Mississippi River shelf, as this is the major pathway of nitrogen loss.

Measurement of the fluxes of N and P from sediments provides a direct means to assess the role of sediment processes on the relative balance of N and P in the overlying water column. There are relatively few NGOM studies where both N and P fluxes from sediments have been determined simultaneously. A compilation of these studies shows a dissolved inorganic nitrogen/dissolved inorganic phosphorus (DIN:DIP) flux ratio that varies from approximately 1:1 to 25:1, with a mean of ~10:1 (Twilley et al., 1999).

Key Findings and Recommendations

The Study Group finds that additional information is needed on internal biogeochemical processes controlling the availability of nutrients to support primary production in the NGOM. The Study Group recommends that research be conducted in the following areas.

- The dynamics of sediment–water exchanges of P on the Louisiana shelf and their relative role in P cycling. Information on both aerobic and anaerobic processes is needed.
- The effects of hypoxia on the rates of denitrification and on long-term burial and regeneration of C, N, and P on the Louisiana shelf.
- N and P biogeochemical processes in sediments that include analysis of oxygen dynamics and the rates of supply of oxygen to the sediment surface.

2.7 Possible Regime Shift in the Gulf of Mexico

Hypoxia can act as a positive feedback to enhance the effects of eutrophication (Vahtera et al., 2007). It has long been known in lakes (Mortimer, 1941) that the internal P loading from sediments during anoxia can sustain eutrophication. In the Baltic Sea, which is one of the largest coastal areas in the world to suffer from eutrophication-induced hypoxia, large internal P loading occurs with hypoxia. The amount of DIP released from sediments in the Baltic is an order of magnitude larger than external inputs from rivers (Conley et al., 2002a). Large sediment–water fluxes of DIP with hypoxia must also occur in the Gulf of Mexico, returning DIP to a partially P-limited water column (Sylvan et al., 2006), stimulating phytoplankton growth and acting as a positive feedback to increase hypoxia severity. As discussed earlier (Section 2.1.6), hypoxia has the potential to reduce rates of denitrification, which would cause less N to be lost from the system, and also act as a positive feedback to increase hypoxia severity.

Recent studies in other coastal marine ecosystems, including Chesapeake Bay (Hagy et al., 2004) and Danish coastal waters (Conley et al., 2007), suggest that repeated hypoxic events can help to sustain hypoxic conditions. Large-scale changes in benthic communities occur with hypoxia, reducing the abundance of large, slow-growing, deeper dwelling animals and facilitating smaller, fast-growing species that can colonize surface sediments rapidly following hypoxia (Diaz and Rosenberg, 1995). Reductions in the abundance and size structure of benthic organisms have been observed in the NGOM with hypoxia (Rabalais and Turner, 2001). These smaller, surface-dwelling species have less capability to irrigate and bring oxygen downward into the sediments, helping to keep the sediments anoxic. The loss of benthic communities and the inability of the communities to recover with repeated hypoxic events (Karlson et al., 2002) may make ecosystems more vulnerable to the development and persistence of hypoxia. In addition, with the loss of sediment buffering capacity through the loss of electron acceptors (NO_3, O_2, Fe^{2+}, Mn^{2+}), there is a change in sediment metabolism from aerobic to anaerobic pathways, changing the production rates and processing of organic matter.

Wiseman et al. (1997) showed that the area of hypoxia along the Louisiana–Texas shelf was correlated to Mississippi River flow. These relationships were similar to those found for Chesapeake Bay (Boicourt, 1992) demonstrating the important role of river inputs in providing both freshwater-induced stratification and adding nutrients stimulating phytoplankton production. However, this apparent relationship has broken down since 1993 (data provided by DiMarco, personal communication). It appears that the Gulf of Mexico hypoxia has worsened following the record breaking 1993 spring floods, e.g., smaller river flows now induce a larger response in hypoxia (see Section 2.1.2). The first large ($>15,000\,km^2$) hypoxic event occurred after the 1993 flood, with large hypoxic areas over $15,000\,km^2$ observed in most following years. This pattern of a more sensitive system is also evident with May–June nitrate loading causing a larger hypoxic area in the NGOM than prior to 1993 (data not shown). A similar pattern of an increasingly sensitive system following the initial occurrence of hypoxia has been observed in Danish coastal waters with worsened hypoxia following the first appearance of large-scale hypoxic events (Conley et al., 2007).

Changes such as those described above suggest that a regime shift has occurred in coastal marine ecosystems that have been affected by large-scale hypoxia (Conley et al., 2007; Turner et al., 2008). Regime shifts are rapid transitions that change the structure and functioning of the ecosystem from one state to another as a consequence of a change in an independent variable. Once a threshold is passed, the ecosystem changes to a new alternative state, with changes in biological variables that can propagate through several trophic levels (Collie et al., 2004; Scheffer et al., 2001). For example, an increase in certain pelagic species (e.g., gelatinous carnivores) can disrupt top-down control of the food web structure causing a regime shift to an alternative stable state. The new stable system may not respond to changes in nutrient levels, a bottom up control, until nutrient input is reduced to a point below which the regime shift occurred. A regime shift due to hypoxia implies that, due to hysteresis in the system, nutrients will need to be reduced below the level at which the threshold occurred in order to reduce hypoxia. The management implications are that nutrients should be reduced as soon as possible before the even larger nutrient reductions are required to reduce the area of hypoxia.

Regime shifts can have large consequences for fisheries (Collie et al., 2004; Oguz and Gilbert, 2007). The Gulf of Mexico ecosystem is a tremendously valuable resource from economic, ecological, and social perspectives. In 2004, the value of commercial fish harvest in the Gulf of Mexico was $670 million (NOAA, 2007). The Gulf of Mexico shrimp fishery is among the most valuable fisheries in the nation, with a total value in 2004 of about $370 million, and about $140 million in Louisiana alone. Additionally, an estimated 24.6 million recreational fishing days occurred in the Gulf of Mexico in 2004, with about 4.8 million of those occurring in Louisiana waters (NOAA, 2007). The Gulf of Mexico also serves as habitat for a host of other species, including endangered sea turtles and marine mammals. Thus, the Gulf of Mexico is a valuable resource that is potentially being threatened by hypoxia.

Earlier studies found it difficult to identify impacts of hypoxia in fisheries landings statistics (Diaz and Solow, 1999; Rabalais and Turner, 2001), although there has been a shift in relative population abundance from benthic to pelagic species (Chesney and Baltz, 2001). A summary of published studies and works in progress on the effects of hypoxia on living resources in the NGOM are mentioned in Appendix A. There is strong scientific evidence that ecosystems in the northern Gulf of Mexico are stressed by hypoxia (Diaz et al., 2003; Diaz and Rosenberg, 2009; Breitburg et al., 2009a). Studies have found impacts ranging from the molecular/genetic level (Brouwer, 2006; Hendon et al., 2006; Perez et al., 2006; Wells et al., 2006), the organismal level (Brouwer, 2006; Zou, 2006; Thronson and Quigg, 2008), and the ecosystem level (Craig et al., 2001; Rabalais, 2006; Rabalais and Turner, 2001; Altieri, 2008; Green et al., 2008; Vaquer-Sunyer and Duarte, 2008). Population effects are indicated as well (Rose et al., 2009). Potential impacts due to displacement from preferred habitat have been identified (Craig et al., 2005; Craig and Crowder, 2005; Switzer et al., 2006). There is also recent evidence that hypoxia has affected the valuable brown shrimp fishery (Zimmerman and Nance, 2001).

There are some indications that the Gulf of Mexico has undergone a regime shift. In the hypoxic/anoxic zone of the Louisiana inner shelf many taxa are lost during the peak of hypoxia. Certain typical marine invertebrates are absent from the fauna, for example, pericaridean crustaceans, bivalves, gastropods, and ophiuroids (Rabalais and Turner, 2001). As noted above, a shift has been observed in the relative abundance of fish species. Changes in benthic and fish communities with the change in frequency of hypoxia are cause for concern (Baustien and Rabalais, 2009; Hazen et al., 2009; Levin et al., 2009). If actions to control hypoxia are not taken, further ecosystem impacts could occur within the NGOM, as has been observed in other ecosystems. The recovery of hypoxic ecosystems may occur only after long time periods (Diaz, 2001) or with further reductions in nutrient inputs. Experience has shown recovery to be greatly delayed, taking years to decades for ecosystems to recover after nutrient inputs are reduced, and with probably less than complete recovery possible (e.g., Diaz, 2001; Diaz et al., 2003; Mee, 2006; Raloff, 2004). Some smaller organisms may respond more rapidly and on annual cycles. For example, in low load years there is less hypoxia, lower phytoplankton biomass and presumably less organic deposition, and lower rates of sediment processes. On the other hand, larger benthic organisms respond more slowly, and resident fish and shellfish populations will require more time to return to previous conditions. One potential concern with regime shifts is that the condition is not always reversible. The system can follow a different path to pre-impact conditions and not return to its former state. This is called a hysteresis effect. However, given that the Gulf of Mexico is an open shelf system, recovery should be more rapid than in enclosed ecosystems. Thus, there are potentially large benefits that justify taking action to control hypoxia and thereby avoiding large-scale changes in the Gulf of Mexico ecosystem.

> **Key Findings and Recommendations**
>
> Hypoxia probably increases sediment–water fluxes of P and may reduce the potential for denitrification and change the degradation of organic matter in sediment from aerobic to anaerobic metabolism. Biological changes have occurred in the benthic communities of the NGOM, and there is evidence that the living resources are impacted by hypoxia. The Gulf of Mexico ecosystem appears to have gone through a regime shift with hypoxia such that today the system is more sensitive to inputs of nutrients than in the past, with nutrient inputs inducing a larger response in hypoxia as shown for other coastal marine ecosystems (Chesapeake Bay, Danish coastal waters). The Study Group therefore provides the following recommendation.
>
> - Nutrients should be reduced as soon as possible before the system reaches a point where even larger reductions are required to reduce the area of hypoxia.

2.8 Single Versus Dual Nutrient Removal Strategies

The *Action Plan* seeks to significantly reduce the size of the Gulf of Mexico hypoxic zone by the year 2015, primarily through reductions in nitrogen (N) loadings from the MARB to the NGOM. Increases in N loads have clearly been occurring throughout the past decades, and there is ample evidence to conclude that N from the

2.8 Single Versus Dual Nutrient Removal Strategies

MARB is a driving force in determining, at least in part, the timing, severity, and extent of the hypoxic zone. Since the mid-1990s, N loadings from the MARB have decreased, although they are still much elevated over historic levels. Total phosphorus loadings, however, have not changed greatly during this period (Battaglin, 2006; Turner et al., 2007; Section 2.1.9 of this book). This trend in nutrient loadings has led to reduced (albeit still very high by "Redfield" standards) N:P ratios. This evidence suggests that P is an additional nutrient of concern, in terms of input reductions. As conveyed in previous sections of this book, a number of investigators (Dagg et al., 2007; Sylvan et al., 2006) have concluded that P is limiting primary production during key periods of high productivity and in zones of high biomass accumulation in the NGOM adjacent to hypoxic waters. Therefore, the role of P in the onset, extent, and duration of the hypoxic zone is worthy of additional consideration.

Many factors influence the cycling and ultimate fate of both N and P. As both play a significant role in driving primary production within the NGOM (and perhaps, in conjunction with Si, in the composition of the primary producers and the likely fate of produced organic carbon), it is logical to consider the potential for removal of either or both elements as a means to reducing hypoxia. The 2001 *Action Plan* focuses on N reductions but does not preclude either P reduction or dual removal strategies. For example, the most recent report of the Mississippi River/Gulf of Mexico Watershed Nutrient Task Force's (MR/GMWNTF's) Management Action Review Team (MART, 2006a) concludes that most load reduction projects developed under the Clean Water Act Section 319 program have targeted both N and P for reduction. Indeed, Howarth et al. (2005) noted that some N control practices utilized in the United States effectively remove P as well, although the reverse is not always the case. However, not all control practices will be effective as a dual nutrient removal strategy; see specific discussion on this topic in Section 4.5.10.

Restoration plans that focus on N alone may not rapidly improve the situation in the MARB where many streams and river segments are degraded by excess P concentrations (MR/GMWNTF, 2001). Given recent discoveries concerning the importance of P in production of organic carbon within significant portions of the NGOM, focusing on N reduction alone may be insufficient to provide the desired reduction in the hypoxic zone. However, some plans being undertaken to reduce nonpoint sources of N (forested buffers, 319 programs, and others [see Section 4.4.2, for example]) will also lead to P reductions, as well. Reductions in P alone will alleviate some of the water quality issues facing freshwater regions of the MARB but are not likely, given our current state of understanding, to significantly address the over-enrichment of the NGOM. Therefore, greater emphasis on a dual nutrient removal strategy is warranted, a conclusion that has been reached in other instances (e.g., Boesch, 2002; Howarth and Marino, 2006; National Research Council, 2000).

Further work is necessary to examine how effectively current reduction strategies target both elements. There may be areas where shifts in removal techniques could improve P reduction. In addition, there is still much to be learned about the response of autotrophic and microbial communities to shifts in nutrient loading and ratios. A better understanding of how these communities have responded to the current loadings and predictions of how they will continue to adapt to nutrient reductions will greatly improve predictions of the likely response in the extent and duration of hypoxia to nutrient reductions in the future.

> **Key Findings and Recommendations**
>
> Recent information clearly indicates that P controls productivity in some portions of the NGOM. The Study Group finds that restoration plans focusing on N alone may not rapidly improve the situation in the MARB and may be insufficient to provide the desired reduction in the hypoxic zone. Reductions in P alone will alleviate some of the water quality issues facing freshwater regions of the basin but are not likely to significantly address the over-enrichment of the NGOM. Therefore the Study Group recommends the following :
>
> - In addition to the N reduction strategy currently in place, reduction strategies for P should be implemented. Section 4.2 provides greater detail on the Study Group's recommended targets for reducing both N and P.

2.9 Current State of Forecasting

There are several types of modeling efforts working toward a better understanding of factors influencing the extent and duration of the Gulf of Mexico hypoxic zone. These vary from the simple to the complex and are based on empirically observed relationships, on mechanistic understanding, or some combination of both.

Empirical models are widely used in the aquatic sciences to establish relationships between variables, with the most well known being the correlation between spring P loading in lakes and summer chlorophyll concentrations (Vollenweider, 1976). This work has been widely used in a management context to justify reductions in anthropogenic phosphorus loading to lakes and to set goals for reductions for particular lakes. Nixon et al. (1996) developed a similar correlation between annual loading of DIN and rates of primary productivity for marine ecosystems. While establishment of empirical models has greatly enhanced understanding of the structure and functioning of aquatic ecosystems (Peters, 1986), the standard criticism of this approach is that correlation does not imply causation. Although correlations between variables exist, they do not explain why variables are correlated or the mechanisms of the relationship. They do, however, provide some very useful predictive capability. In addition, when ecosystem production is greatly different from that predicted, controls on productivity other than nutrients may be dominating, such as light limitation or limitation from rapid flushing (Howarth et al., 2006a).

2.9 Current State of Forecasting

Some new forecast modeling work has been completed since the *Integrated Assessment*. Turner et al. (2006) developed simple linear and multiple regression models to examine hypoxia in the NGOM. Empirical models require important decisions regarding the choice of variables and of the timescales of model operation. Turner et al. (2006) tested many different nutrient loading lag times and concluded that the best relationship was obtained 2 months (May) prior to the maximum observed extent of hypoxia (July), with significant correlations for nitrate+nitrite, total nitrogen (TN), ortho-P, and total phosphorus (TP) (r^2 values of 0.50, 0.27, 0.54, and 0.60, respectively). A multiple regression analysis was also developed incorporating nutrient load and a new variable "Year" to account for the increase in carbon in surface sediments after the 1970s causing significantly more sediment oxygen demand. A lag of 2 months of nutrient loading was, again, the most significant variable to describe hypoxic area with r^2 values of 0.82, 0.80, 0.69, and 0.64 obtained with nitrate+nitrite, TN, ortho-P, and TP, respectively. Turner et al. (2006) then used the nitrate+nitrite model to extrapolate beyond the data range used to construct their models to predict hypoxic area prior to available measurements. When the hindcasted values became negative, they were plotted as zero values. In general, it is considered incorrect to extrapolate model results in this manner beyond the range of the data supporting the model, as other mechanisms and relationships may exist that may not be included in the regression analysis. Further, the Study Group believes that the addition of the variable "Year" in the multiple regression analysis is inappropriate as the addition of one more year will cause prediction of a positive increase in hypoxia with time.

Among models that address Gulf of Mexico hypoxia and include some consideration of processes and mechanisms, that of Scavia et al. (2003) is one of the simplest. Their model uses a relationship between the nitrogen loading from the MARB and the decay of oxygen "downstream" (i.e., in the NGOM – within the plume and the near-shore reaches to the west of the Mississippi and Atchafalaya River outflows). When used in a forecast mode, this model is able to explain only approximately 45–55% of the variability in hypoxic length and area. This model explicitly addressed uncertainty in prediction. The Study Group found this approach to be very useful. Recently, in combination with a watershed model, the model of Scavia et al. (2003) has been used to address how climatic variability and change may affect Gulf hypoxia (Donner and Scavia, 2007). A similar model has also been applied very successfully to understand hypoxia and anoxia in Chesapeake Bay (Scavia et al., 2006). The Scavia et al. (2003) model focused on N loading and did not consider P. Consideration of P would seem to be a timely addition to the model, as was recently discussed by Scavia and Donnelly (2007). This model approach, and the modeling efforts of Bierman and colleagues and Justić and colleagues (see below) all provide reasonably consistent guidance and suggest similar levels of N reduction that might be required to reduce the extent of the hypoxic zone.

Other process-based models are more complex and attempt to model both physical and biological controls occurring in the hypoxic region. Examples include those of Bierman et al. (1994), Justić et al. (1996, 2002), and Green et al. (2006b). The Bierman et al. (1994) model is the most complex of these approaches and simulates the steady-state summertime conditions for the hypoxic area using three-dimensional modeling of the physics as well as interactions between food web processes, nutrients, and oxygen. The model of Justić et al. (1996, 2002) simulates oxygen dynamics at one location within the hypoxic zone using a simple model that has two vertical layers and meteorological conditions and nitrogen loads as drivers. The Green et al. (2006b) surface mixed layer model is based on food web dynamics and relatively simple two-dimensional physics (no vertical dimensionality) of the Mississippi River plume. This model predicts, among other things, the relationship between carbon sources and bottom water oxygen depletion; the model does not include changes to either N or P inputs or dynamics. None of these more complex models explicitly presented analysis of uncertainty or sensitivity analysis of potential biasing terms. As with the Scavia et al. (2003) model, Bierman et al. (1994) and Justić et al. (1996, 2002) do not consider P loads or dynamics.

It should be pointed out that complex water quality models that could be very useful in the NGOM have been developed and used in other environmentally stressed regions like the Chesapeake Bay system (Cerco and Cole, 1993), the New York/New Jersey Harbor/New York Bight complex (Landeck-Miller and St. John, 2006), and the Massachusetts/Cape Cod Bays system (Besiktepe et al., 2003). These models include a coupling to three-dimensional and time-dependent hydrodynamics, a water column eutrophication submodel and a sediment diagenesis/nutrient flux submodel. [The water-column eutrophication submodel includes state variables for three functional phytoplankton groups; dissolved inorganic nutrients (ammonium, nitrate+nitrite, ortho-phosphate, and silica); labile and refractory forms of dissolved and particulate organic nitrogen and phosphorus; biogenic silica; labile and refractory forms of particulate and dissolved organic carbon; and dissolved oxygen.] The sediment nutrient flux submodel includes state variables for labile, refractory, and inert organic carbon, nitrogen, and phosphorus as well as biogenic silica. Inorganic substances tracked include ammonium, nitrate+nitrite, ortho-phosphate, silica, sulfide, and methane. Processes tracked in the sediment flux model include organic matter deposition; sediment diagenesis; burial; the flux of inorganic nutrients between the water column and the sediment bed; and the generation of sediment oxygen demand (SOD).

There is an inherent tradeoff between model simplicity (where many potentially important factors are not considered) and complexity (where many coefficients and a great amount of data are required). More complex models may have value to help devise effective management strategies, especially if N reductions alone will not be sufficient to control hypoxia and if the more complex models can reasonably capture the importance of P. However, with complexity comes greater numbers of estimated parameters and the uncertainty associated with them. Hence this type of model may not improve forecasting capabilities dramatically. The development of more complex models is likely to prove extremely valuable for understanding the physical factors controlling water and carbon (C) transport, the dynamics of nutrient interactions with primary producers, and the recycling and loss of C and nutrients from the system. There is also great value in refining and further developing simple models, which may, in the end, prove most valuable for making management decisions. Scavia et al. (2004) explicitly compared the models of Scavia et al. (2003), Biermann et al. (1994), and Justić et al. (1996, 2002) for use in managing Gulf of Mexico hypoxia and showed that all three models gave broadly consistent guidance.

The physics of the NGOM region is complex, and there is clear value in developing more complex models of physical processes for this region. Improved three-dimensional models with finer grid structure than present models would have many uses. These uses include assisting the interpretation of monitoring data and serving as platforms upon which improved models of biogeochemistry and ecological response could be built. However, the level of complexity in the biogeochemistry and ecology need not match the complexity of the physical models (Hetland and DiMarco, 2007). Complex physical models could be very valuable in constructing simple box mass balance accounting models for C, N, P, Si, and O, for example. The importance of developing such budget-based models is discussed further below.

In addition to statistical and simulation models, another modeling format that should be considered involves construction and evaluation of material budgets or mass balance models. These are basically quantitative input–output budgets with additional complexity added by consideration of internal processes of production, recycling, and loss. These relatively simple budgets provide a quantitative mass balance framework to test the understanding of how the systems work. These budgets should be developed on a seasonal basis (e.g., summer hypoxic season) and evaluated for distinctive areas (e.g., Mississippi River Plume). These budgets are largely based on empirical observations and are not simulated through time, although data used in a budget analysis are needed in simulation models for both calibration and verification. As an example, an oxygen budget (Eq. 2.1) would involve DO inputs/outputs from air–sea diffusion, horizontal advective/dispersive transport, and vertical transport between euphotic and sub-pycnocline zones. In addition, DO is added through daytime photosynthesis and lost through water column and sediment respiration. Evaluation of these pathways indicates especially important processes, and imbalances in the budget point to areas where understanding or measurements are inadequate. We suggest that conceptual mass balance models also be used to provide a checklist of needed measurements for future NGOM hypoxia research/monitoring.

Other general points regarding modeling efforts are summarized in Section 3.4 of this chapter. An important conclusion for both models of the response of the NGOM to nutrient inputs and watershed models generating estimates of nutrient loads is that a diverse ensemble of models is needed, including both relatively simple and more complex ones. No one best approach to modeling can be identified, and management of Gulf hypoxia is best served by having multiple models with multiple outputs. The Study Group suggests that modeling efforts, ranging from the simple to complex, be conducted in parallel wherein there is the opportunity for cross-testing of results among model formats. When predictions tend to agree, managers can have more confidence in deciding upon courses of action. When models do not agree, dissecting the reasons for divergence can lead to better understanding and, ultimately, better management.

Key Findings and Recommendations

Since the *Integrated Assessment*, a number of modeling approaches have been employed to characterize the onset, volume, extent, and duration of the hypoxic zone. Models have been able to explain approximately 45–55% of the variability in hypoxic length and area. However, the Study Group finds that model development, calibration, and verification are hampered by the relative paucity of data on the duration and extent of hypoxia and on rates of important biogeochemical and physical processes that regulate hypoxia. In addition, the Study Group finds that a diverse ensemble of models is needed, including both relatively simple and more complex ones. No one best approach to modeling can be identified, and management of Gulf hypoxia is best served by having multiple models with multiple outputs. The Study Group provides the following recommendations to advance the science for characterizing the onset, volume, extent, and duration of the hypoxic zone.

- To the extent reasonable, future models (particularly more complex models that rely on accurate representation of ecological and biogeochemical processes) of hypoxia in the Gulf should consider nitrogen, phosphorus, and their interactions. However, this is a significant challenge since these interactions are so poorly studied in the NGOM at present.
- The development of more comprehensive monitoring should be coordinated with model development. For example, the more complex physical models of the NGOM should be used to aid in interpretation of monitoring data on extent and duration of hypoxia. These models can also feed into both simple and complex biogeochemical and ecological models.
- Because there is great value in developing simple mass balance models in the NGOM for organic C, dissolved oxygen, and nutrients, mass balance models should be used to provide a checklist of needed measurements for future NGOM hypoxia research/monitoring.
- Gulf hypoxia models should be designed so that they can be compatible with watershed models. That is, there must be compatibility in (1) the time step between a Gulf hypoxia model and a watershed model, and (2) the form of key variables that serve as outputs from a watershed model and inputs for a Gulf hypoxia model (e.g., a watershed model that predicts total nitrogen is not compatible with a Gulf hypoxia model that requires specific forms of nitrogen).

Chapter 3
Nutrient Fate, Transport, and Sources

The Study Group was asked to review the available literature and information, especially that developed since 2000, that would allow them to assess any changes and improvements in the understanding of nutrient sources and flux estimates within the Mississippi and Atchafalaya River basins (MARB) (see Fig. 1.2) and the current ability to use watershed models to route and predict nutrient delivery to the Gulf of Mexico. The following sections discuss the current levels of understanding and provide brief summaries of the Study Group's key findings and recommendations.

3.1 Temporal Characteristics of Streamflow and Nutrient Flux

The research needs identified in the *Integrated Assessment* to understand and document the temporal characteristics of MARB riverine nutrient loads included (1) studies on small watersheds to better document nutrient export on the short timescales needed; (2) detailed information on tile drainage intensity; (3) increased monitoring of stream sites; and (4) measurements of point source discharges rather than estimates from permits. Only a limited number of these needs have been met.

However, more recent estimates of agricultural drainage appear to be more representative than those used in the original assessment (e.g., see Sands et al., 2008), and new procedures for load calculations have resulted in changes in estimates of nutrient fluxes. A brief discussion of each of the improvements follows.

Current extent and patterns of agricultural drainage. The *Integrated Assessment* relied largely on the 1987 USDA-ERS report (Pavelis, 1987), which based estimates of agricultural drainage on land capability class and crop information from the 1982 Natural Resources Inventory (NRI). NRI estimates were dropped after 1992, and NRI is statistically valid only at a watershed or county level. Based on the USDA surveys, some degree of subsurface drainage is present on 13 million hectares (over 32 million acres) in the Midwest states. However, there is considerable uncertainty with respect to the actual extent and distribution of drainage of cultivated cropland. In the absence of additional survey data, more recent estimates of the extent of drained agricultural land have been developed based on land use and soil class/characteristics (Jaynes and James, 2007; Sugg, 2007). This general approach needs further development and validation but seems to provide the best current estimate of the extent of agricultural drainage. The approach takes advantage of the now extensive and detailed GIS coverages and provides a considerably finer level of spatial resolution than previously available.

In the following example, USDA STATSGO soil data were used to estimate the extent of agricultural drainage based on the distribution of row crops (primarily corn and soybean) on soils with a drainage class of poorly drained soils and slopes 2% or less (Fig. 3.1, per D. Jaynes, National Soil Tilth Lab, Ames, IA). These patterns of agricultural drainage predicted using this approach are generally similar to patterns in land use (Fig. 3.2) and in-stream nitrate concentration estimated from STORET data selected to exclude point source influences (Fig. 3.3). Drainage estimates could be further refined by using improved land-use data and by using SSURGO rather than STATSGO data.

The relationship between nitrate concentration and land use is further illustrated in Fig. 3.4 for 52 NASQAN stations (Alexander et al., 1998) in the upper Mississippi and Ohio River basins selected to exclude sites with large upstream reservoirs or extensive upstream urban areas (Crumpton et al., 2006). See Section 4.5.7 for further discussion on urban nonpoint sources. Percent cropland (corn or soybean) accounts for 90% of the observed variation in the average of 1980–1993 annual flow-weighted average nitrate concentrations for the 52 stations examined. Reduced nitrogen (calculated as total nitrogen minus nitrate) shows a slight, but statistically significant, increase with percent crop land.

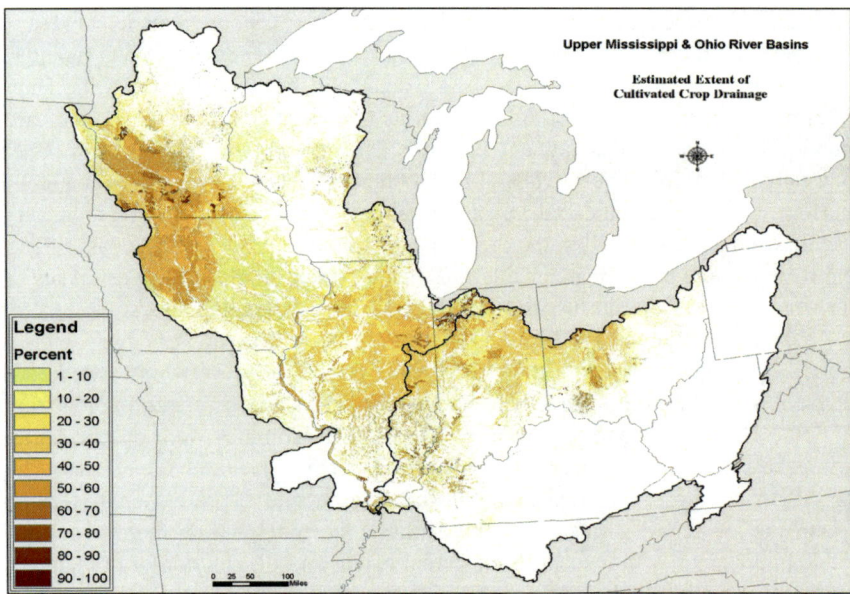

Fig. 3.1 Estimated extent of agricultural drainage based on the distribution of row crops, largely corn and soybean, and poorly drained soils (per D. Jaynes, National Soil Tilth Lab, Ames, IA)

3.1 Temporal Characteristics of Streamflow and Nutrient Flux

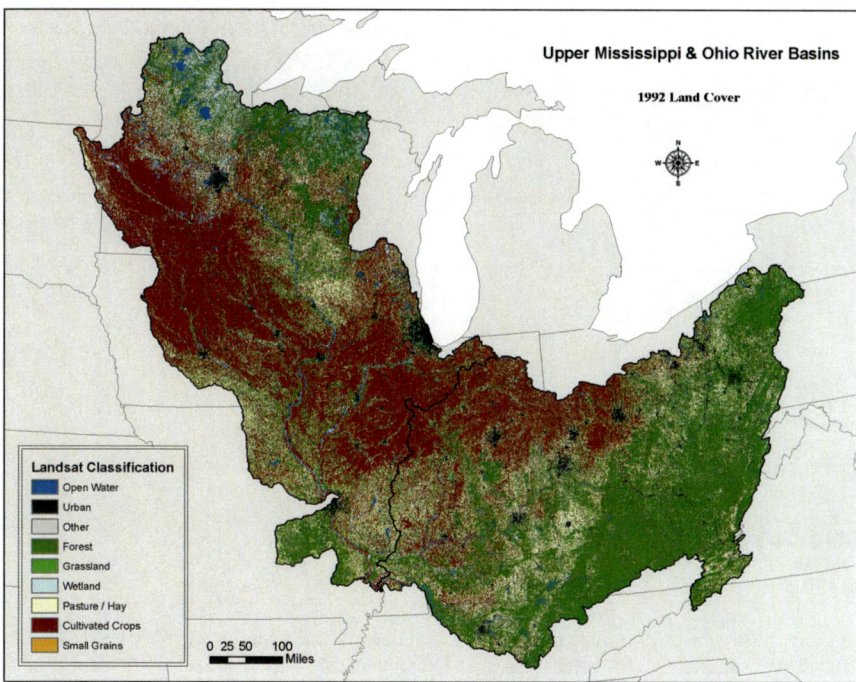

Fig. 3.2 Land cover based on Landsat data (adapted from Crumpton et al., 2006)

Flow-weighted average nitrate concentrations estimated by applying the regression for NASQAN sites to 1992 Landsat land cover data for UMR and Ohio River basins are similar to those estimated from STORET data. The relationship between nitrate concentration and the estimated extent of agricultural drainage was also examined, and for these 52 stations, nitrate concentrations were more closely related to land use than to STATSGO-derived estimates of drainage. There is certainly more error in estimates of drainage than in estimates of cropland distribution, and this error could degrade the fit of nitrate concentration with drainage. However, much of the cropland not directly drained by field tile still contributes to nitrate discharged through drainage networks, and at some spatial scale, nitrate concentrations might depend more on cropland distribution than on artificial drainage (i.e., if the land is successfully cropped, then some combination of natural and artificial drainage can be implied).

It is clear that agricultural drainage in the Corn Belt is extensive; the general distributions of drainage and cropland are correlated; and nitrate concentrations are correlated with patterns of cropland and drainage. Additional research is needed to better define the extent, pattern, and intensity of agricultural drainage, including cropland drained by field tile as well as cropland not directly drained by field tile but contributing to drainage networks.

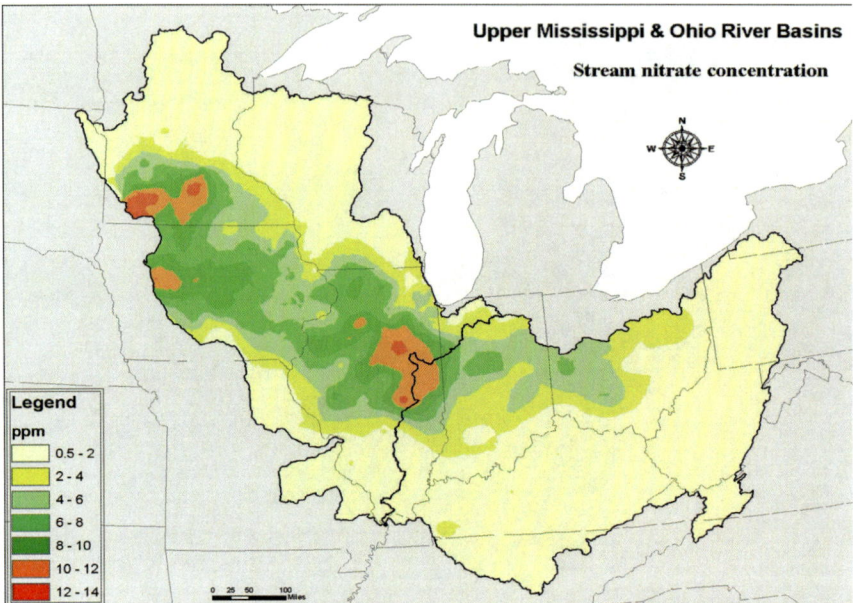

Fig. 3.3 Flow-weighted average nitrate concentrations estimated from STORET data selected to exclude point source influences (adapted from Crumpton et al., 2006)

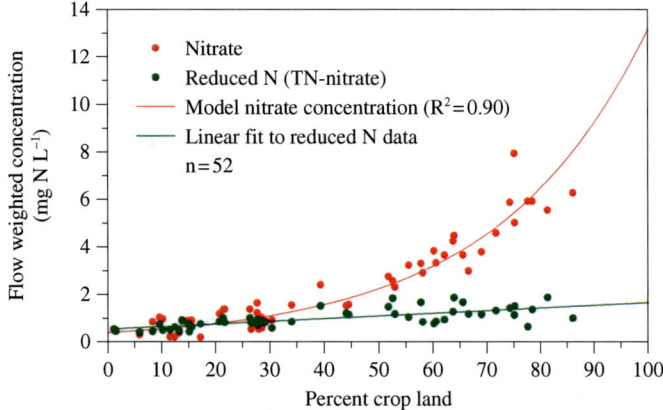

Fig. 3.4 Flow-weighted average nitrate and reduced N versus percent cropland (adapted from Crumpton et al., 2006)

Change in the flux estimation method. Riverine loads can be calculated with many different methods; the method chosen is dependent on sampling frequency as well as river size, which determines how quickly the concentration changes. A comparison of the estimates of annual N flux for the combined Mississippi and Atchafalaya

3.1 Temporal Characteristics of Streamflow and Nutrient Flux

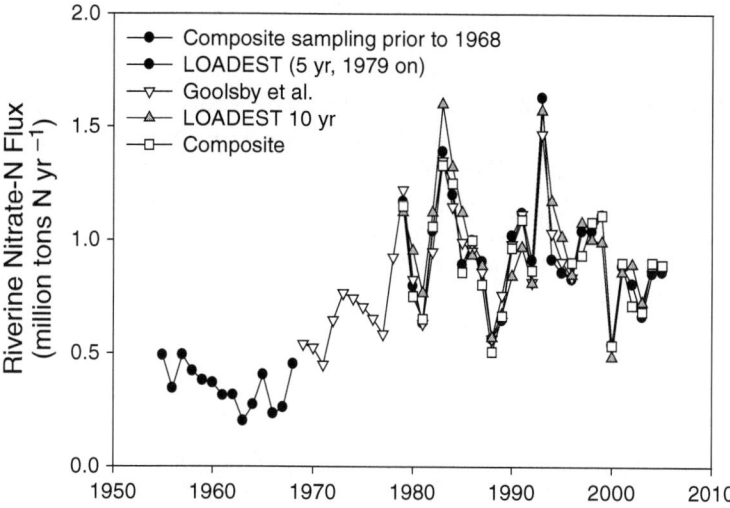

Fig. 3.5 MARB nitrate-N fluxes for 1955 through 2005 water years comparing estimates from various methods for 1979–2005. Based on USGS data from Battaglin (2006) and Aulenbach et al. (2007)

Rivers using five different methods is shown in Fig. 3.5. Goolsby et al. (1999) presented nitrate-N loads to the Gulf from 1955 through 1996. For the years prior to 1968, loads were calculated from daily samples composited at 10- to 30-day intervals for analysis. For the period 1968–1996, they used a multiple regression approach to calculate daily concentrations based on about 10–15 samples per year (or less) and daily flow [shown as Goolsby et al. (1999)]. Goolsby et al. (1999) calibrated one model (using a minimum variance unbiased estimator, MVUE) for 1968–1975, and one model for 1976–1997. This type of regression equation provides a good measure of the overall flux of a nutrient for the entire period of fitting but is less accurate for a given year. Since the *Integrated Assessment*, USGS has modified load estimation procedures to reduce the bias in the regression models. These modified procedures are all based on the rating-curve method but differ in the form of the equation and/or calibration periods. In July 2002, USGS posted load estimates for the entire period of record using ESTIMATOR (Cohn et al., 1992; Gilroy et al., 1990), a regression model method using the same MVUE technique used by Goolsby et al. (1999) with a 10-year moving window calibration period, and provided updated annual estimates through June 2002, followed by annual updates through June 2005 (shown as LOADEST 10 years). In this case the MVUE procedure used was equivalent to the adjusted maximum likelihood estimate (AMLE, discussed below) used in later estimates because there were no censused nitrate values in the calibration data sets. In 2006, the USGS posted new estimates for the entire period of record using Load Estimator (LOADEST) (Runkel et al., 2004) with the AMLE procedure and a 5-year moving window (shown as LOADEST 5 years). In addition to a shorter calibration period, the

AMLE procedure modifies the rating-curve equation in an attempt to correct for transformation bias. However, the AMLE procedure can still suffer from serial correlation in the residuals; so when sufficient data are available, the USGS applies a period-weighted interpolation to correct the AMLE estimate for the serial structure in the residuals (Aulenbach and Hooper, 2006). Results from this composite method for the mainstem Mississippi and Atchafalaya Rivers are nearly the same as just using a period-weighted (or linear interpolation) approach for nitrate-N (shown as composite). This suggests that the regression model in the composite method adds little when at least 10 samples are available for a given year, as well as demonstrating that concentrations of nitrate-N change slowly in these large rivers. (For additional information on methods used to estimate nutrient fluxes see http://toxics.usgs.gov/pubs/of-2007-1080/methods.html.)

Although the overall year-to-year pattern of N flux is consistent across the various methods, there is considerable variability among the estimates of each annual N flux. Figure 3.6 shows the percent difference between three of the methods and the current LOADEST 5-year method in both percent and metric tons for the entire period of record. The LOADEST 10-year method estimated N fluxes that ranged from as much as about 18% less (1990) to 28% more (1994) than the N fluxes estimated by the LOADEST 5-year method. That translates into an underestimate of about 180,000 metric tons or 198,000 tons of N that was delivered to the Gulf in 1990 and an overestimate of about 260,000 metric tons of N (287,000 tons of N) in 1994. Research published since 2003 would have used the LOADEST 10-year fluxes in models predicting the Gulf hypoxic zone in which case they likely used the more recent estimates (2003 and 2004 in Fig. 3.5), which ranged from only 3 to 10% or 25–50,000 metric ton of N (28–55,000 tons of N) more than the estimated flux using the current LOADEST 5-year method. The flux estimates presented in the following sections of this book are based on the new LOADEST 5-year method.

3.1.1 MARB Annual and Seasonal Fluxes

The following analysis is based on US Geological Survey streamflow and water quality monitoring data described in Aulenbach et al. (2007) and available on the Internet at http://toxics.usgs.gov/pubs/of-2007-1080/. The nutrient flux estimates were calculated as the combined fluxes at the Mississippi River near St. Francisville, LA, and the Atchafalaya River at Melville, LA (Fig. 3.7), using the LOADEST 5-year method discussed in the previous section.

3.1.1.1 Annual Patterns

Nitrogen: During the past 5 years (2001–2005 water years), an average of 813,000 metric tons (896,000 tons) of nitrate-N and 429,000 metrics tons (473,000 tons) of total Kjeldahl N (TKN) were transported annually to the Gulf.

3.1 Temporal Characteristics of Streamflow and Nutrient Flux 57

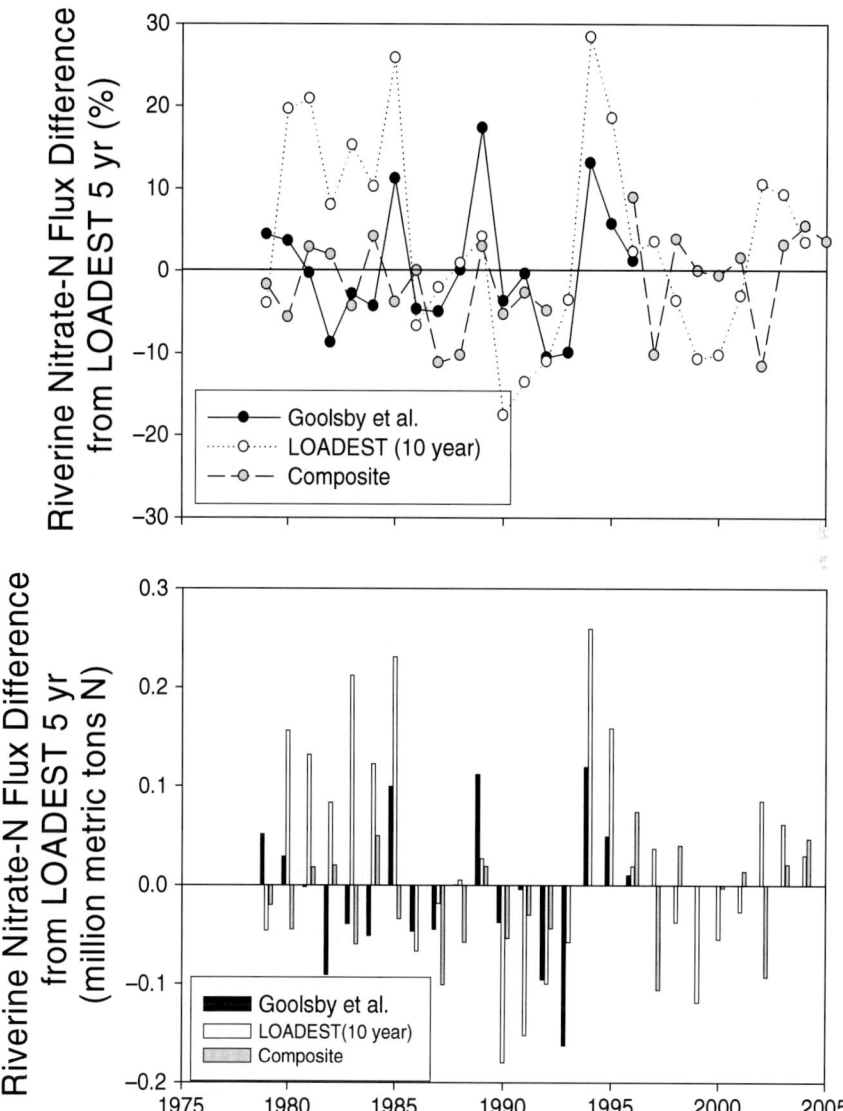

Fig. 3.6 Comparison (percent and absolute basis) of MARB nitrate-N fluxes to LOADEST 5-year method for 1979 through 2005 water years. Based on USGS data from Battaglin (2006) and Aulenbach et al. (2007)

There is considerable interannual variability in these flux values, driven primarily by precipitation patterns and resulting streamflow (Fig. 3.8), which appears to have increased slightly since the 1950s. Since the mid-1990s, annual nitrate-N flux has steadily decreased, which is more clearly shown by the 5-year running average. In addition, TKN has also shown a steady decline since the mid-1980s, so

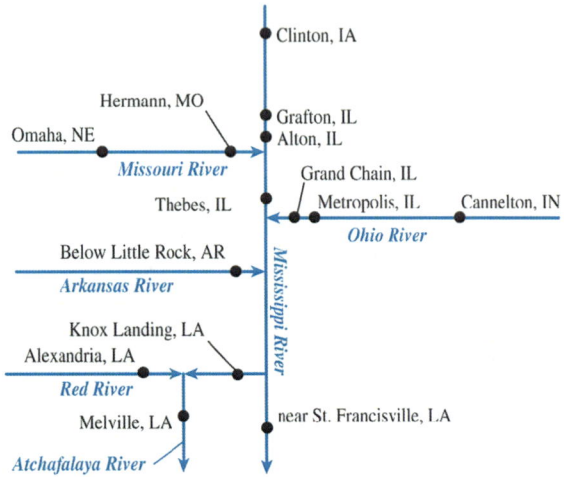

Fig. 3.7 Schematic showing locations of MARB monitoring sites (Aulenbach et al., 2007)

the total N flux, although highly variable from year to year, shows a very striking decline. The annual NH_4-N flux also decreased during the monitoring period (from 77,000 metric tons N/year [85,000 tons N/year] in 1980–1984 to 12,000 metric tons N/year [13,000 tons N/year] for 2001–2005) but was not the primary reason for the decline in TKN, as particulate and organic N declined. The decline in NH_4-N is likely due to improvements in sewage treatment as is at least part of the decline in particulate and organic N (Larson, 2001; Metropolitan Council, 2004). In addition, reduced sediment loads, because of a reduction in soil erosion, may also be a driving factor in reducing particulate N losses (Richards and Baker, 2002).

Phosphorus and silicate: Temporal trends in total P, soluble reactive P (SRP), and dissolved silicate fluxes for the combined rivers are less striking than the trends in N flux. The average annual total P flux (Fig. 3.9) was 154,000 metric tons P/year (170,000 tons P/year) for the water years 2001–2005, with SRP flux 24% of total P flux. Battaglin (2006) reported that total P flux increased during that period, but this was in comparison to the average flux during the period 1980–1996. When total P flux is viewed during the entire period of 1980–2005 and a LOWESS curve fit to the data set, there appears to be a slight increasing trend since the mid-1990s. The annual flux of dissolved silicate appears to have declined slightly since the early 1990s.

Nutrient ratios: Ratios of N to P and Si to N can be important in determining the growth of various phytoplankton species in the Gulf. The Si:DIN (dissolved inorganic N) ratio ranged from about 2 to 4.5 during the 1950s and 1960s but then greatly decreased as silicate concentrations declined by about 50% between the 1950s and 1980s (Rabalais et al., 1999b; Turner and Rabalais, 1991). Ratios since 1980 of Si:DIN have been just above 1 annually (Fig. 3.10), averaging 1.08 for 2001–2005 water years. Nitrogen to P ratios averaged 18 for 2001–2005 have shown little variability since the early 1990s, with perhaps a declining trend. These

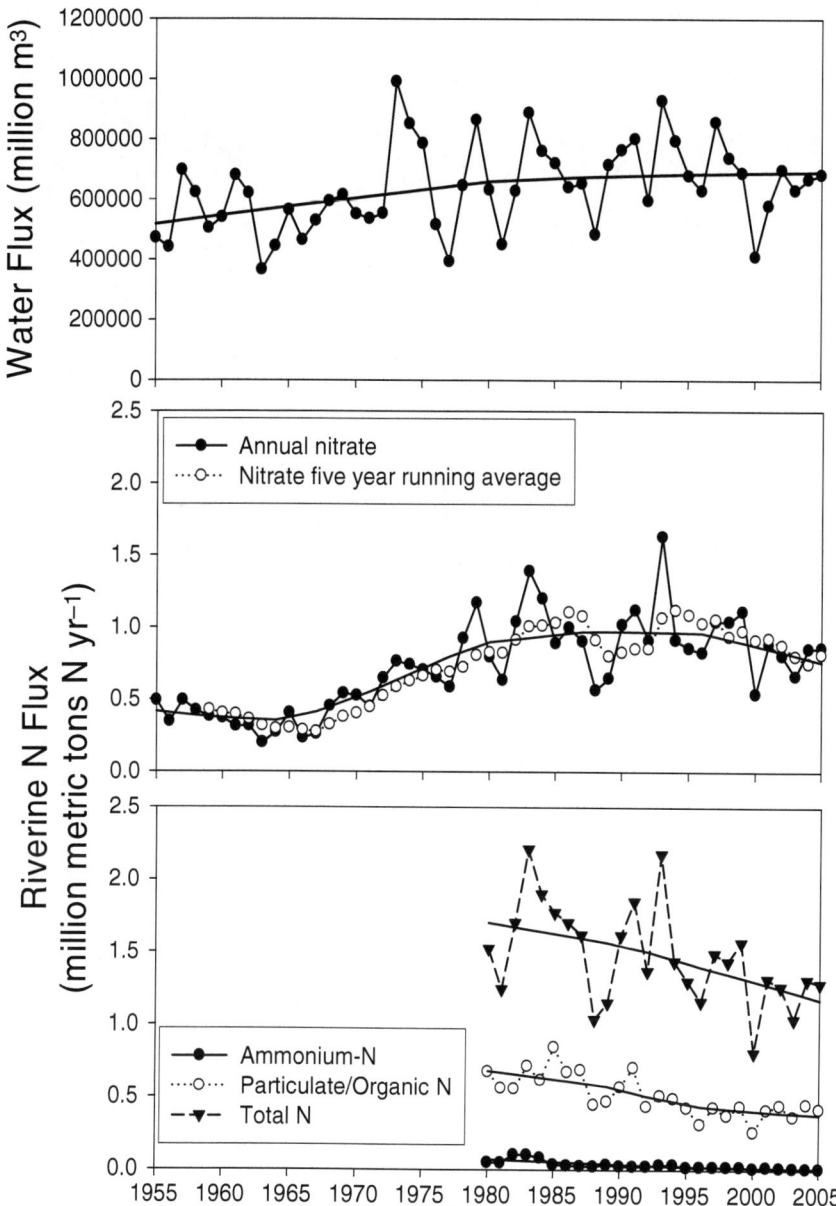

Fig. 3.8 Flow and available nitrogen monitoring data for the MARB for 1955 through 2005 water years (LOWESS, locally weighted scatterplot smooth, curves shown as a *solid line*). LOWESS describes the relationship between Y and X without assuming linearity or normality of residuals and is a robust description of the data pattern (Helsel and Hirsch, 2002)

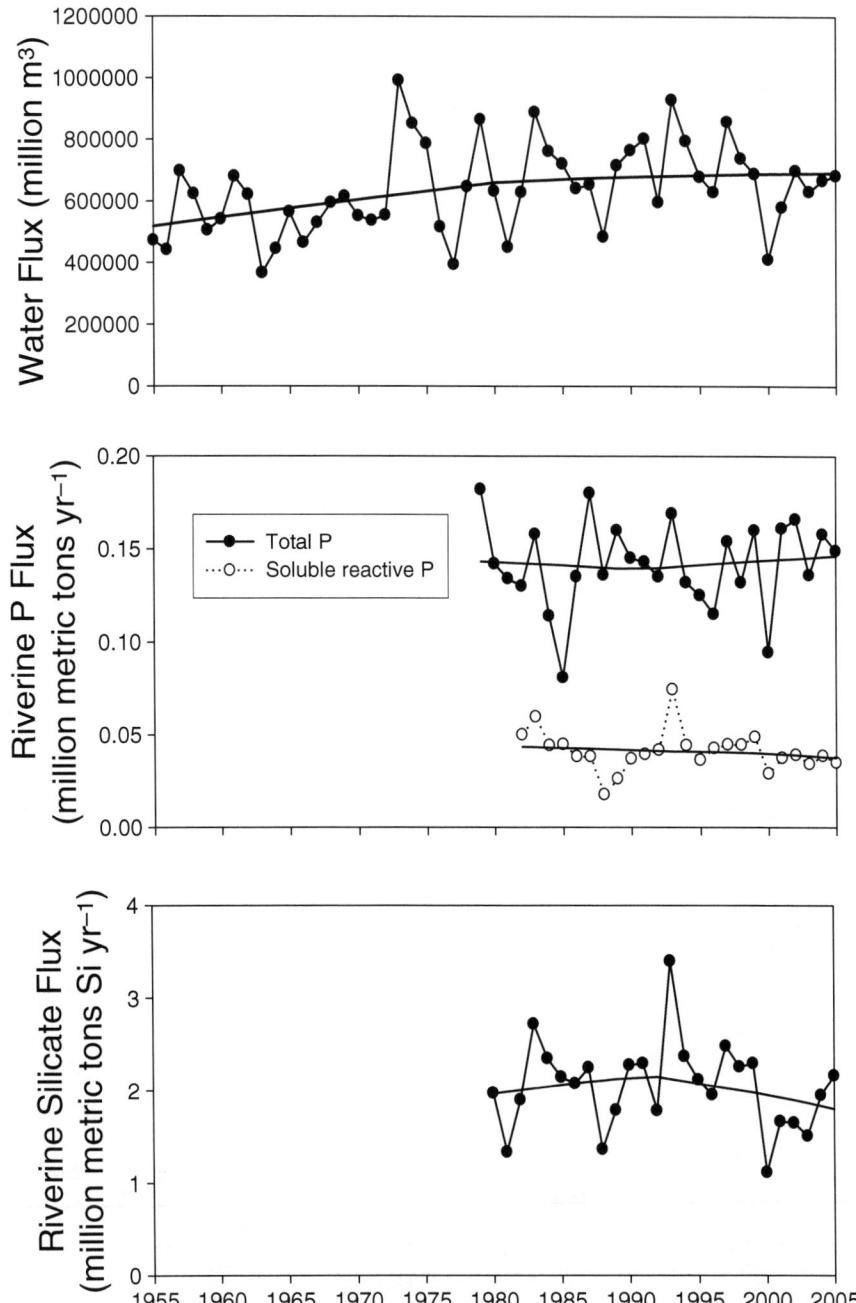

Fig. 3.9 Flow, available phosphorus, and available silicate monitoring data for the MARB for 1955 through 2005 water years (LOWESS curves shown as a *solid line*). Based on USGS data from Battaglin (2006) and Aulenbach et al. (2007)

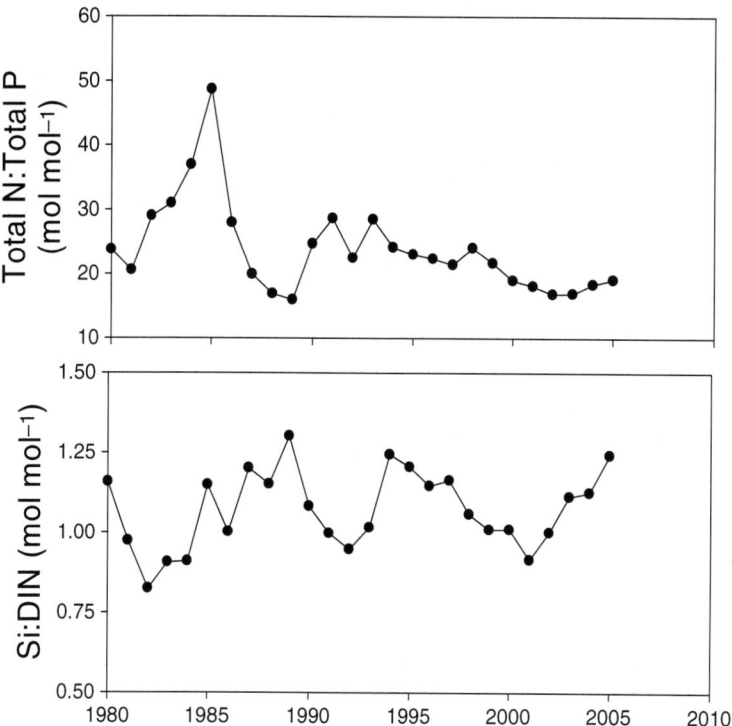

Fig. 3.10 Ratio of total N to total P and dissolved silicate to dissolved inorganic N for MARB for the 1980 through 2005 water years. Based on USGS data from Battaglin (2006) and Aulenbach et al. (2007)

ratios are useful to compare to the Redfield ratio (Si:N:P = 16:16:1) and suggest, as Rabalais et al. (1999) concluded, that annual nutrient fluxes to the Gulf are quite close to this ratio. However, spring ratios, discussed later, are somewhat different and may have a more important effect on Gulf phytoplankton growth.

3.1.1.2 Seasonal Patterns

Nitrogen: Since the *Integrated Assessment*, greater emphasis has been placed on the spring flux of nutrients (sum of April, May, and June fluxes) as a possible important regulator of hypoxia, and, therefore, fluxes for this period were examined using the available data for the period 1979–2006. Whereas the annual water flux showed a slightly increasing trend since 1990 (Fig. 3.8), the spring water flux, although highly variable, appears to show a decreasing trend (Fig. 3.11). Spring nitrate-N flux also has declined, with even larger decreases in TKN flux and, therefore, total N flux.

Phosphorus and silicate: Spring P flux (both total and SRP) has changed relatively little, with perhaps a small decrease in total P flux (Fig. 3.12). The spring

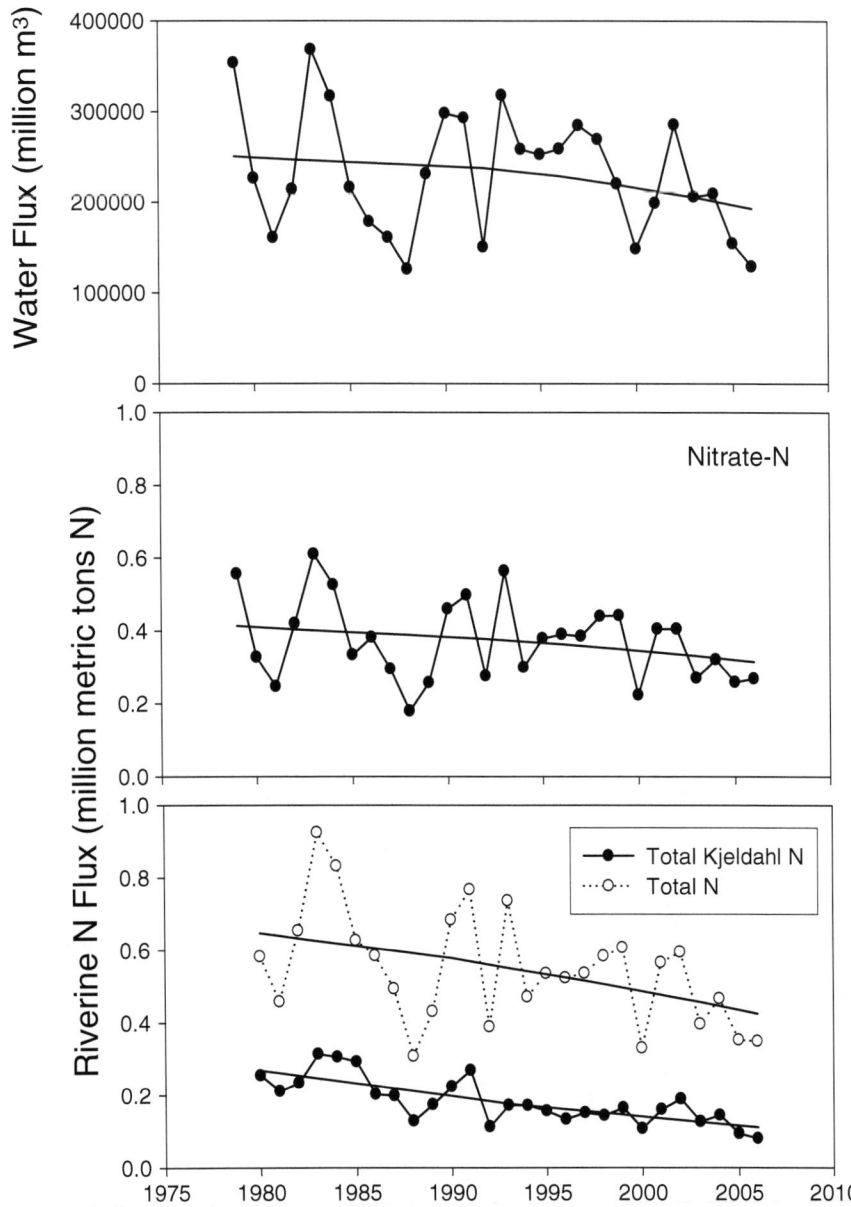

Fig. 3.11 Flow and nitrogen flux for the MARB during spring (April, May, and June) for the period 1979–2005 (LOWESS curve shown as a *solid line*). Based on USGS data from Battaglin (2006) and Aulenbach et al. (2007)

3.1 Temporal Characteristics of Streamflow and Nutrient Flux

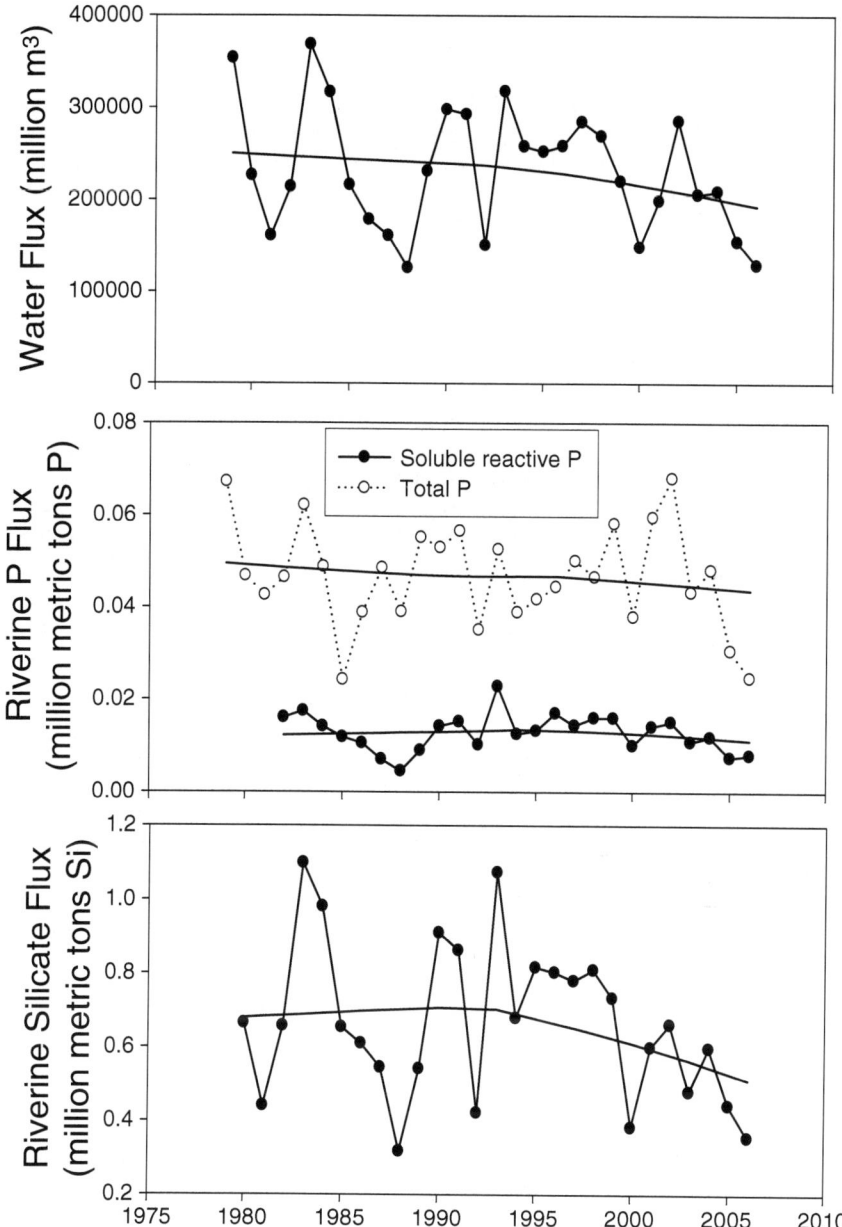

Fig. 3.12 Flow, phosphorus, and silicate flux for the MARB during spring (April, May, and June) for the period 1979–2006 (LOWESS curve shown as a *solid line*). Based on USGS data from Battaglin (2006) and Aulenbach et al. (2007)

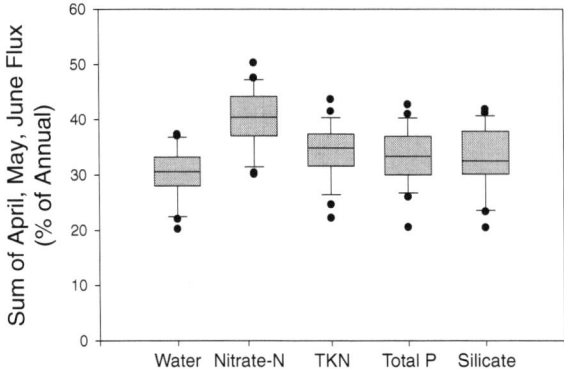

Fig. 3.13 Sum of April, May and June fluxes as a percent of annual (water year basis) for combined Mississippi mainstem and Atchafalaya River. *Box* plots show median (*line in center* of box), 25th and 75th percentiles (*bottom and top* of box, respectively), 10th and 90th percentiles (*bottom and top error bars*, respectively), and values <10th percentile and >90th percentile (*solid circles* below and above *error bars*, respectively). Based on USGS data from Battaglin (2006) and Aulenbach et al. (2007)

dissolved silicate flux has shown a pronounced decline since 1990s, greater than the decline in water flux. The reason for this decline is not known.

Figure 3.13 shows the spring fluxes (sums of April, May, and June fluxes) as a percentage of the annual fluxes. There is considerable interannual variability in the annual fluxes that occurs during spring, as indicated by the whiskers on the box plots. Spring water flux was, on an average, 30% of annual flux, whereas nitrate-N was 40%, TKN 34%, and total P 34% of their annual fluxes. Therefore, the river is disproportionately enriched with all nutrients during the spring but particularly with nitrate. This result further substantiates the conclusion drawn earlier that tile-drained fields are a primary source of N, which is released beginning in winter (Ohio into central Illinois) to spring (northern Illinois, Iowa and Minnesota). This influence was very evident in 2002, when 50% of the nitrate-N flux occurred during the three spring months. Royer et al. (2006) pointed out how most of the N and P flux from tile-drained watersheds occurred during a few months during winter and spring each year, further supporting the trends at this larger scale.

Nutrient ratios: N-to-P ratios during spring flow to the Gulf averaged 22 for 2001–2005 (Fig. 3.14), greater than the annual value of 18 for the same time period. As discussed previously, nitrate transport is greater during this period than is P transport. The Si:DIN ratio was also lower during the spring compared to the annual mean for 2001–2005 (spring ratio 0.84, annual ratio 1.08), reflecting greater transport of nitrate compared to silicate. Turner et al. (1999) concluded that decreasing Si:DIN ratios to less than 1.1 could greatly alter Gulf food web dynamics because the proportion of diatoms in the phytoplankton community would be reduced, which would impact zooplankton and higher trophic levels.

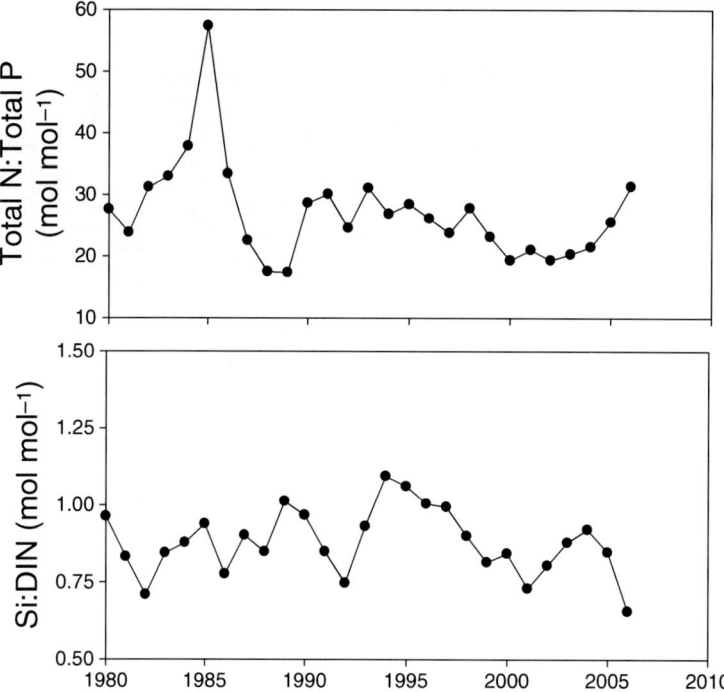

Fig. 3.14 Ratio of total N to total P and silicate to dissolved inorganic N for the MARB during spring (April, May, and June) for the period 1980–2006. Based on USGS data from Battaglin (2006) and Aulenbach et al. (2007)

3.1.2 Subbasin Annual and Seasonal Flux

3.1.2.1 Annual Patterns

USGS estimates (Aulenbach et al., 2007) were used to examine nutrient fluxes within subbasins of the MARB. Annual nutrient fluxes were calculated with an adjusted maximum likelihood estimate (AMLE), a type of regression model method, with a 5-year moving average calibration period (composite method estimates were not made for subbasin data). Figure 3.15 shows the location of nine subbasins comprising the MARB. Figure 3.7 shows a schematic of the MARB sampling stations to assist with the following analyses. The initial analysis discusses the cumulative fluxes of five major subbasins: (1) Upper Mississippi (upstream of Thebes, IL, minus the inflow from the Missouri River), (2) Ohio-Tennessee (upstream of Grand Chain, IL), (3) Missouri (upstream of Hermann, MO), (4) Arkansas-Red (combined flux from the Arkansas and Red Rivers), and (5) Lower Mississippi.

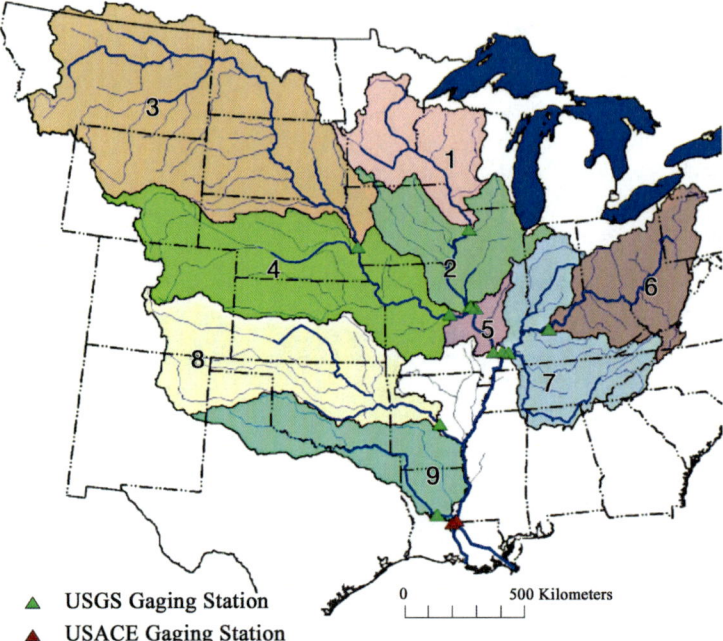

Fig. 3.15 Location of nine large subbasins comprising the MARB that are used for estimating nutrient fluxes (from Aulenbach et al., 2007)

3.1.2.2 Annual Flux Estimates

The flux estimates from the five subbasins are listed in Table 3.1. During the past five water years, most of the nitrate-N flux (84%) and TKN flux (73%) was from the Upper Mississippi and Ohio-Tennessee subbasins. The Missouri subbasin contributed 9.8% of the nitrate-N flux to the Gulf, with much smaller fluxes coming from the Arkansas-Red and lower Mississippi River subbasins. These data clearly illustrate that the source of both nitrate-N and TKN is from the upper Mississippi River basin before the Missouri River enters. For total P flux, the Missouri subbasin was more important and contributed 20% of the flux, compared to 26 and 38% for the upper Mississippi and the Ohio-Tennessee subbasins, respectively.

To further examine source areas of N, P, and silicate, the nutrient fluxes in the MARB were divided into ten smaller subbasins (see Fig. 3.15 and Table 3.2), with some of the values calculated as the difference between an upstream and a downstream monitoring station. The lower Mississippi River subbasin is again calculated by difference and is the same in both the five- and ten-subbasin analyses (this subbasin is not shown in Fig. 3.1 but was included in the Table 3.1 analysis). These results are listed in Table 3.2. For nitrate-N, this further breakdown of the basin indicates that the largest sources are the upper Mississippi and Ohio-Tennessee River subbasins. These subbasins represent about 31% of the total land area within

3.1 Temporal Characteristics of Streamflow and Nutrient Flux

the MARB, yet they contribute about 82% of the nitrate-N flux, 69% of the total Kjeldahl N, and 58% of the total P flux. Furthermore, when the subbasins are further divided, the subbasin feeding into the upper Mississippi River between Clinton, IA, and Grafton, IL, contributes about 29% of the nitrate-N flux, while representing only 7% of the drainage area. The Missouri River at Hermann also was a relatively large contributor of total P (14% of total flux). For dissolved silicate, percentages did not include the Red River because estimates were not available. Again, most of the silicate flux was from the upper Mississippi River and the Ohio-Tennessee River, similar in proportion to water flux.

Table 3.1 Average annual nutrient fluxes in 1000 metric tons for the five large subbasins in the MARB for the 2001–2005 water years. (Percent of total basin flux shown in parentheses)

Subbasin	Area (km^2)	Flow (M m^3/year)	Nitrate-N (1000 MT)	TKN (1000 MT)	Total P (1000 MT)
Upper Mississippi[a]	493,900	116,200	349 (43%)	136 (32%)	40.4 (26%)
Ohio–Tennessee	525,800	279,800	335 (41%)	175 (41%)	58.7 (38%)
Missouri	1,353,300	60,080	78.6 (9.8%)	83.8 (20%)	30.4 (20%)
Arkansas-Red	584,100	67,200	28.7 (3.5%)	43.9 (10%)	8.7 (6%)
Lower Mississippi[a]	183,200	129,550	22.1 (2.7%)	−8.4 (−2%)	16.1 (10%)

[a]Nutrient fluxes calculated by difference. Negative values occur where downstream site had a lower flux than upstream site, the result of either error in the flux estimates or a real net loss of nutrients within the subbasin (Aulenbach et al., 2007).

3.1.2.3 Annual Yield Estimates

Similarly, the nitrate-N and TKN yields were dominated by the Upper Mississippi and Ohio-Tennessee River subbasins, with nitrate-N values of 7.1 and 6.4 kg N/ha-year (6.3 and 5.7 lb N/ac-year) and TKN values of 2.7 and 3.3 kg N/ha-year (2.4 and 2.9 lb N/ac-year) for the upper Mississippi and Ohio-Tennessee River subbasins, respectively (Table 3.3). The Missouri and Arkansas-Red River subbasins had much lower nitrate-N yields of 0.6 and 0.5 kg N/ha-year (0.53 and 0.44 lb N/ac-year) for this 5-year period. Similar to N, yield of total P was much greater in the upper Mississippi and Ohio-Tennessee River subbasins when compared to the Missouri River. The greater yields from the upper Mississippi and Ohio-Tennessee River basins no doubt reflect the relative sizes of the basins when compared to the Missouri River but also the importance of point sources in the basins, as well as more intensive agricultural inputs.

When nutrient yields from the nine smaller subbasins are examined, the yields from the upper Mississippi River between Clinton and Grafton and the entire Ohio

Table 3.2 Average annual nutrient fluxes for 10 subbasins in the MARB for the 2001–2005 water years. Some subbasin fluxes are calculated as the difference between the upstream and the downstream monitoring station. (Percent of total basin flux shown in parentheses)

Subbasin	Area (km^2)	Flow (M m^3/year)	Nitrate-N	TKN	Total P	Si
			1,000 metric tons			
Mississippi–Clinton	222,000	48,300	88.3 (11%)	50.1 (12%)	8.5 (6%)	219 (12%)
Mississippi–Grafton[a]	221,700	52,100	237 (29%)	71.7 (17%)	21.2 (14%)	162 (9%)
Missouri–Omaha	836,000	23,900	24.1 (3%)	25.4 (5.9%)	8.1 (5%)	102 (6%)
Missouri–Hermann[a]	517,000	36,100	54.6 (7%)	58.4 (14%)	22.3 (14%)	161 (9%)
Mississippi–Thebes[a]	50,300	15,800	23.8 (3%)	13.9 (3%)	10.8 (7%)	8.5 (0.5%)
Ohio–Cannelton	251,000	133,400	160 (20%)	92.1 (21%)	35.2 (23%)	355 (20%)
Ohio–Grand Chain[a]	275,000	146,400	175 (22%)	82.7 (19%)	23.5 (15%)	320 (18%)
Arkansas–Little Rock	409,300	33,900	21.9 (3%)	19.5 (5%)	4.4 (3%)	102 (6%)
Red River–Alexandra	175,000	33,200	6.8 (1%)	24.3 (6%)	4.3 (3%)	757 (20%)[b]
Lower Mississippi[a]	183,200	129,550	22.1 (2.7)	−8.4 (−2)	16.1 (10%)	

[a]For these basins, fluxes were calculated as the difference between upstream and downstream stations.
[b]For these two subbasins, fluxes were calculated by difference from overall basin flux minus eight subbasins where Si flux was estimated.

Table 3.3 Average annual nutrient yields in kg/ha-year for the five large subbasins in the MARB for water years 2001–2005

Subbasin	Nitrate-N	TKN	Total P
Upper Mississippi	7.1	2.7	0.8
Ohio–Tennessee	6.4	3.3	1.1
Missouri	0.6	0.6	0.2
Arkansas-Red	0.5	0.8	0.1
Lower Mississippi	1.2	−0.5	0.9

River basin were 10.7 and 6.4 kg N/ha-year (9.6 and 5.7 lb N/ac-year), respectively (Table 3.4). The largest total P yield (2.1 kg P/ha-year or 1.9 lb P/ac-year) was from the subbasin measured on the Mississippi River at Thebes, which would include row crop lands of Missouri River and southern Illinois River along with sewage effluent from St. Louis. Greatest dissolved silicate yields were from the Ohio River, followed by the upper and lower Mississippi River, again reflecting water flux.

3.1 Temporal Characteristics of Streamflow and Nutrient Flux

Table 3.4 Average annual nutrient yields for nine subbasins in the MARB for the 2001–2005 water years. Some subbasin yields are calculated as the difference between the upstream and the downstream monitoring stations

Subbasin	Nitrate-N	TKN	Total P	Silicate
	(kg/ha-year)			
Mississippi–Clinton	4.0	2.3	0.4	9.9
Mississippi–Grafton	10.7	3.2	1.0	7.3
Missouri–Omaha	0.3	0.3	0.1	1.2
Missouri–Hermann	1.1	1.1	0.4	3.1
Mississippi–Thebes	4.7	2.8	2.1	1.7
Ohio–Cannelton	6.4	3.7	1.4	14.1
Ohio–Grand Chain	6.4	3.0	0.9	11.6
Arkansas–Little Rock	0.5	0.5	0.1	2.5
Red River–Alexandra	0.4	1.4	0.2	9.9[a]
Lower Mississippi	1.2	−0.5	0.9	

[a]Flux calculation available only for the sum of two subbasins.

Subbasin nitrate-N yield compared to net N inputs: The complete time series records were examined to better understand longer term patterns in subbasins contributing the largest N and P fluxes. At the five-subbasin level, the trend lines for flow and N fluxes for the Ohio River basin have been relatively flat since the early 1980s (Fig. 3.16).

However, the upper Mississippi River subbasin has experienced a decreasing trend in annual flow since the mid-1990s (Fig. 3.17). What appears to be only a slight decrease in nitrate-N yield in the upper Mississippi subbasin in response to what the Study Group thinks are greatly decreasing net N inputs demonstrates the difficulty in predicting riverine nutrient yields in tile-drained agricultural lands. Many interacting factors are at work, which are difficult to estimate and/or measure. For example, there are uncertainties in some of the estimates, such as biological N_2 fixation (primarily soybean), as well as our assumption that large soil N pools are in a steady state. The predominant soil types in the upper Mississippi subbasin are Mollisols, which are high in organic matter with large soil organic N pools (much larger than the Ohio River subbasin). As fertilizer rates have stayed constant and yields have increased, several possibilities may account for the lack of riverine response. These include increasing soybean N_2 fixation percentages, net N mineralization of soil organic N (David et al., 2001), long lag times due to a buildup of relatively easily degradable organic N (amino sugar N, Mulvaney et al., 2001) that is now being released, or perhaps increasing tile drainage and loss of fall-applied N. Figure 3.17 includes a recalculation of net N inputs for 1998–2005, increasing soybean fixation rates from 50 to 70%, and assuming a corn acre net soil mineralization rate of 10 kg N/ha-year (8.9 lb N/ac-year). These two changes greatly alter the net inputs, pushing the value back up to where it was during the 1980s.

Soybean production is a net depletion to soil N pools, and the fixation rate is a function of available inorganic N (nitrate) in the soil (Gentry et al., 2001). When

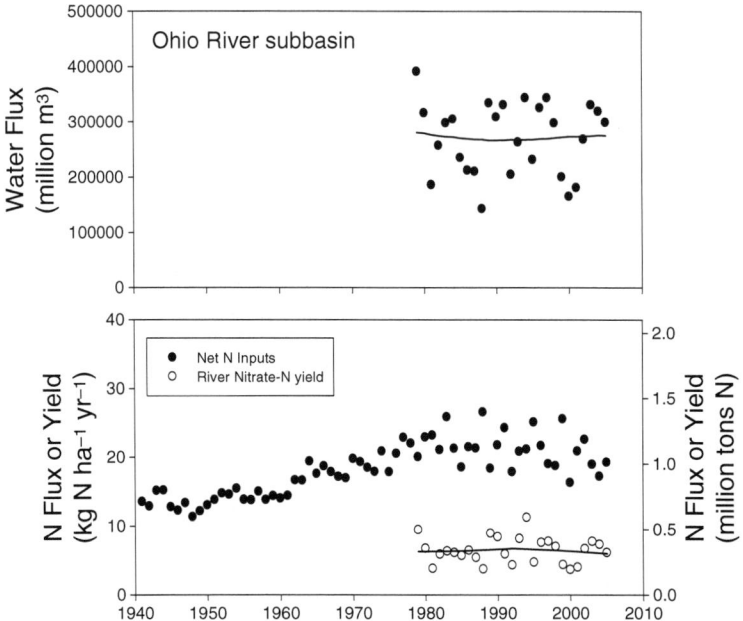

Fig. 3.16 Net N inputs and annual nitrate-N fluxes and yields for the Ohio River subbasin. (LOWESS curves for riverine nitrate-N shown with *solid lines*.) Based on USGS data from Battaglin (2006) and Aulenbach et al. (2007)

there was more inorganic N left from corn production prior to the late 1990s, soybeans would have fixed less N compared to recent growing seasons when corn yields have set records and little residual soil nitrate would be expected. This could be leading to increasing soybean N_2 fixation rates, which are not accounted for in typical net N input calculations.

A second factor is soil mineralization. Net N input calculations assume that the soil organic N pool is at a steady state (McIsaac et al., 2002), with mineralization rates over a year balanced by immobilization (both microbial and crop residue inputs). It is possible that with greater corn production and steady fertilizer rates, increased mineralization rates occur, so that there is a net depletion of soil organic N (one component of soil organic matter, which is discussed further in Section 4.5.6). This depletion, as discussed earlier, may be small (about 10 kg N/ha-year or 8.9 lb N/ac-year) but over many acres would be an important additional input.

Finally, another factor may be an increase in tile drainage intensity in the region, combined with increasing fall fertilization and warmer winter temperatures. New and replacement tile drainage is added every year to this region, although no data are available to quantify the increase. Fall application of anhydrous ammonia in much of the region has increased greatly since the 1980s (see later discussion in Section 4.5.6 for supporting sales and USDA ARMS data). The four states of the upper Mississippi River basin (Minnesota, Wisconsin, Iowa, and Illinois) all show

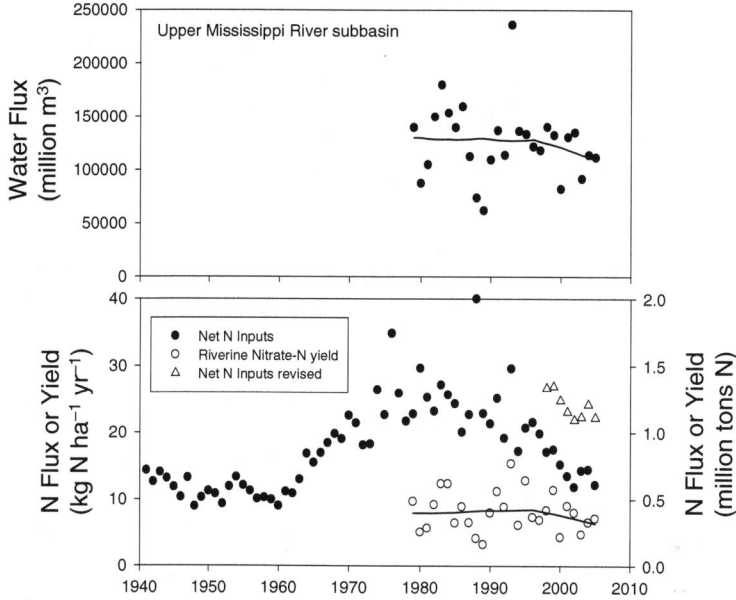

Fig. 3.17 Net N inputs and annual nitrate-N fluxes and yields for the upper Mississippi River subbasin. (LOWESS curves for riverine nitrate-N shown with *solid lines*.) Shown in triangles is a recalculated net N input for the upper Mississippi River basin, increasing soybean N_2 fixation from 50 to 70% of above ground N, and a soil net N mineralization rate from 0 to 10 kg N/ha-year. Based on USGS data from Battaglin (2006) and Aulenbach et al. (2007)

an increasing winter (November through March) temperature (for the months following fall application of anhydrous ammonia, all show strong increasing trends in winter temperatures during the past 30 years; data not shown). Warmer soils would increase nitrification rates and lead to higher concentrations of soil nitrate that could be lost with late winter and spring precipitation. Therefore, fall-applied anhydrous ammonia could be a more important source of spring nitrate-N flux in this subbasin during recent years and, when combined with changing N input and output patterns, may be keeping the flux steady despite the reduction in annual net N inputs.

Changes in subbasin P: As discussed previously, total P flux for the MARB has increased during the monitoring period. Most of this increase was found to have occurred in the Ohio River subbasin, particularly during the 2001–2005 time period (Fig. 3.18). In comparing the 2001–2005 period with 1980–1996, Ohio River total P increased by 51%, while water flux increased only by 6%, and reactive P decreased by 20%. This led to a large increase in particulate/organic P of 89% between these two time periods. Because TKN decreased by 3% during this period, it does not seem that increased erosion can explain this pattern (all indications are that erosion has decreased). The 89% increase in particulate/organic P represents most of the increase in total P flux to the NGOM between 1980–1996 and 2001–2005. Unfortunately, data are not available because of monitoring limitations for smaller

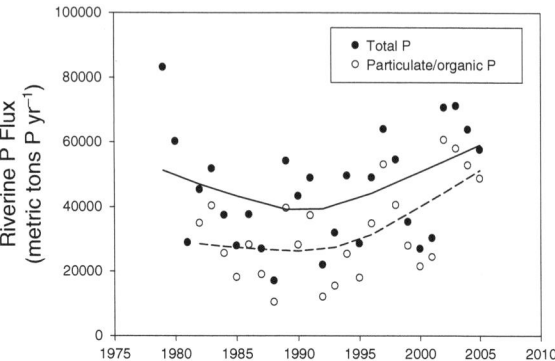

Fig. 3.18 Total P and particulate/organic P fluxes for the Ohio River near Grand Chain, Illinois (LOWESS curves shown in *solid* and *dashed lines*). Based on USGS data from Battaglin (2006) and Aulenbach et al. (2007)

basins within the Ohio River subbasin to further determine the source of this P flux. However, this trend seems to be more widespread than just the Ohio subbasin.

The Missouri and Upper Mississippi River subbasins are following a similar trend as the Ohio River, although their absolute increase in total P is much less than the Ohio River. In both of these subbasins flow has decreased (by 10 and 31% for the Upper Mississippi and Missouri River subbasins, respectively, for 1980–1996 compared to 2001–2005), while total P flux has increased (about 10% in each subbasin). Again, TKN flux has decreased. Therefore, in the Missouri, Upper Mississippi, and Ohio River subbasins flow-weighted total P concentrations have increased greatly during the past 15 years.

These observations are not consistent with overall TKN riverine fluxes in the MARB, and at this time the Study Group has no explanation for this large, yet potentially very important, change in total P concentrations and flux for these subbasins that could influence management decisions.

3.1.2.4 Seasonal Patterns

Spring fluxes (sum of April, May, and June) were examined for the Mississippi River at Grafton and the Ohio River at Grand Chain, and little change in water flux was detected (Fig. 3.19). However, for nitrate-N, there seems to be a slight increasing pattern of spring flux based on LOWESS curves.

When the sum of the upper Mississippi River at Grafton and Ohio River at Grand Chain spring nitrate-N flux is plotted against the flux for the entire basin, an interesting pattern emerges (Fig. 3.20). During the 1980s into the early 1990s, some of the spring flux was from other subbasins, mostly the Missouri River. However, the Missouri River flux has greatly decreased so that now the upper Mississippi River above Grafton and the Ohio River contribute nearly all of the spring flux. Sprague et al. (2006) discuss the riverine fluxes in the Missouri River basin (due to decreasing flow and management practice changes) in a recent report that supports this observation.

3.1 Temporal Characteristics of Streamflow and Nutrient Flux

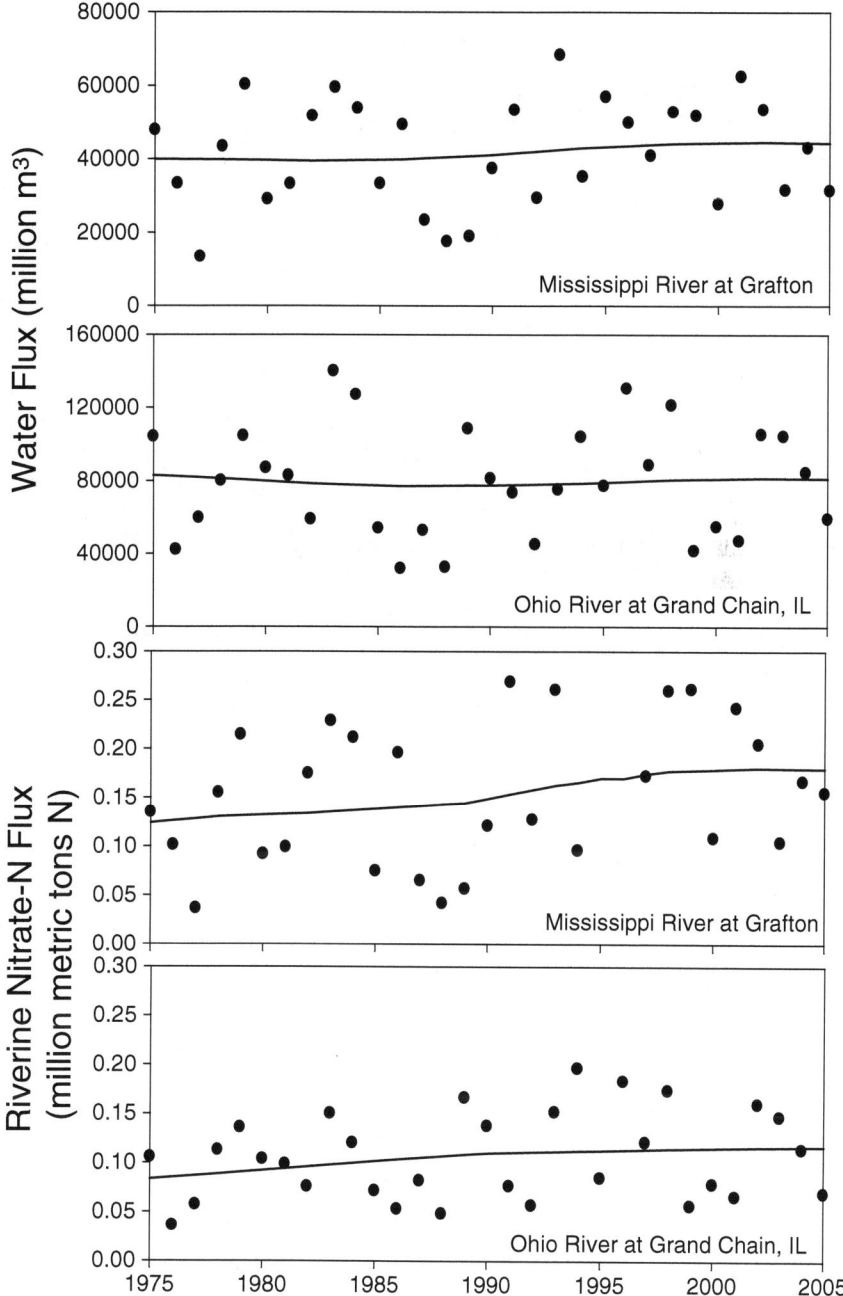

Fig. 3.19 Spring water flux and nitrate-N flux for the Mississippi River at Grafton and the Ohio River at Grand Chain, IL, for water years 1975–2005 (LOWESS curves shown with *solid lines*.) Based on USGS data from Battaglin (2006) and Aulenbach et al. (2007)

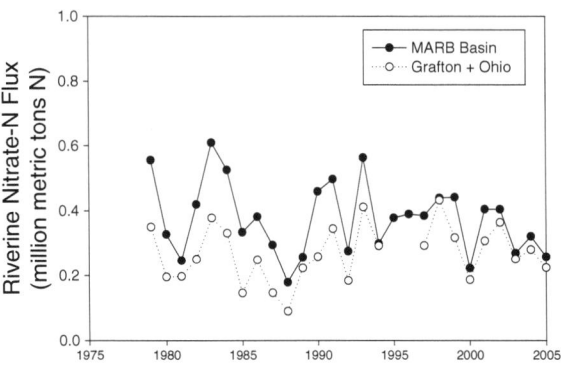

Fig. 3.20 Spring nitrate-N flux (sum of April, May, and June) for the Mississippi River at Grafton plus Ohio River at Grand Chain subbasins compared to the combined Mississippi and Atchafalaya River for 1979 through 2005. Based on USGS data from Battaglin (2006) and Aulenbach et al. (2007)

Key Findings and Recommendations

Most of the research needs identified in the *Integrated Assessment* have not been met, and fewer rivers and streams are monitored today than in 2000. Data continue to be available for the large river sites, but many intermediate and smaller river monitoring sites have been dropped from monitoring programs. Recently, the USGS has initiated real-time (every 2 h) monitoring of three large river sites with field nitrate-N measurement. These types of new efforts to provide expanded monitoring data are critically needed. To more fully assess the response of the entire suite of management programs and changes at the subbasin and large river scale in the MARB, we need more robust monitoring programs that have adequate sampling intensities to allow the composite method (the preferred one) of estimating stream loads to be utilized. At the small-watershed (1,000–50,000 hectares or about 2,500–125,000 acres) scale, there have been many studies, but they provide data for only the period of funding, which is often short. A monitoring network is needed throughout the MARB focused on small watersheds with larger N and P loads and that provides intensive, long-term data. This network will allow determination of how effective particular individual or suites of management programs are in reducing nutrient loads. However, because of year-to-year weather patterns and the often slow response of changes in outputs, these programs will need to be in place for decades. Finally, there is a critical need for the ability to document tile drainage intensity, which requires that new techniques be developed and applied.

Changes in USGS flux calculation methods have altered estimates of nutrient flux as reported in the *Integrated Assessment*. LOADEST 5-year and a new COMPOSITE method seem to be the best estimation methods. Although water flux for the MARB has increased slightly during the past 25 years, total N, primarily nitrate-N and particulate/organic N, has decreased. The total N flux averaged 1.24 million metric tons/year (1.37 million tons/year) from 2001 to 2005 (65% of the flux is nitrate), and the total P flux averaged 154,000 metric tons/year (170,000 tons/year). During the spring (April–June), water flux for the MARB appears to have decreased slightly, causing similar decreases in total N (nitrate-N and TKN). Spring dissolved silicate flux has declined more than water flux. Neither total P nor SRP fluxes show major annual or seasonal trends during the full period of record.

The subbasin analysis provides clear evidence that while the upper Mississippi and Ohio-Tennessee River subbasins represent about 31% of the total drainage area of the MARB, they contribute about 82% of the nitrate-N flux, 69% of the TKN flux, and 58% of the total P flux to the Gulf. Furthermore, when the subbasins are further divided, the subbasin feeding into the upper Mississippi River between Clinton, IA, and Grafton, IL, contributes about 29% of the nitrate-N flux while representing only 7% of the drainage area. Perhaps more importantly, the upper Mississippi and Ohio-Tennessee River subbasins currently represent nearly all of the spring N flux to the Gulf. These subbasins represent the tile-drained, corn-soybean landscape of Iowa, Illinois, Indiana, and Ohio and illustrate that corn–soybean agriculture with tile drainage leaks considerable N under the current management system. The source of riverine P is more diffuse, although these subbasins are also the largest sources of P. A large increase in the Ohio River subbasin particulate/organic P flux occurred during the 2001–2005 time period, which was the source of nearly all of the increase in total P to the NGOM. At the same time flow-weighted total P concentrations increased in the Upper Mississippi and Missouri River subbasins as well, although increases in flux were smaller than the Ohio River due to decreased water flux. The Study Group has no explanation for this striking change in P concentrations in these subbasins.

Based on these findings, the Study Group recommends the following:

- Establishment of a monitoring network (20–100) of small watersheds will provide long term (tens of years), intensive flux data to determine the response of management programs and decisions in the MARB.
- More intensive monitoring of larger rivers at the subbasin and entire MARB scale is needed to allow for monthly calculation of fluxes using the composite estimation method, the most accurate method estimating fluxes.
- Further research is needed to determine why riverine spring nitrate-N fluxes are not declining in response to annual net N input decreases, which will inform management decisions for corn–soybean agriculture.
- The increase in riverine total P concentrations needs to be fully explored to verify the increase and to further document the source, potentially considering management implications for control of P in the MARB.

- The tile-drained Corn Belt region of the MARB is an important target for reductions in both N and P, focusing on both surface (P) and subsurface losses (N).
- Additional research is needed to better define the extent, pattern, and intensity of agricultural drainage, including cropland drained by field tile as well as cropland not directly drained by field tile but contributing to drainage networks.

3.2 Mass Balance of Nutrients

Mass balance can be used to better understand sources, sinks, and transformations of nutrients in ecosystems, although losses to stream water are not specifically determined. Goolsby et al. (1999) constructed a detailed annual N mass balance for 1960–1996 and a P mass balance for 1992. Improving flux estimates was identified as a research need. In particular, better estimates are needed for soil N mineralization, soil immobilization, plant N volatilization, denitrification, and biological N_2 fixation.

3.2.1 Cropping Patterns

Mass balances reflect the types and areas of crops grown across the MARB. There were large changes in these crops over the past half-century (Fig. 3.21). Earlier cropping systems had more diverse rotations, including corn, wheat, hay, and oats. With the onset of modern agriculture and large fertilizer inputs, much of the MARB is now in a corn and soybean rotation. By the late 1990s, corn and soybean areas were equal but more recently corn acreage has increased and soybean has decreased, with this trend very apparent in 2007. This trend is expected to continue as demand for corn increases due to expanding ethanol production, the implications of which are discussed in detail in Section 4.5.9.

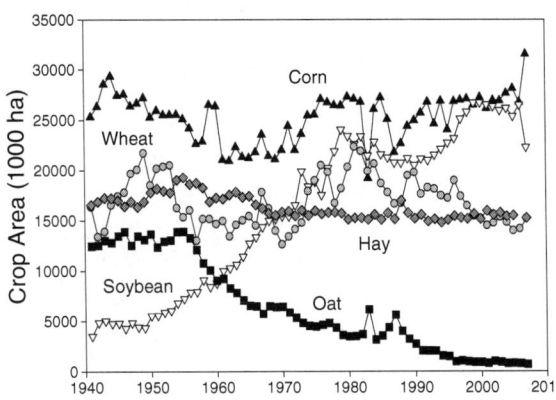

Fig. 3.21 Area of major crops planted in the MARB from 1941 through 2007. Adapted from McIsaac (2006)

3.2.2 Nonpoint Sources

Nitrogen: The N mass balance described in the *Integrated Assessment* indicated that there was a greater surplus of N during the 1950s than during the 1980s and 1990s (Goolsby et al., 1999). McIsaac et al. (2001, 2002) used the same data set to determine the N mass balance using a method described by Howarth et al. (1996) that also has been used by many others (e.g., David et al., 2001; David and Gentry, 2000; McIsaac and Hu, 2004 for Illinois). Net anthropogenic N inputs (NANI) were calculated (sum of fertilizer, NOy deposition, N_2 fixation, minus net food and feed imports) from existing MARB data bases, assuming that the large soil organic N pool is in a steady state. Manure is included in this calculation as part of the feed imports, where grain consumed and excreted as a part of animal agriculture is estimated. NANI is N that should be available for denitrification, loss to groundwater, or leaching and transport in streams.

The recalculated NANI for the MARB showed a clear increase from about 9 kg N/ha-year (8 lb N/ac-year) in the 1940s to about 16 kg N/ha-year (14 lb N/ac-year) from the early 1980s to present, with a maximum value of 20.9 kg N/ha-year (18.7 lb N/ac-year) in 1988 (Fig. 3.22). This increase was due to increasing fertilizer N inputs (from 0 to ~20 kg N/ha-year or 17.9 lb N/ac-year) and higher N_2 fixation from the increased soybean production (from about 8–14 kg N/ha-year or about 7–12.5 lb N/ac-year). Atmospheric deposition appears to be the greatest in the Ohio River basin (about 16% of NANI) and shows a slight increase basin-wide but generally is a small component of the NANI (for a more detailed discussion see Appendix B). Manure shows a slight decrease across the MARB, as extensive animal production has moved to feedlots further west, but represents only about 16% of the total inputs. However, animal production has become concentrated in specific regions of the MARB, creating localized nutrient surpluses compared with crop needs and offtake (US Department of Agriculture, 2003). Up to now, this has led to water quality impairment at a local rather than MARB scale because the animal operations have become concentrated (for more information on distribution see Section 4.5.5 and Appendix C). Therefore, the major changes in inputs were due to fertilizer and N_2 fixation. However, when compared to the amount of N removed during crop harvest, which has dramatically increased since 1940, the increase in N inputs from fertilizer and N_2 fixation do not appear to have increased proportionately. In fact, this rapid increase in crop production has led to a small decrease in NANI from about 17 kg N/ha-year (15 lb N/ac-year) in 2000 to net N inputs of 14 kg N/ha-year (12.5 lb N/ac-year) in 2004 and 2005 (McIsaac, 2006).

The subbasins that contribute the greatest N flux to the Gulf are the upper Mississippi and Ohio River basins, due largely to the intensity of agriculture with concomitant large inputs of N from fertilizer and fixation combined with the system of tile drains. Therefore, when the nitrogen balance is presented by subbasin (Fig. 3.23), the highest net nitrogen inputs are to those subbasins.

However, a closer look at the inputs to the upper Mississippi River basin shows that, even though N inputs from fertilizer and N_2 fixation appear to be fairly level during recent years, the amount of N removed during harvest continues to

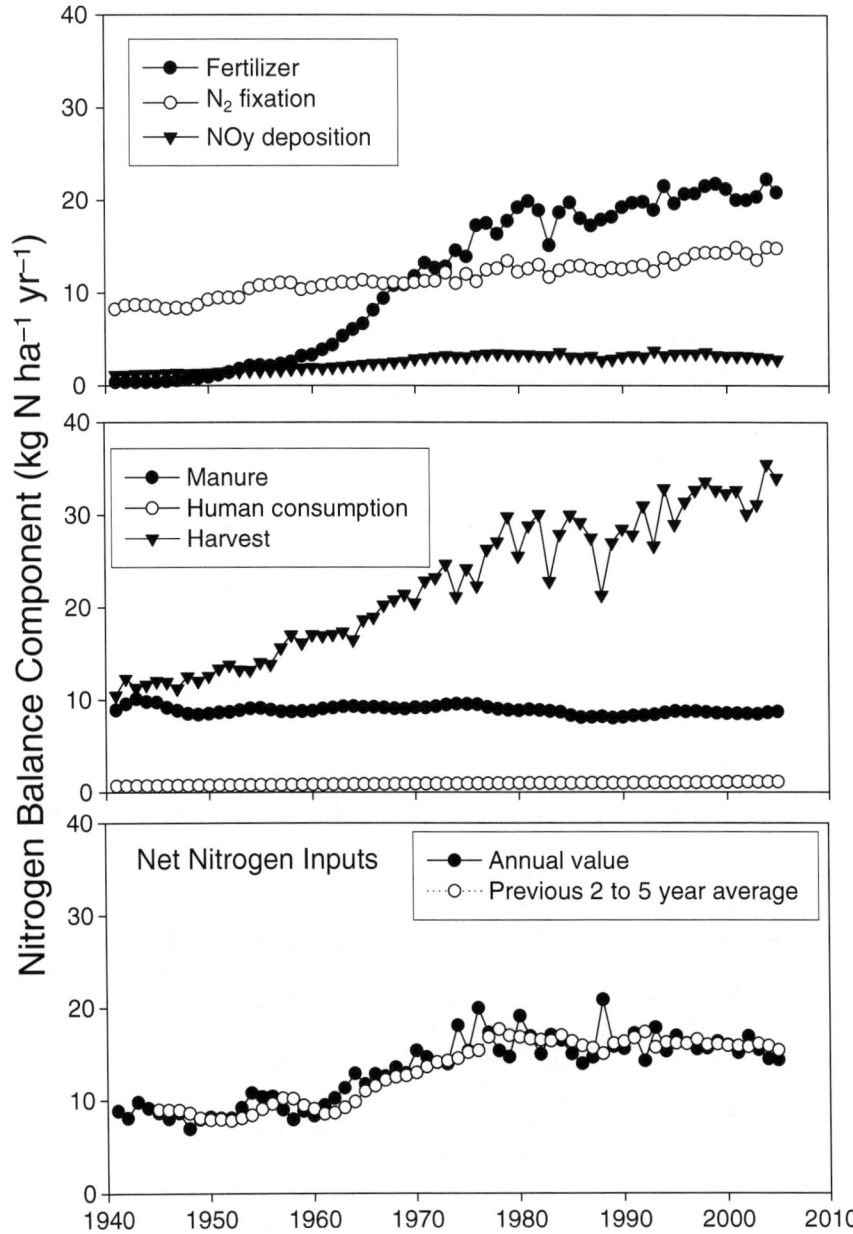

Fig. 3.22 Nitrogen mass balance components and net N inputs for the MARB, as calculated by McIsaac et al. (2002) and updated through 2005 by McIsaac (2006)

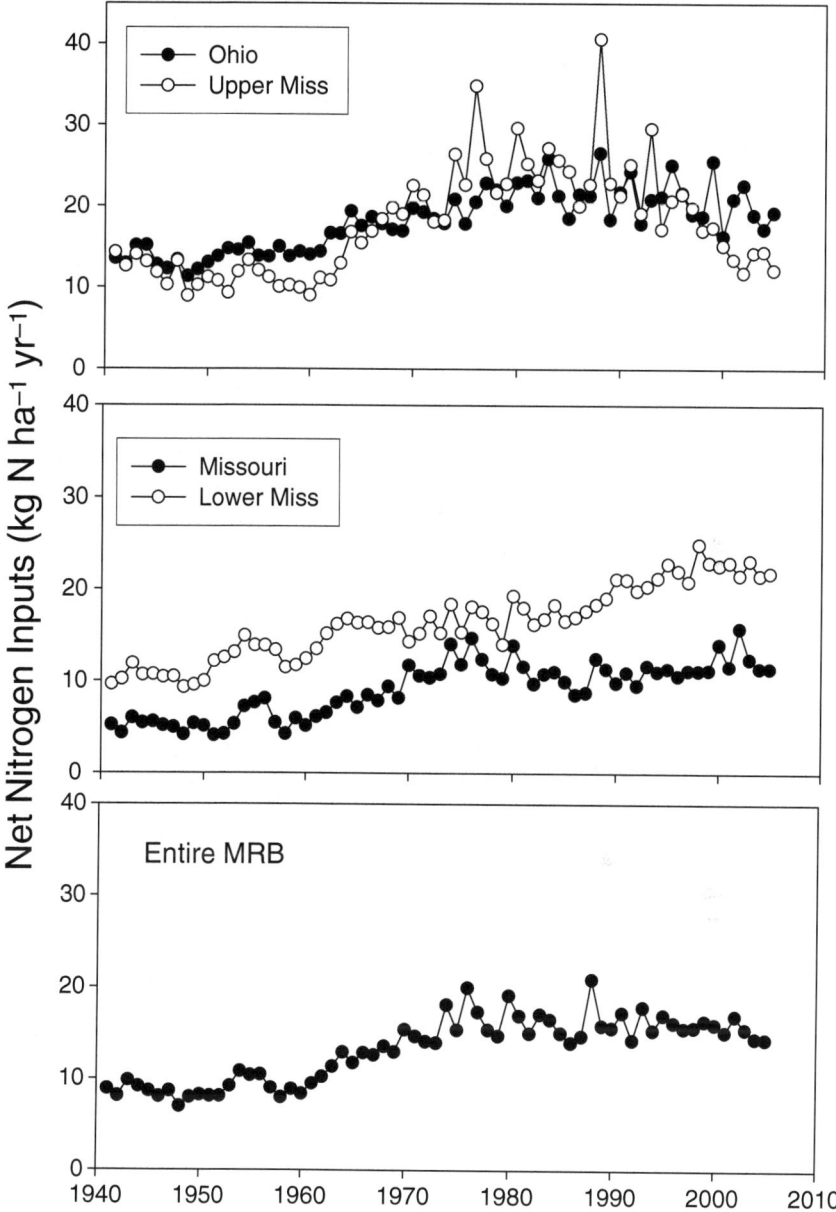

Fig. 3.23 Net N inputs for the four major regions of the MARB through 2005. Adapted from McIsaac (2006)

increase, resulting in a substantial decline in NANI (Fig. 3.24). These changes are not reflected in the other subbasins, which lead to a small decline in NANI to the overall basin. However, given the importance of the upper basin as a source of nitrate-N, it might be expected that the riverine flux of N would start to decrease.

McIsaac et al. (2001, 2002) showed that net N inputs could be used, in combination with riverine water flux, to predict export of nitrate-N to the Gulf. They found that a 2–5 year lagged net N input explained the most variation in nitrate-N export, with 6–9 year lagged net N inputs explaining less, but a significant amount of the variation. Therefore, given the large decrease in net N inputs in the upper Mississippi River subbasin, it is reasonable to expect riverine export of nitrate should decrease. However, there is a factor that is not assessed in the net N input mass balance that may be important.

McIsaac and Hu (2004) showed that, for tile and nontile-drained regions of Illinois, net N inputs were similar but that riverine export of N was much greater in the tile-drained watersheds. They found that during the 1990s net N inputs were equal to riverine N flux, about 27 kg N/ha-year (24 lb N/ha-year). This would leave no N available for other fluxes that are thought to be important, such as terrestrial and aquatic denitrification. More recent net N inputs in these same tile-drained watersheds are about zero, yet riverine N export has continued. Given that there are denitrification losses (that are unmeasured), this result indicates that N must be coming from a depletion of soil N pools, as suggested by Jaynes et al. (2001). With steady fertilizer N rates, high corn and soybean yields, and high stream N export, the only source available to supply N would be the large soil N pool (often 10,000–15,000 kg N/ha or 8,930–13,400 lb N/ac) in the Mollisols of the upper Midwest. Techniques are not yet available to document the small change that would be occurring in this N pool from a small annual depletion of 25–50 kg N/ha-year (22–45 lb N/ac-year); however, this possibility has critical implications for the sustainability of production.

Another possibility raised by McIsaac et al. (2002) is that estimates of crop harvest N, N_2 fixation, or animal consumption of N and manure production could be inaccurate. Although Goolsby et al. (1999) recommended improvement in estimates of the N mass balance, there has been no progress in methods or data available to calculate individual fluxes of N. Manure is an important component of the mass balance and can be thought of as N that is not exported in grain (or forage that is consumed) or, therefore, the N that is returned to the landscape in the MARB. There are many assumptions in calculating the manure flux that could also alter our interpretation of the overall mass balance.

Phosphorus: A P mass balance for 1992 was included in the *Integrated Assessment* that incorporated fertilizer, manure, grain harvest, hay harvest, and pasture grazing (Goolsby et al., 1999). Small but potentially important changes in the large soil pool were not included because methods are not available for making this estimate for short time spans.

A P mass balance was calculated using the extended N mass balance (McIsaac, 2006) for 1951–2005 for each state, and these values were then summed for the MARB (Fig. 3.25). P fertilizer inputs have decreased since the 1970s such that the

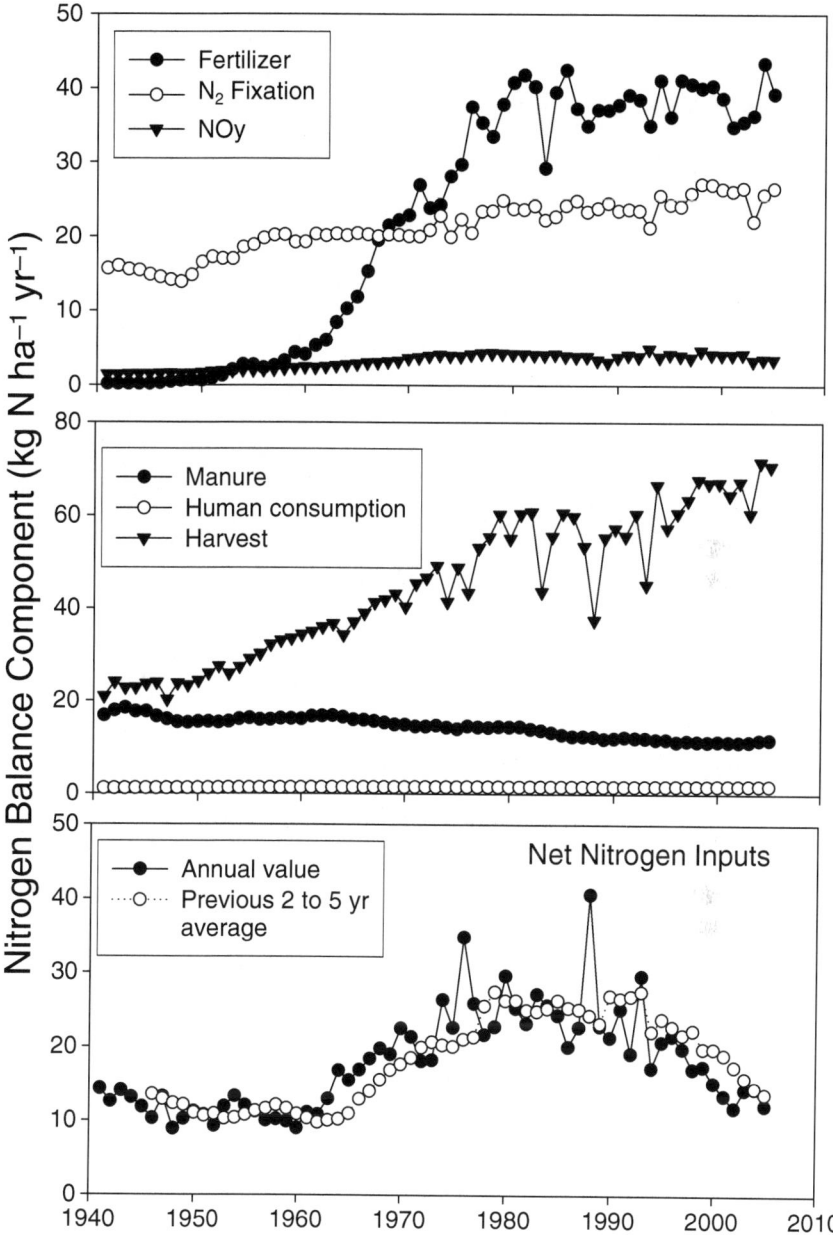

Fig. 3.24 Nitrogen mass balance components and net N inputs for the upper Mississippi River basin, as calculated by McIsaac et al. (2002) and updated through 2005 by McIsaac (2006)

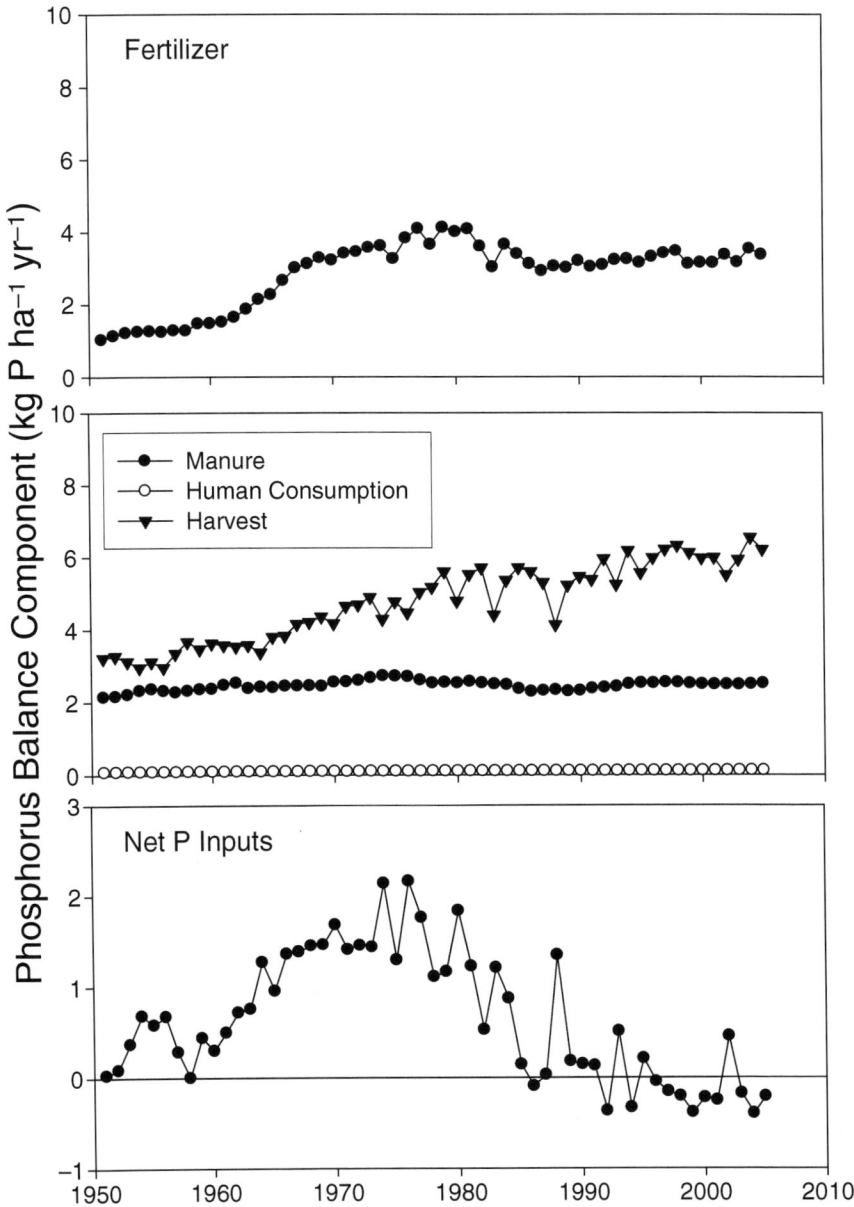

Fig. 3.25 Phosphorus mass balance components and net P inputs for the MARB. Adapted from McIsaac (2006)

increased harvest now exceeds fertilizer inputs (and manure retention) most years, so large soil P pools are being utilized by crops. The large buildup of soil P in the 1970s and 1980s led to a large positive net P balance, but decreased fertilizer inputs and high crop yields result in the current negative balance.

3.2 Mass Balance of Nutrients

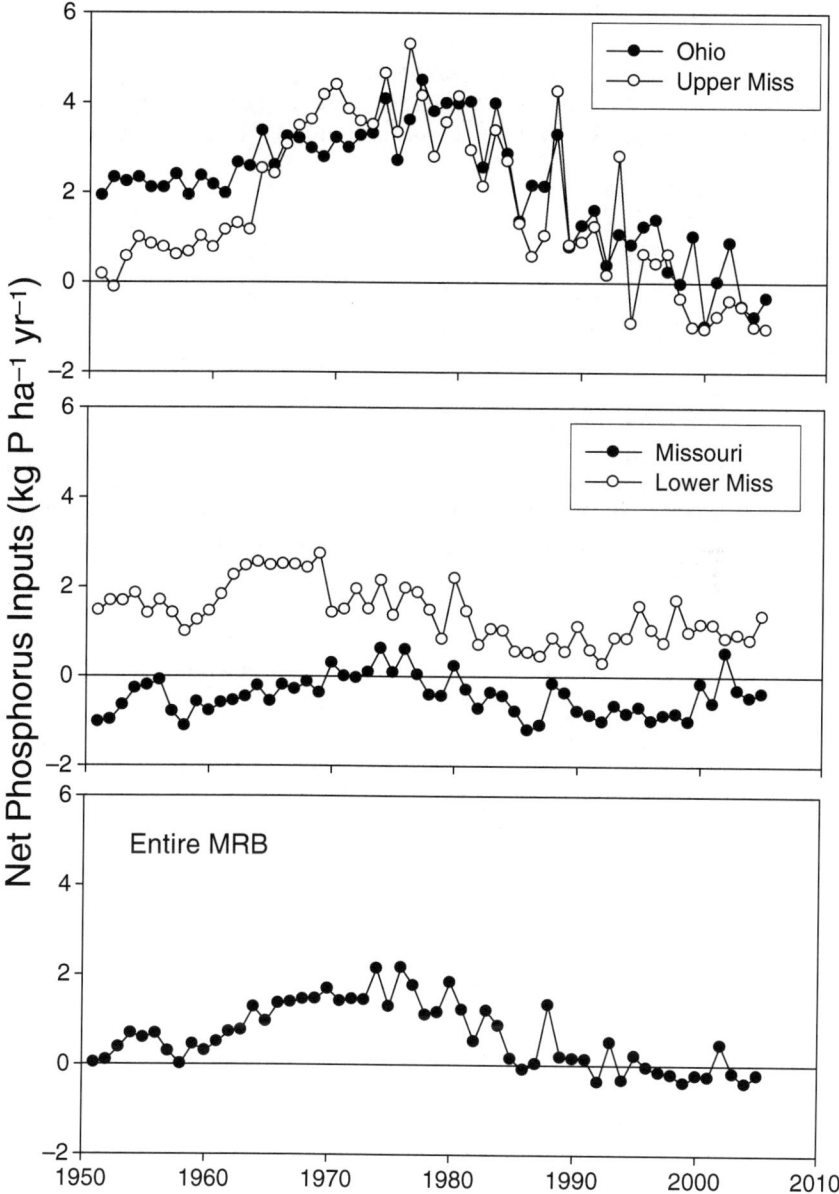

Fig. 3.26 Net P inputs for the four major subbasins of the MARB through 2005. Adaptive from McIsaac (2006)

When P mass balance is calculated for major subbasins, only the lower MARB still has a positive P balance (Fig. 3.26). The Missouri River P balance has shown little change, while the Ohio and Upper Mississippi River have a negative P balance. In contrast to N, the amount of P lost to streams and exported by rivers is small relative to agronomic fluxes; hence it is not expected that these changes in P mass balances will cause short term (or even relatively long term) changes in stream P concentrations and loads (David and Gentry, 2000).

A closer look at the upper Mississippi River basin (Fig. 3.27) shows an even larger decline in P from fertilizer and a steady decline in P from manure.

3.2.3 Point Sources

In the *Integrated Assessment*, point sources were estimated to contribute about 11% of the total nitrogen and an undefined, though likely somewhat lower, total phosphorus flux to the MARB. This assessment (Tetra Tech, Inc., 1998) was based on 1996 information, and it estimated fluxes at 321,000 metric tons N/year (354,000 tons N/year) and 91,500 metric tons P/year (101,000 tons P/year).

A reassessment (MART, 2006b) was based on 2004 permit information, adjusted assumptions, evaluated more facilities, and revised estimated fluxes downward to 233,000 metric tons N/year (257,000 tons N/year) and 39,500 metric tons P/year (43,500 tons P/year). Municipal treatment plants (STP) were thought to account for about 65% of the total point source fluxes for both N and P. However, few permits have suitable data for direct flux calculations, and only 11.1% of the mass flux was directly calculated from the permit information. The rest of the mass flux was estimated using "typical pollutant concentrations" (TPC) and estimated daily water flows from point sources. The TPCs used in the MART (2006b) estimates are lower than those used by other water quality programs; therefore, the Study Group has recalculated the contribution of N and P from municipal sewage treatment plants based on effluent concentrations that better reflect measured nutrient concentrations from point sources during 2004. These calculations also assume that the point source load is delivered to the NGOM without any in-stream losses. Therefore, they are the upper estimate for the contribution of point sources to the total N and total P riverine load. The Study Group's calculation indicates that load estimates would need to be revised upward to 267,000 metric tons N/year (294,000 tons N/year) (72% from STPs and 28% from industrial sources) and 53,000 metric tons P/year (58,500 tons P/year) (77% from STPs and 23% from industrial sources). (See Appendix D for a more detailed discussion of the Study Group's estimates.) When the contributions from all point sources are compared to the average annual N and P fluxes for the period 2001–2005, these new estimates indicate that point sources contribute to the Gulf about 22 and 34% of the average annual N and P flux, respectively. When compared to 2004 N and P fluxes (slightly higher than average fluxes), the percentage of the N flux contributed by point sources drops to about 20%, and the P flux remains constant at about 34%. Fluxes from point sources are equally distributed throughout the year, but spring flux is critical to the Gulf. Assuming equal monthly loads from point sources, the Study Group's estimates indicate that point sources are responsible for approximately 14% of spring N flux and 27% of spring P flux for 2001–2005. Again, the Study Group emphasizes that these are rough estimates, as measured data are not available at this time to make more accurate determination of point source contributions.

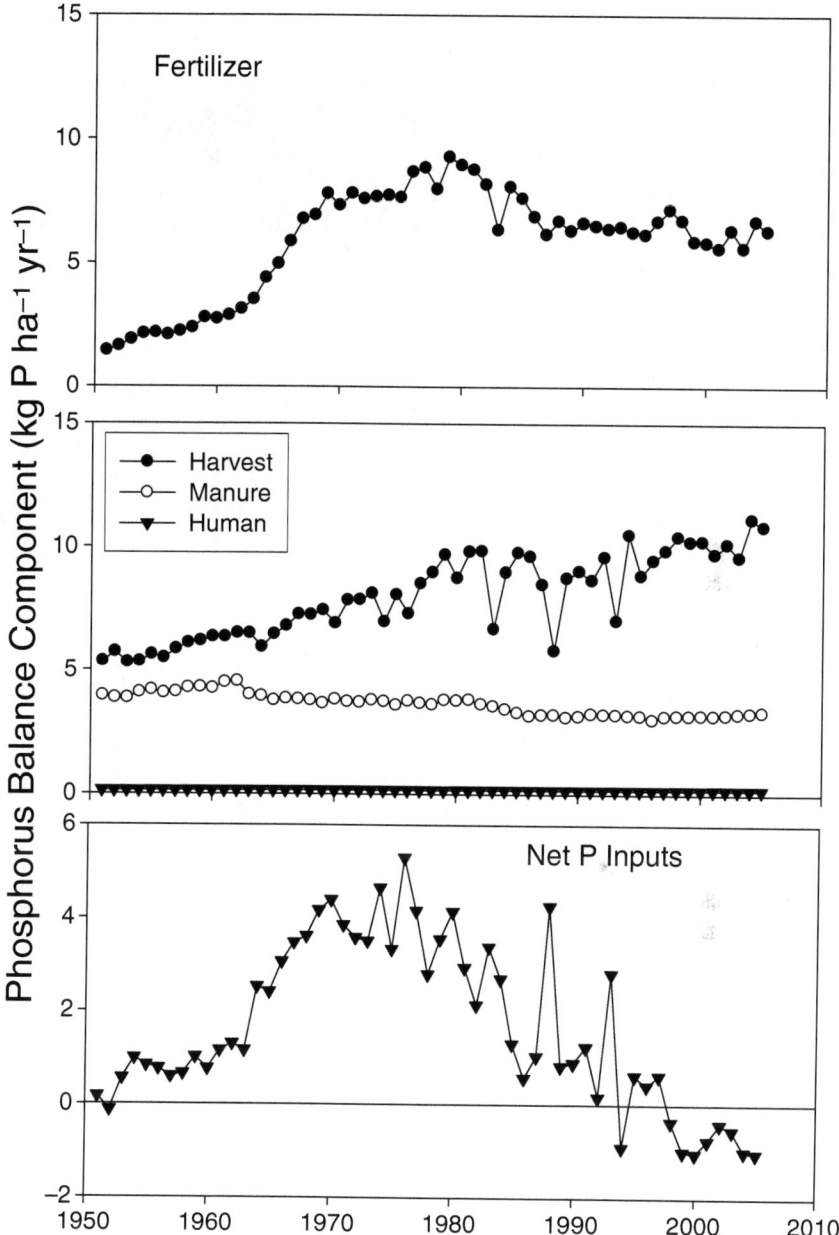

Fig. 3.27 Phosphorus mass balance components and net N inputs for the upper Mississippi River basin. Adapted from McIsaac (2006)

Fig. 3.28 Total phosphorus point source fluxes as a percent of total flux for the MARB for 2004 by hydrologic region

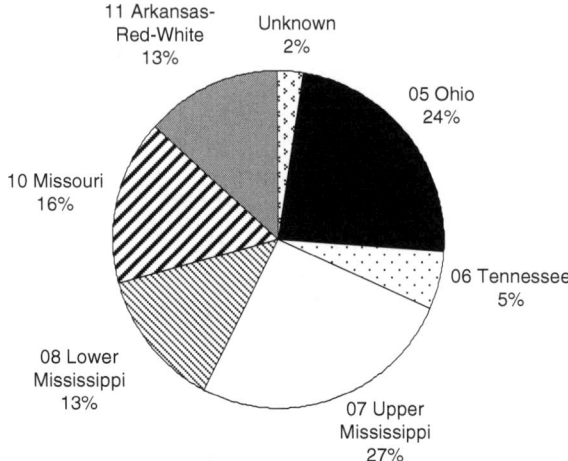

A summary of the percent of P fluxes by major hydrologic region, based on the new estimates, is shown in Fig. 3.28. Collectively, the upper Mississippi and Ohio River basins account for about half the P flux from point sources in the MARB.

This analysis suggests that point source P fluxes are a significant source of both annual and spring fluxes to the MARB and the Gulf and that substantial reductions in P fluxes in the MARB are likely if P fluxes from point sources are reduced. Point sources are a less important source of spring and annual N flux; however, reduction in N fluxes from point sources may offer a certain and cost-effective means of achieving some of the N reductions needed in the MARB. It is important to emphasize that the differences in assumptions used to estimate fluxes based on TPC have a major impact on annual and seasonal flux estimates for the MARB and would likely affect the estimated cost-effectiveness of requiring N or P removal from point sources in the MARB (discussed further in Section 4.5.88).

Key Findings and Recommendations

Although N mass balances have been recalculated since the *Integrated Assessment*, the research needs described in that report remain. Components of the N mass balance such as denitrification, N_2 fixation, manure N, and soil N pool processes such as mineralization and immobilization are not measured each year. Only N_2 fixation and manure N can even be estimated, with the other fluxes having little data available to make calculations. Point sources export N and P directly to rivers, yet their contributions continue to be estimated from permits.

New methods have been used to calculate N mass balances in this book (net anthropogenic N inputs, NANI). NANI and net P inputs for MARB have increased greatly since the 1950s but have decreased in the past decade because of steady fertilizer applications and increased crop yields for N and reduced fertilizer applications and increased crop yields for P. Mass balances in the upper Mississippi River subbasin suggest that, under the current tile-drained corn and soybean management system, depletion of soil organic N pools may be occurring. From a sustainability viewpoint, this needs to be fully documented and decreased as new systems are put in place to reduce N export in rivers. Point sources represented 22% of riverine N flux and 34% of P flux delivered to the Gulf. Manure is a more significant source of P than N; and where riverine N flux is greatest, excess manure N tends to be a less important input. Manure is likely more important basin wide to local water quality problems, rather than a large component of MARB export of N or P, because of which concentrated animal production has relocated. The greatest decrease in net N and P inputs was seen in the upper Mississippi River basin. From 1999 to 2005, 54% of N inputs were from fertilizer, 37% from fixation, and 9% from deposition for the entire basin. Deposition was most important in Ohio basin (16% of inputs). Based on these findings, the Study Group offers these recommendations.

- Continue and expand research to more accurately and fully measure the N mass balance in the MARB by developing methods and gathering data for improving the estimates of critical fluxes such as N_2 fixation, manure, denitrification, and soil N pool changes.
- Sustainability of soils in the MARB must be fully addressed by research to improve measurement of changes in soil N pools as a result of new management systems, with changes in soil N pools incorporated into more complete N mass balances. Section 4 discusses the need for research on changes in N pools associated with different management practices, e.g., tillage systems and other practices.
- N and P from point sources should be estimated from direct measurements, rather than relying on estimated values based on permits, so that more accurate calculations can be made of their contributions to the nutrient fluxes.

3.3 Nutrient Transport Processes

3.3.1 Aquatic Processes

Studies conducted since the *Integrated Assessment* have addressed many of the research needs that were identified for nutrient transport processes: quantification of in-stream processes such as denitrification (particularly in small streams), research in small watersheds to identify dynamics and timing of N transport and to better understand the impact of drainage practices on nutrient flux, and development of a better understanding of N behavior during floods. We review these advances for nitrogen, phosphorus, and silicate transport and transformation.

Nitrogen. In-stream nitrogen removal in river networks is variable, but it can be substantial, particularly in river networks with relatively low nitrogen concentrations. In 16 river networks in the northeastern United States, the Riv-N model predicted that 37–76% of nitrogen inputs were removed within streams (Seitzinger et al., 2002), and the SPARROW model predicted that 7–54% of nitrogen inputs were removed (Alexander et al., 2002b). Estimates of the percentage of annual N inputs removed by in-stream processes in regional drainages in the Mississippi River basin range from 20 to 55% (SPARROW model, Fig. 3.29). The Ohio and White River basins removed the lowest percentage and the Arkansas and Missouri River basins the highest. Although these are estimates of the role of in-stream processes on an annual basis, the SPARROW model results strongly reflect the effects of seasonal pulses, especially the high spring values, because the mean annual flux is a flow-weighted estimate (Alexander, personal communication).

In-stream N removal accounts for a much smaller fraction of annual N export in tile-drained agricultural regions and other areas where stream water nitrogen concentrations are extremely high and water residence time is short. The proportion of the nitrate flux that was denitrified was highest in forested systems, lowest in

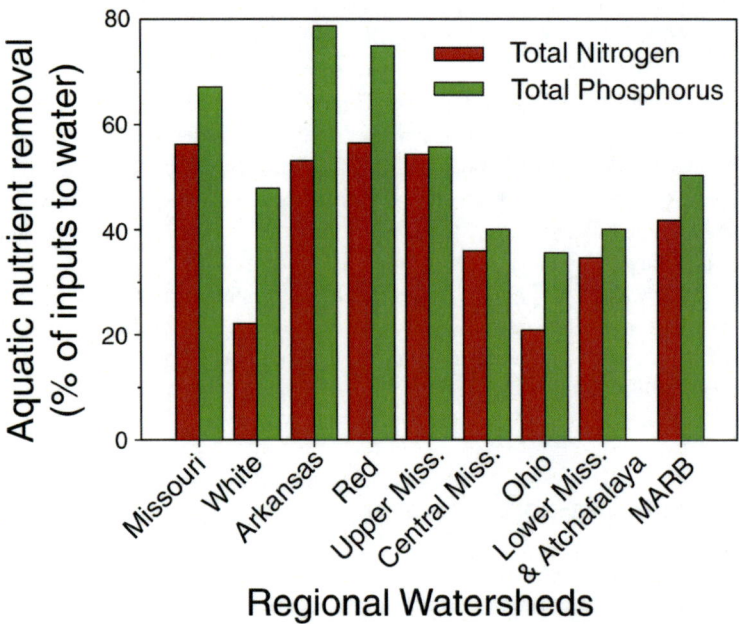

Fig. 3.29 Percentage of nutrient inputs to streams that are removed by in-stream and reservoir processes as predicted by the SPARROW model (Alexander et al., 2008)

urban, and intermediate in agricultural streams in Michigan (Inwood et al., 2005). Denitrification removed a greater fraction of N in meandering than in channelized reaches, but removal never exceeded 15% per day except during periods of low flow and warm temperature (Opdyke et al., 2006). Denitrification is a significant pathway for N removal in midwestern tile-drained streams during low flow, warm periods (summer and autumn), which improved local water quality at those times (Royer et al., 2004; Schaller et al., 2004). However, most of the nitrate is exported to the Gulf during high flows from January to June (Royer et al., 2006), and denitrification removes an insignificant fraction of this flux (Royer et al., 2004, 2006). Because in-stream removal is a small fraction of total flux at high flows, enhancing N removal by 50% during low flows (Q < median) would reduce annual N export only by less than 2% in Illinois agricultural streams, whereas enhancing removal by 25% during high flows (greater than 75th percentile flows) would reduce annual N export by 21% (Royer et al., 2006).

Recent research on streams in predominantly forested watersheds has shown that, in comparison to larger rivers, small streams remove a higher proportion of their incoming nitrogen per unit of water travel time (Alexander et al., 2000), per stream reach (Seitzinger et al., 2002), and per unit length (Helton, 2006; Wollheim et al., 2006). However, larger streams remove larger masses of nitrogen because more nitrogen passes through them (Helton, 2006; Seitzinger et al., 2002; Wollheim et al., 2006). Small streams receive and transport a significant amount of N to larger rivers, e.g., N loads to headwaters account for 45% of the load delivered to the entire river network in the northeastern United States (Alexander et al., 2007). Similar calculations have not yet been done for the Mississippi River basin (Alexander, R.B., 2007, personal communication: U.S. Geological Survey). Enhancing nutrient removal in small streams by restoring stream length that has been lost to straightening or burial could improve local water quality and decrease both N and P load to larger rivers (Bernot and Dodds, 2005); however, these reductions would be greatest at low stream flows and less effective at high discharges when the bulk of nutrient load is being transported to the Gulf.

Denitrification is not the only pathway for N removal in streams, although it is the most permanent. Removal of nitrate from stream water and its assimilation into biological tissues transforms N from dissolved to particulate form, which reduces the rate at which it is transported downstream. Particulate N can be deposited and stored in sediments, where it can be mineralized and potentially denitrified.

Effectiveness of N removal in aquatic systems increases with water residence time, so reservoirs can make a significant contribution to N removal in river networks. Denitrification in an Illinois reservoir reduced average annual N export by 58%, but the percent reduction in annual export over a 23-year period varied from 31 to 91% as retention time increased (David et al., 2006). N retention in Illinois reservoirs is higher than observed from rivers and reservoirs with lower nitrate concentrations (Fig. 3.30). The difference can be attributed to lower removal efficiencies in natural lakes than in reservoirs where elevated inputs of N support high rates of denitrification in the sediments (David et al., 2006). Denitrification in aquatic sediments (80% in reservoirs in the tile-drained part of Illinois and 20% in streams) was estimated to reduce N export from Illinois by 25% (David et al., 2006). Existing

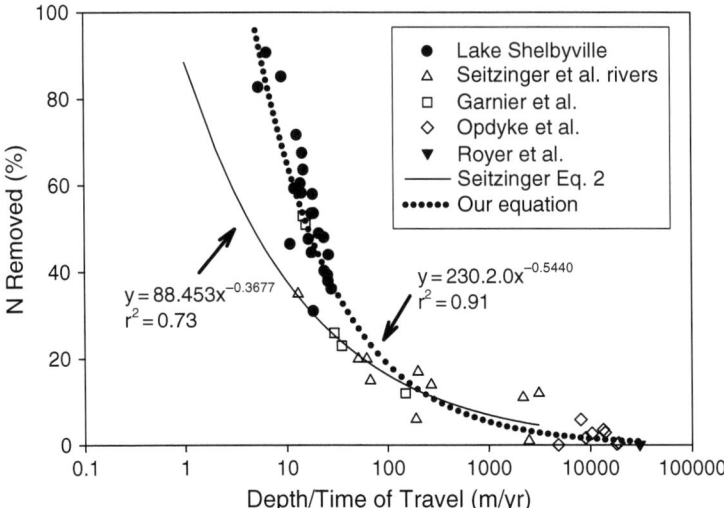

Fig. 3.30 N removed in aquatic ecosystems (as a % of inputs) as a function of ecosystem depth/water travel time (modified from David et al., 2006). Values shown are for 23 years in an Illinois reservoir (David et al., 2006), French reservoirs (Garnier et al., 1999), Illinois streams (an average from Royer et al., 2004), agricultural streams (Opdyke et al., 2006), and rivers (Seitzinger et al., 2002). The curve from Seitzinger et al. (2002) is not as steep as the curve that includes information from reservoirs in an agricultural region

floodplain backwaters on the upper Mississippi River basin are limited in their effectiveness in N removal by denitrification because of short water retention times and a lack of hydrologic connectivity with the main stem (David et al., 2006; Richardson et al., 2004). Enhancing connectivity and water residence time on floodplains during periods of high discharge and high nitrate concentrations in the spring has been suggested as an effective way to reduce N loading to the Gulf (David et al., 2006).

Because N_2O is a potent greenhouse gas, whether the end product of denitrification is N_2O or N_2 is of importance. The IPCC estimates that 1.25 and 0.75% of N that enters agricultural soils and rivers, respectively, is converted to N_2O (Mosier et al., 1998). However, that fraction includes N_2O production via both nitrification and denitrification. IPCC assumes that only 0.5% of N that is denitrified in rivers is converted to N_2O (Mosier et al., 1998), but they do not estimate this fraction for soils. A review of 32 studies of terrestrial denitrification reported the fraction of denitrified N converted to N_2O to be highly variable (0–100%) with a mean of 27% (Stevens and Laughlin, 1998). Thus available data suggest that denitrification in aquatic systems produces less N_2O as a fraction of denitrified N than terrestrial systems. Therefore, where denitrification occurs on the landscape will influence its contribution to greenhouse gases. However, enhancing denitrification to reduce water quality impacts of leached nitrogen will increase greenhouse gas emissions if nitrogen leaching rates remain high.

Phosphorus. An understanding of P transport and transformation in streams and rivers has developed in parallel with the studies on N just described. Stream networks alter the timing, magnitude, and bioavailability of edge-of-field P loss during transport to the Gulf via geochemical and biological processes: sediment sorption and desorption, precipitation and dissolution, microbial and algal uptake, and riparian floodplain and wetland retention. Many of the geochemical processes are mediated by biota; e.g., co-precipitation of dissolved P with calcite may be biologically mediated during active photosynthesis (Neal, 2001), and aquatic biota accounted for 30–40% of sediment P uptake and release in wetland (Khoshmanesh et al., 1999) and stream sediments (McDowell and Sharpley, 2003).

Fluvial sediments come from overland flow and erosion of stream channels and banks. High discharge events that generate overland flow in agricultural regions commonly account for most of the annual phosphorus load (e.g., Gentry et al., 2007). Soils eroding from stream banks may be subsoils poor in P, which is less available for release to water; hence the subsoils will likely represent a net sink for P (McDowell and Sharpley, 2001). Land-disturbing activities (e.g., urban development and mining) can be a significant source of sediment P, particularly when eroded sediments are rich in nutrients because of past agricultural practices. For example, construction of one side channel on the Missouri River floodplain has been calculated to contribute ~4,000 metric tons P (4,400 tons P) to the river (Kristin Perry, Missouri Clean Water Commission, personal communication, June 2007).

Regardless of sediment source, particulate P is the predominant form in transport. Both fluvial hydraulics and adjacent land use influence the properties of sediment within river systems (McDowell et al., 2002). To link P loss from the landscape to channel processes, variability in flow, local sources of P, sediment properties, and changes in P forms and loads should be simulated in models that estimate P loss from catchments, although this is rarely done.

In tile-drained agricultural regions, P is transported to streams by both overland flow and the artificial drainage systems, which have been associated with elevated dissolved reactive P (DRP) concentrations (Xue et al., 1998). DRP concentrations remained high in successive tile flow events, suggesting a pool of soil P that is readily desorbed (Gentry et al., 2007). In a tile-drained Illinois watershed, P loss via tiles represented 45% of total P loss in 1 year and 91% during a wetter year (Gentry et al., 2007). One rain-on-snow event transported about 40% of the annual P load in 1 week, 80% of which was DRP (Gentry et al., 2007). Clearly artificial drainage alters both the amount and the form of P exports, and the amount exported is dependent on both the magnitude and the timing of storms.

In fluvial systems with good hydraulic mixing (e.g., shallow streams), P availability in sediments can be estimated by the equilibrium P concentration (EPC_0). At low flow, EPC_0 will have a major influence on soluble P concentration, for P will desorb from sediments if P concentration in water is less than the sediment's EPC_0 or P will adsorb to sediments if P concentration is greater than EPC_0. P desorbed from sediments will be available for biological uptake. Bioavailable P from desorption is likely to be most significant as salinity increases in the estuary (Sutula et al., 2004).

Although cellular uptake and growth rates are generally saturated at low P concentrations, maximum biomass accrual in streams often occurs at somewhat greater concentrations (0.015–0.050 mg PO_4-P/L, Popova et al., 2006). This range of dissolved P concentrations might be more typical of streams draining agricultural catchments, and therefore, algal and microbial uptake likely plays a significant role in dissolved P retention, especially at low flow. Dissolved P uptake rates of algae vary with light, water velocity, temperature, grazing, and time following in-stream disturbances (Mulholland et al., 1994).

Estimates of the percentage of total P inputs removed by these in-stream processes in regional drainages in the Mississippi River basin range from 20 to 75% (SPARROW model, Fig. 3.29). The Ohio River basin removed the lowest percentage and the Arkansas River basin the highest. These percentages are considerably higher than what was used in the *Integrated Assessment* (28–37% in small streams and negligible in the mainstem).

P concentrations and loads generally increase with increasing discharge and are greatest on the rising limb of the hydrograph (e.g., Green and Haggard, 2001; Novak et al., 2003; Richards et al., 2001). Although P concentrations are greater during high flows, the importance of in-stream P retention is minimized at those times because of sediment resuspension and scouring within the channel. However, P deposition on floodplains may be a significant P sink during storms. Many streams export most of their P loads during episodic storm events; for example, in Illinois agricultural watersheds, extreme discharges (>90th percentile) are responsible for 84% of P export, and 98% of P export occurred at discharges that were greater than the median (Royer et al., 2006). This export is primarily particulate P; in contrast, over half of dissolved P export can occur during base flow conditions (Novak et al., 2003). Dissolved P constitutes a larger proportion of P export in watersheds with extensive tile drainage (Royer et al., 2006). Because most P transport occurs at high flows, models from Illinois agricultural watersheds suggest that enhancing in-stream P removal by 50% during low flows (e.g., less than the median) would reduce P export by less than 1%, whereas enhancing P removal by 25% during high flows (more than the 75th percentile) would increase P removal by 24% (Royer et al., 2006).

Silicate. Understanding of Si transport and transformations in rivers and streams lags far behind that of N and P. Although first-generation models for Si transport and transformations are available (Garnier et al., 2006; Sferratore et al., 2006), there are currently no models in the Mississippi River basin to predict the transport of dissolved silicate or biogenic Si (amorphous Si contained in diatoms and phytoliths). Once dissolved silicate is weathered, there are a number of transformations that occur, including inorganic transformations (such as new clay formation and precipitation as amorphous Si in soils) and biological transformations (such as the uptake and deposition in terrestrial plants, uptake and deposition in diatoms in aquatic systems) (Conley, 2002). Unlike models developed for N and P, there are no models that describe the complexity of biological transformations that occur with Si. In addition, significant reductions in the transport of Si have occurred with the building of dams along the Mississipi River leading to potentially significant changes in food webs on the Mississippi River shelf (Turner and Rabalais, 1994).

3.3.2 Freshwater Wetlands

The *Integrated Assessment* recognized the historical loss of many freshwater wetlands as one of the primary land-use changes contributing to excess nutrient loads in the Mississippi River basin. Mitsch et al. (1999) suggested the creation and restoration of wetlands for the specific purpose of controlling nonpoint source nutrient loads and emphasized the importance of targeting wetland creation and restoration in areas where nitrogen concentrations and loads were highest. They estimated that restoring about 2 million hectares (5 million acres) of wetlands would reduce N loads to the Gulf of Mexico by 20%, assuming a denitrification rate of 150 kg N/ha (134 lb N/ac) of wetland per year. Subsequent research (Section 4.5.2 of this book) suggests that wetlands can achieve substantially higher N removal rates in areas with elevated nitrate concentrations (Fig. 3.3), underscoring the importance of targeting restorations.

Wetland restoration is a particularly promising approach for heavily tile-drained areas like the Corn Belt (Fig. 3.1). This region was historically rich in wetlands, and in many areas, farming was made possible only as a result of extensive wetland drainage (Dahl, 1990; Pavelis, 1987). There are widespread opportunities for wetland restoration in the Mississippi River basin, and since the CENR reports, approximately 570,000 hectares (1.4 million acres) of wetlands have been restored, created, or enhanced within the basin under the Wetland Reserve Program (WRP), Conservation Reserve Program (CRP), Conservation Reserve Enhancement Program (CREP), Environmental Quality Incentive Program (EQIP), and Conservation Technical Assistance (CTA) (Table 3.5). However, the vast majority of wetland restorations have been motivated primarily by concern over habitat loss, and site selection criteria for wetland restorations have not primarily considered water quality functions. This past emphasis does not lessen the promise of wetlands for water quality improvement but rather underscores the need for programs focused on restoring wetlands explicitly for the purpose of reducing nonpoint source nutrient loads.

Table 3.5 Acres of wetlands created, restored, or enhanced in major subbasins of the Mississippi River from 2000 to 2006 under the Wetland Reserve Program (WRP), Conservation Reserve Program (CRP), Conservation Reserve Enhancement Program (CREP), Environmental Quality Incentive Program (EQIP), and Conservation Technical Assistance (CTA). (Personal communication, Mike Sullivan, USDA)

2-Digit Watershed	Hectares
Ohio River basin	33,300
Tennessee River basin	2,130
Upper Mississippi River basin	133,227
Lower Mississippi River basin	241,868
Missouri River basin	93,108
Arkansas, White, and Red River basins	68,161
Total	571,794

3.3.3 Nutrient Sources and Sinks in Coastal Wetlands

The general conclusion in the *Integrated Assessment* was that coastal wetlands are of secondary importance as nutrient sinks in comparison to other sources and sinks. Their role as a source of organic matter was discussed in an earlier section, and a more detailed review of that subject is in Section 2.1.5. Mitsch et al. (1999) assessed the utility of wetlands as nutrient and sediment sinks and concluded that 1) potential NO_3 reduction by coastal wetlands was likely less than 10–15% of the total river load; 2) water passage through coastal wetlands would likely decrease water column N:P and N:Si ratios; 3) the concept of coastal wetlands as net nutrient sinks remains controversial (e.g., Turner, 1999) so more large-scale measurements are needed; 4) deltaic systems might become N-saturated or begin to release N in forms other than NO_3; and 5) research and modeling was needed to better understand relationships between land subsidence, river diversions into wetlands, and N uptake in the coastal wetland/delta area. The *Integrated Assessment* concluded that although coastal denitrification rates were substantial (10–25 g N/m^2-year) relative to many shallow estuarine areas, diversion of river water into coastal wetlands might lead to N removal rates of 50–100 metric tons N/year (55–110 tons N/year), which is a relatively small fraction of N reduction goals.

A number of papers have been produced concerning nutrient sources and sinks in coastal wetlands since the *Integrated Assessment*. Lane et al. (2002) reported large decreases in nitrate as river water passed through an estuarine/wetland complex (Fourleague Bay); this estuarine–marsh complex appears to buffer the impact of the Atchafalaya River on coastal waters by causing an estimated 41–47% reduction in river nitrate concentrations. Denitrification rates in coastal wetlands ranged from 30 to 40 g N/m^2-year (larger than rates typically measured in adjacent estuaries), accretion rates of 8–11 mm/year or about 2,300 g dry sediments m^2/year (approximating sea level rise), and N burial rates of about 7 g N/m^2-year. Day et al. (2003) and others argued for river diversions to wetlands to prevent land losses and remove nutrients via denitrification, burial, and plant uptake. Nitrogen reductions of about 4 g N/m^2-year and 10–20 g N/m^2-year have been recorded for forests and wetlands, respectively. Particulate N burial rates of 13–23 g N/m^2-year have been measured in some wetlands. These are substantial rates by estuarine standards but modest relative to wetlands/reservoirs in the upper MARB connected to or adjacent to agricultural drainage. However, Turner (1999) reported very small N concentration reductions and modest TSS, POC, and particulate P concentration reductions in waters flowing through the Atchafalaya system and hence concluded river diversions would remove small amounts of nutrients relative to nutrient input loads.

The recent literature supports the importance of forested and other types of coastal wetlands for nutrient uptake and sediment accretion, both of which would lead to reductions in loads to the GOM. Rates appear to be substantial compared to most sub-tidal estuarine locations (excluding areas like the Mississippi River plume) and moderate to small relative to many freshwater natural and created wetlands. Rates lower than those observed in more northern wetlands of the MARB may be due to the generally lower nutrient loading rates to these coastal wetland systems

(<10 g N/m^2-year). However, given the data currently available, it is doubtful that a predictive model of nutrient losses in these coastal wetlands can be developed following the general form of statistical models used for predicting nutrient losses in freshwater wetlands (Saunders and Kalff, 2001; Spieles and Mitsch, 2000).

Missing from the GOM hypoxia analysis is a regional scale (i.e., larger spatial scale) analysis of both nutrient (N and P) and OM losses associated with coastal wetlands. It would appear that sufficient information is currently available to delineate the spatial extent of various coastal wetland habitats. It seems less certain that essential nutrient and OM loss rate estimates (e.g., long-term burial of C, N, and P or denitrification) are available to achieve this goal.

There appear to be few nutrient, sediment, or organic matter budgets available for these coastal wetlands that can be used to judge the effectiveness of wetlands as either sinks or sources. For example, nutrient sink behavior of wetlands has been inferred from nutrient concentration reductions with distance from a nutrient source. While this approach has appeal, it would be more convincing if nutrient loads (i.e., concentrations coupled to water flows) entering and leaving wetland systems were compared in a mass balance format. Additionally, more emphasis on process measurements (e.g., burial, denitrification, and plant uptake rate) would allow for better understanding of observed differences between wetland inputs and outputs. It appears that process measurements in these coastal wetlands lag behind those made in natural and created wetlands in other parts of the MARB.

Key Findings and Recommendations

The percentage of annual N and P inputs removed by in-stream processes varies by MARB subbasin and ranges from 20 to 55% for N and 20–75% for P based on model estimates. There currently are no models to predict the transport of dissolved silicate. Denitrification can be a significant pathway for N removal in small streams during low-flow, warm periods, thereby enhancing local water quality. However, most nitrate is exported to the Gulf during high flows in the period from January to June, when denitrification is not effective in removal. Since the effectiveness of N removal in aquatic systems increases with water residence time, enhancing the connectivity and water residence time in floodplains and backwater areas on the upper Mississippi River during periods of high flow and high nitrate concentrations could increase the effectiveness of N removal by denitrification and, therefore, reduce the N flux to the Gulf. Likewise, since high-flow events that generate overland flow in agricultural regions generally account for most of the annual P flux (as much as 84%, primarily as particulate P), deposition on floodplains and in backwater areas could represent a significant P sink. However, in tile-drained areas, dissolved reactive P represents a much larger percentage of P flux (45–91%), and deposition is a less significant sink.

There has been substantial wetland restoration within the MARB since the CENR reports, but restorations have not been targeted for water quality benefits. The greatest water quality benefits will be realized in areas of the Corn Belt with highest nitrate concentrations and loads.

Although current estimates of denitrification rates in coastal wetlands are higher than the estimates used in the *Integrated Assessment*, current studies still conclude that river diversions to coastal wetlands would remove only small amounts of nutrients relative to the total fluxes. However, better estimates of nutrient and organic matter loss rates (denitrification; long-term burial of C, N, and P; and plant uptake) are needed to better understand observed differences between wetland inputs and outputs in coastal areas. Based on these findings, the Study Group offers these recommendations.

- Removal of both N and P can be increased by implementing management strategies that include enhancing hydrologic exchange and retention on floodplains and in backwater habitats when discharge, total P, and nitrate concentrations are high (e.g., during spring), particularly in rivers of intermediate size.
- More reliable and process-driven models that simulate fluvial processes and estimate N and P transfer to stream channels need to be developed to more accurately predict land management or best-management-practice (BMP) impacts on nutrient inputs to receiving waters.
- First-generation models need to be developed to describe the transport and transformations of Si in the MARB.
- Programs focused on restoring wetlands explicitly for the purpose of reducing nonpoint source nutrient loads need to be implemented and targeted in areas of the Corn Belt with highest nitrate concentrations and loads.
- Better measurements of key processes (e.g., burial, denitrification, plant uptake rate) are needed in coastal wetlands to provide a better understanding of observed differences between inputs and outputs.
- Regional scale studies of coastal wetlands are needed to develop nutrient, sediment, and organic matter budgets that can be used to better evaluate the effectiveness of coastal wetlands as sinks or sources.

3.4 Ability to Route and Predict Nutrient Delivery to the Gulf

The Study Group concurs with the *Integrated Assessment's* identification of modeling as a critical component of an adaptive management approach to improving Gulf hypoxia. Along with monitoring, interpretation, and research, modeling can improve the scientific understanding of the impacts of land and nutrient management actions

3.4 Ability to Route and Predict Nutrient Delivery to the Gulf

on watershed and Gulf of Mexico environmental quality. The *Integrated Assessment* used only limited modeling results at the field-scale and none at the watershed scale in its assessment.

Research was proposed in the *Integrated Assessment* to develop an effective modeling framework, including improved watershed and basin-scale simulation of nutrient transport and transformations from natural, urban, and agricultural landscapes, improved estimates of nutrient mass balances throughout the landscape, and improved understanding of biogeochemical cycling within the basin. Within this modeling framework, further research was called upon to assist in four areas: (1) to characterize the dynamics and timing of nutrient movement from the edge of the field in agricultural landscapes to small streams and tributaries, particularly from agricultural tile drainage systems; (2) to scale up from experimental plots to watershed/farm-scale studies of on-farm practices and edge-of-field strategies to reduce and intercept nutrients; (3) to assess the effects on nutrient loads and hypoxia of long-term change in climate, hydrology, and population; (4) to evaluate the role of flood events and the potential role of flood prevention strategies on nutrient transport to the NGOM; and (5) to improve understanding of the social and economic trade-offs and impacts of various management and policy alternative strategies.

Numerous models have been used to describe sources, transport, and delivery of nutrients at various spatial and temporal scales within the MARB (Table 3.6). Several of these studies address needs identified in the *Integrated Assessment*, including improved understanding of basin-scale nutrient transport, nutrient cycling processes, tile drainage nutrient transport, watershed-scale simulation of in-field and edge-of-field practices, climate effects, and loading from high-flow events. Issues associated with social and economic trade-offs of alternative strategies are discussed in Sections 4.3 and 4.4.

While each of the models listed in Table 3.6 may prove useful for developing adaptive management to mitigate hypoxia, we single out three models for further discussion based on the fact that they have been applied at the basin-wide scale within the MARB. In so doing, we do not intend to suggest that only these three models should be relied upon for insight into the processes of nutrient fate and delivery.

3.4.1 SPARROW Model

The SPARROW (SPAtially Referenced Regressions On Watershed attributes) model is a hybrid mechanistic/empirical, basin-scale simulation model developed by the US Geological Survey (Alexander et al., 2007; Smith et al., 1997). The model uses spatially distributed data on nutrient sources, climate, soils, topography, and natural and artificial drainage densities to estimate N and P delivery to streams and removal processes in streams and reservoirs under long-term steady-state conditions. Nutrient sources include atmospheric deposition (N only); urban/human sources; agricultural runoff and subsurface drainage; and natural sources from forest, barren, and shrub lands.

Table 3.6 Attributes of models used to estimate sources, transport, and/or delivery of nutrients to the Gulf of Mexico

Model[a]	Type[b]	Time[c]	Space[d]	Components[e]	Inputs[f]	Outputs[g]	Predicts[h]	Strength[i]	MARB Refs[j]
ADAPT	E,M	D	F,W	R,D,S,E,N,P,U	C,F,P,A,S	F,S,N,P	C,L,M	Overland/drainage	1
AnnAGNPS	E,M	D	H,W	R,E,N,P,U	C,F,P,A,S	F,S,N,P	C,L,M	In-field sediment	2
DAFLOW/BLTM	M	D	R	R,N,Q	C,G	F,N	C,L	River routing	3
DRAINMOD	E,M	D	F	R,D,N	C,F,L	F,N	C,L,M	Drainage	4
EPIC	E,M	D	F	R,S,E,N	C,F,A,L	F,S,N,P	C,L,M	In-field practices	5
GLEAMS	E,M	D	F	R,E,N	C,F,A,L	F,S,N,P	C,L,M	Overland	6
HSPF, LSPC	E	D	W	R,D,S,E,N,Q,U	C,F,P,A,S	F,S,N,P	C,L,M,R	Overland/stream	–
IBIS/THMB	E,M	D	W,B	R,D,S,E,N,Q,P	C,F,P,A,S	F,S,N,C	C,L,M	Ecosystem	8
L-THIA	C,E	A	F,W	R,E,N,U	C,P,N,F	F,S,N,P	L	W-s land-use change	–
NANI	B	A	W,B	N,P	F,P,A,S	N	C,L,M	Process accounting	10
PLOAD	C	M	W	R,E,N,U	F,C,P,S	F,S,N,P	C,L,M	Distribute w-s loads	–
REMM	M	D	F	R,D,E,N	C,F,L	F,S,N,P,C	C,L,R	Riparian ecosystem	12
RZWQM	E,M	S	F	R,D,S,N	C,F,A,L	F,S,N,P	C,L,M	Subsurface/plants	–
SPARROW	S,M	A	W,B	R,D,N,Q,P,U	C,F,P,A,S,G	F,N,P	C,L	Data-driven	14
SWAT	E,M	S,D	H,W,B	R,D,S,E,N,Q,P,U	C,F,P,A,S	F,S,N,P,C	C,L,M,W,R	Overland/w-s	15
WARMF	E,M	D	W	R,E,N,P,U	C,F,P,A,N	F,S,N,P	C,L,M	TMDL study	–
WEPP	M	S	F,W	R,D,S,E,N	C,F,P,A,S	F,S,N,P	C,L,M,R	Hillslope	17

[a]Although all models cited above (as well as numerous other models not included) are relevant to MARB and Gulf of Mexico hypoxia issues, not all of the above models have been discussed within this book. In this section, the discussion was focused on models considered most applicable to MARB basin-scale processes. More details on applications of these models within the MARB can be found in cited references.
[b]**Type**—*model classification*: S=statistical/stochastic, C=export coefficient, B=mass balance, E=empirical/process-based, M=mechanistic/process based

3.4 Ability to Route and Predict Nutrient Delivery to the Gulf

Table 3.6 (Continued)

[c]**Time**—*smallest time-scale for output*: S=subdaily, D=daily, M=monthly, A=annually
[d]**Space**—*organizing spatial scale*: F=field, H=hydrologic resource unit, W=watershed, B=basin, R=river network
[e]**Components**—*model features*: R=runoff, D=drainage, S=snowmelt, E=erosion, N=nutrients, Q=stream processes, P=ponds/reservoirs, U=urban
[f]**Inputs**—*input types*: C=climate (temperature, precipitation), F=fertilizers, P=point sources, A=Atmospheric deposition, S= spatial land-use, L=single-field land use, G=stream gage data
[g]**Outputs**—*constituents modeled*: F=flow, S=sediment, N=nitrogen, P=phosphorous, C=carbon
[h]**Predicts**—*predictive capability*: C=climate change, L=land-use change, M=land-management change, W=wetland change, R=riparian change
[i]**Strength**—*application for which model tends to be well suited* (w-s=watershed)
[j]**MARB Refs**—*Recent and key references from Mississippi/Atchafalaya River basin model applications*:

1—ADAPT: Dalzell, B.J., P.H. Gowda, D.J. Mulla. 2004
Gowda, P.H., D.J. Mulla. 2006.
Gowda, P.H., B.J. Dalzell, D.J. Mulla. 2007
Sogbedji, J.M., G.F. McIsaac. 2006.
2—AGNPS:
Yuan, Y., R.L. Bingner, R.A. Rebich. 2001
Yuan, Y., R.L. Bingner, F.D. Theurer. 2006
Yuan, Y. R.L. Bingner, F.D. Theurer, R.A. Rebich, P.A. Moore. 2005
3—DAFLOW/BLTM:
Broshears, R.E., G.M. Clark, H.E. Jobson. 2001
4—DRAINMOD:
Northcott, W.J., R.A. Cooke, S.E. Walker, J.K. Mitchell, M.C. Hirschi. 2001
5—EPIC:
Atwood, J.D., V.W. Benson, R. Srinivasan, C. Walker, E. Schmid. 2001
Chung, S.W., P.W. Gassman, R. Gu, R.S. Kanwar. 2002
6—GLEAMS:
Wedwick, S., B. Lakhani, J. Stone, P. Waller, J. Artiola. 2001
8—IBIS/THMB:
Donner, S.D., M.T. Coe, J.D. Lenters, T.E. Twine, J.A. Foley. 2002
Donner, S.D., C.J. Kucharik. 2003a

Table 3.6 (Continued)

Donner, S.D., C.J. Kucharik, J.A. Foley. 2004.
Donner, S.D. 2006
10—NANI:
McIsaac, G.F., M.B. David, G.Z. Gertner, D.A. Goolsby. 2002
Howarth, R. W., G. Billen, D. Swaney, A. Townsend, N. Jarworski, K. Lajtha, J. A. Downing, R. Elmgren, N. Caraco, T. Jordan, F. Berendse, J. Freney, V. Kueyarov, P. Murdoch, and Zhu Zhao-liang. 1996.
12—REMM:
Graff, C.D., A.M. Sadeghi, R.R. Lowrance, R.G. Williams. 2005
14—SPARROW:
Alexander et al., 2008.
Alexander, R.B., R.A. Smith, G.E. Schwarz., 2004.
Smith, R.A., G.E. Schwarz, R.B. Alexander, 1997
Alexander, R.B., R.A. Smith, G.E. Schwarz., 2000
15—SWAT:
Anand, S., K.R. Mankin, K.A. McVay, K.A. Janssen, P.L. Barnes, G.M. Pierzynski. 2007
Du, B., A. Saleh, D.B. Jaynes, J.G. Arnold. 2006
Gassman, P.W., M.R. Reyes, C.H. Green, J.G. Arnold. 2007
Green, C.H., M.D. Tomer, M. DiLuzio, J.G. Arnold. 2006a
Hu, X., G.F. McIsaac, M.B. David, C.A. Louwers. 2007
Jha, M., J.G. Arnold, P.W. Gassman, F. Giorgi, R.R. Gu. 2006
Kirsch, K., A. Kirsch, J.G. Arnold. 2002.
Santhi, C., J.G. Arnold, J.R. Williams, W.A. Dugas, R. Srinivasan, L.M. Hauck. 2001
Shirmohammadi A., I. Chaubey, R. D. Harmel, D.D. Bosch, R. Muñoz-Carpena, C. Dharmasri, A. Sexton, M. Arabi, M.L. Wolfe, J. Frankenberger, C. Graff, T.M. Sohrabi. 2006..
Stone, M.C., R.C. Hotchkiss, C.M. Hubbard, T.A. Fontaine, L.O. Mearnes, J.G. Arnold. 2001
Vache, K.B., J.M. Eilers, M.V. Santelman. 2002
VanLiew, M.W., T.L. Veith, D.D. Bosch, J.G. Arnold. 2006
Wang, X., A.M. Melesse. 2005
17—WEPP:
Tiwari, A.K., L.M. Risse, M.A. Nearing. 2000

Wet-deposition data are the basis for atmospheric deposition N source, which assumes that dry deposition and ammonium deposition are spatially correlated to wet deposition. Urban nutrient sources include all human-population-dependent nutrient sources: municipal and septic-system wastewater, stormwater runoff, and other sources that are spatially correlated to human population data (such as wet and dry deposition from vehicles, power plants). Agricultural nutrient sources include commercial fertilizers, livestock manure, and biological N_2 fixation. Soil transformations of N and P are not considered and assumed to be in equilibrium between immobilization and mineralization.

SPARROW simulates N and P fluxes (mass) and yields (mass per unit area) within sub-catchments using three first-order attenuation terms, with mass balance constraints, to represent nutrient losses in overland transport, riverine processes, and reservoir trapping. Model parameters are calibrated based on the source conditions for a base year and the flow-adjusted, long-term mean annual loads of total N and total P estimated using rating curves fit to stream monitoring data. The calibrated SPARROW model can be used to assign these loads to specified nutrient sources and sub-catchments with quantifiable uncertainty.

SPARROW has been applied to diverse watersheds including the MARB (Alexander et al., 2000; 2007; Smith et al., 1997), the Chesapeake (Preston and Brakebill, 1999), the Neuse (McMahon et al., 2003) in North Carolina, and the Waikato (Alexander et al., 2002a) in New Zealand. The most recent application of SPARROW to the MARB includes more and better data for model parameter estimation and greater detail in the model specification (Alexander et al., 2008). The result is an increased number of nutrient source terms and 20% less model error compared to previous applications of SPARROW (Alexander et al., 2000; Smith et al., 1997). Several important assumptions are embedded in the modeling approach, however, and these must be considered in interpretation of model results for the MARB.

SPARROW does not assess flow-related changes in nutrient loads, which are important to Gulf hypoxia extent and severity. In comparing SPARROW predictions with observed loads for a particular period or location, SPARROW does not estimate nutrient load for any particular year, but rather a flow-adjusted or flow-independent load. This is the load predicted under long-term average flow conditions for the source input conditions of a particular year (the base year). For example, the SPARROW estimates for the 1992 base year in Alexander et al. (2008) are the mean annual loads that would be predicted under the source conditions of 1992 and the mean annual flow of the period from 1975 to 2000. These represent the loads that would have been expected in 1992 *if 1992 had had the mean annual flow of the period 1975–2000*. They are not an estimate of the nutrient loads *in 1992*. As a result, the comparisons between 1992 and 2002 in Alexander et al. (2008) are not based on the loads predicted for 1992 and 2002 but rather the loads that would have been expected in 1992 and 2002 under the different source inputs for those 2 years, but assuming both years had had exactly the same flow patterns, i.e., the mean annual flow of the period 1975–2000.

Model input coefficients for each nutrient source are statistically estimated by the model, and, as such, are influenced by all sources that are spatially correlated to these sources (whether these correlated sources are in the model or not). For example, wet-deposition N was spatially characterized from monitoring data, but no data for dry deposition were used; and urban sources were modeled assuming that all sources were correlated and spatially distributed similar to the model input of population. Particularly in these two cases, coefficients may be artificially high as they include the effects of other spatially correlated sources that are not in the fitted model. The model does not account for soil storage of nutrients but assumes that stream inputs are correlated to the agricultural nutrient source inputs. The lack of a soil storage term may ignore nutrient carry-over effects that are often important in determining stream export (David et al., 1997; David and Gentry, 2000; McIsaac et al., 2001; Mulvaney et al., 2001). These limitations and others discussed in Alexander et al. (2008) should be considered in interpretation of the SPARROW model results.

The most recent version of SPARROW (Alexander et al., 2008) is thought to have improved many aspects of this statistical model. For example, the percentage of N and P that enters streams and is actually delivered to the NGOM has increased, as in-stream removal terms have been reduced. SPARROW can be used to examine source inputs for the nutrients being transported by streams, and with each new version these can change. Source areas are quite dependent on land-to-water transfer coefficients and the way the model represents inputs and their availability to be transferred to a stream. As expected, agriculture was found to be the major source of nutrients to the NGOM in this recent application (Alexander et al., 2008). Important nonagricultural nutrient contributions were from atmospheric deposition and urban sources. The largest source of N is attributed to fertilizer inputs to corn and soybean fields (52%) followed by atmospheric deposition (16%). In contrast, the largest source of P is attributed to animal manure on pasture and rangelands (37%) followed by corn and soybeans (25%), other crops (18%), and urban sources (12%) (Alexander et al., 2008). It is important to note that in the model structure, manure is the only source of P that is available for transport from pasture and rangelands. These lands are otherwise assumed to be in steady state. Similarly, fertilizer, N fixation, and manure N are the only sources of N that are available for transport from corn and soybean. These lands are assumed to be in long-term steady state, and there is assumed to be no net soil mineralization. Statistical coefficients for agricultural sources suggested N delivery to streams ranging from 6% of applied nutrients for pasture/rangeland to 16% for corn and soybeans, and the opposite trend for P delivery, ranging from 2% for corn and soybeans to 14% for pasture/rangeland. These results give a very different picture of important inputs to the basin and their effects on riverine N and P fluxes. For example, atmospheric deposition (for N) and manure (for P) are thought to much more important, and point sources and corn and soybean production much less important than mass balance-type calculations would suggest (see Section 3.2). SPARROW is continually being developed and improved. However, the Study Group cautions against the sole use of SPARROW (or any model) for making decisions about where to target management efforts given the current stage of development of this approach.

3.4.2 SWAT Model

The Soil and Water Assessment Tool (SWAT) model is a physically based, deterministic, continuous, watershed-scale simulation model developed by the USDA Agricultural Research Service (Arnold et al., 1998; Neitsch et al., 2004). It uses spatially distributed data on topography, soils, land cover, land management, and weather to predict water, sediment, nutrient, and pesticide yields. A modeled watershed is divided spatially into subwatersheds using digital elevation data according to the density specified by the user. Subwatersheds are modeled as having uniform slope and climatic conditions, and they are further subdivided into lumped, nonspatial hydrologic response units (HRUs) consisting of all areas within the subwatershed having similar soil, land-use, and land-management characteristics. The use of HRUs allows soil and land-use heterogeneity to be simulated within each subwatershed but ignores pollutant attenuation between the source area and the stream and limits spatial representation of wetlands, buffers, and other BMPs within a subwatershed.

The model includes subbasin, reservoir, and channel-routing components. The subbasin component simulates runoff and erosion processes, soil water movement, evapotranspiration, crop growth and yield, soil nutrient and carbon cycling, and pesticide and bacteria degradation and transport. It allows simulation of a wide array of agricultural structures and practices, including tillage, fertilizer and manure application, subsurface drainage, irrigation, ponds and wetlands, and edge-of-field buffers. The reservoir component detains water, sediments, and pollutants, and degrades nutrients, pesticides, and bacteria during detention. The channel component routes flows, settles and entrains sediment, and degrades nutrients, pesticides, and bacteria during transport. SWAT typically produces daily results for every subwatershed outlet, each of which can be summed to provide monthly and annual load estimates.

The SWAT model has been tested for a wide range of regions, conditions, practices, and timescales (Gassman et al., 2007). Evaluation of monthly and annual streamflow and pollutant outputs indicate that SWAT functioned well in a wide range of watersheds. Relatively poor results in some cases, particularly for daily flow and pollutant outputs, were attributed partly to input and calibration data uncertainty and partly to model limitations. In general, the model had more difficulty simulating wet years than dry years and tended to overestimate soil water in dry soil conditions and underestimate in wet soil conditions.

Numerous studies have applied the SWAT model in the Mississippi River basin. Several recent studies have addressed issues identified in the *Integrated Assessment*, such as application of field-scale hydrologic processes to large watershed scale (Anand et al., 2007; Arnold et al., 1999), effectiveness of various nutrient-reduction strategies in agricultural watersheds (Hu et al., 2007; Santhi et al., 2001; Vache et al., 2002), model enhancements to address tile-drained cropland (Du et al., 2006; Green et al., 2006a), and assessment of the impacts of climate change on large-basin hydrology and nutrient export (Jha et al., 2006). These studies are discussed below.

Studies from field scale (Anand et al., 2007) to basin scale (Arnold et al., 1999) in the MARB have demonstrated the ability of SWAT to scale up processes to the large watershed scale. Arnold et al. (1999) validated the water balance component of SWAT in a large-scale modeling study of the conterminous United States and concluded that it would be useful in studying the effects of climate and BMPs on annual and seasonal runoff. The long-term effects of various BMPs was assessed in a 4,277 km^2 (1,651 mi^2) pasture-rangeland-dominated watershed experiencing urban growth in Texas (Santhi et al., 2001). They found future (2020) loads could be reduced by about 50% by implementing a combination of practices, including a 1 mg/L limit for wastewater treatment plant P effluent, limiting dairy manure land applications to the P rate, exporting 38% of manure from the watershed, and reducing P in livestock diet. Vache et al. (2002) explored the impact of traditional and alternative agricultural practices on water quality in two agricultural watersheds in Iowa. Continuing current trends in Midwestern agricultural production to 2025 (including increased conservation tillage, increased farm size and total acres, and current BMPs) resulted in simulated increase of nitrate export and decrease of sediment export relative to present. Two other scenarios representing different combinations of practices, such as complete conversion of cropland to no-till, implementation of riparian buffers on all streams, and increased use of perennial cover (CRP, pasture, and alfalfa), resulted in reductions of nitrate loads by 54–75% and sediment load by 37–67% simulated by SWAT. In a tile-drained watershed in east-central Illinois, reductions in N fertilizer resulted in 10–43% decrease in riverine nitrate export (Hu et al., 2007). However, SWAT overestimated nitrate export during major wet periods and had several other unrealistic aspects of N cycle components. Recent enhancements have been made to allow better simulation of tile drainage in agricultural fields by SWAT (Du et al., 2006; Green et al., 2006a). This change indicates that previous modeling results by SWAT in heavily tile-drained watersheds should be reassessed using the revised model.

Shifts in future precipitation and climate may impact flow and nutrient loads from the MARB. Jha et al. (2006) used SWAT to assess the effects of future climate change on UMRB flows. They found a doubling in CO_2 (to 660 ppmv) to result in a 36% increase in average annual streamflow and a 20% increase in precipitation to increase streamflow by 58%. Similar increases were found in average monthly streamflow in the April–May period, which is in the critical period for hypoxia development. Mean annual streamflow changes in response to six general circulation model scenarios ranged from –6 to +51%. Results indicated increases in rainfall and snowmelt in January and February and large increases in spring stream flow.

3.4.3 IBIS/THMB Model

The Integrated Biosphere Simulator (IBIS) land-surface and terrestrial ecosystem model and the Terrestrial Hydrology Model with Biogeochemistry (THMB, an enhanced version of Hydrological Routing Algorithm, HYDRA) are two physically based models that have been linked to model large basin-scale hydrological, carbon, and nutrient processes (only nitrogen at this time) (Donner, 2006; Donner et al., 2002). The IBIS model represents phenomena such as land-surface biophysical processes, canopy physiology, vegetation phenology, and long-term ecosystem dynamics at different time steps, ranging from 60 min to 1 year, to simulate time-transient surface and subsurface hydrological fate and transport processes. IBIS requires spatially distributed inputs of climate, soil texture, vegetation type, and associated management information. It uses these inputs to simulate terrestrial processes at a user-defined grid-cell scale. The terrestrial model is coupled with THMB, which represents phenomena such as solute transport, surface and subsurface leaching, point source inputs, and in-stream chemical and biological transformations to simulate river, wetland, lake, and reservoir flow and storage of water and nutrients.

The IBIS/HYDRA model, with a simplified nitrogen leaching algorithm, was found to represent much of the spatial and temporal variability in stream discharge and nitrate export within the MARB (Donner et al., 2002). A study of 29 stations in the MARB from 1965 to 1994 found interannual errors in simulated river discharge were less than 20% for the majority of the data (76%), although the seasonal errors were greater than 20 for 65% of the station months and particularly underestimated the magnitude of spring discharge. A similar analysis found simulated annual mean nitrate export of the Mississippi River at St. Francisville was within 1% of the USGS estimate, but annual errors at various stations varied widely.

Results of the IBIS/HYDRA modeling study indicated that nitrate export from the MARB was significantly greater during the latter half of the 1955-to-1994 period, largely due to the increase in N fertilizer application, with greatest contribution from the central and eastern subbasins (Donner et al., 2002). This analysis made many simplifying assumptions about nitrogen inputs, fate, and transport in order to isolate the impact of hydrology on nitrate export variability. Donner et al. (2002) concluded that the observed increase in river discharge was responsible for about 25% of the increase in nitrate export between 1966 and 1994, with an error of 7%. The remainder of the increase was inferred to arise predominantly from an increase in fertilizer N inputs. In the Upper MARB (1974–1994), Donner and Kucharik (2003a) found that a ±30% change in N fertilizer application resulted in little change in corn yields (+4%/−10%) but greater sensitivity in dissolved inorganic N subsurface drainage (+53%/−37%). They noted that soil N storage resulting from the +30% fertilizer N case appeared to lead to almost 60% increase in nitrate export after 20 years and that this effect was greatest during wet years.

Further work for the entire MARB based on IBIS/HYDRA results led Donner et al. (2004) to conclude that the doubling of nitrate export to the Gulf of Mexico over the 1960-to-1994 period resulted largely from an increase in fertilizer application rates, particularly to corn; an increase in runoff across the basin; and the expansion of soybean cultivation. Their results indicated that by the 1990s, fertilized cropland (particularly in Corn Belt hot spots across Iowa, Illinois, and Indiana) became the overwhelming nitrate source in the river system, contributing almost 90% of the nitrate from just 20% of the watershed area. Changes in MARB crop production systems associated with a shift away from meat production were simulated by IBIS/THMB (Donner, 2006). Results indicated a reduction in total land and fertilizer demands by over 50% and N export by 49–54% without any change in total production of human food protein.

3.4.4 Discussion and Comparison of Models

The Study Group found only one study that compared any of the three focus models. A study comparing SWAT and a statistical approach based on SPARROW within the Great Ouse watershed in the United Kingdom found similar total oxidized nitrogen load estimations and similar statistical reliability of the two models (Grizzetti et al., 2005). They suggested using SPARROW as a screening tool for identifying sources and using SWAT for testing management practice scenarios but found that both models demonstrated utility for nitrogen load estimation.

Different modeling approaches resulted in different assessments of nutrient sources and distribution within the MARB among the models, where comparisons were possible. Cropland was found to contribute 90% of nitrate within the MARB by IBIS/HYDRA (Donner et al., 2004) compared to 66% of total N for all crops by SPARROW (Alexander et al., 2008). Both models suggested the major nutrient source yields (mass per unit area) originated from the central Mississippi and Ohio River basins. Future work with SWAT (J. Arnold, personal communication) applied to the entire MARB will provide another estimate of nutrient sources and distribution to assist with watershed planning and management decisions.

3.4.5 Targeting

The models cited in Table 3.6 vary considerably in type, scale, and approach. The Gulf hypoxia issue requires a diversity of model types, scales, and approaches. Models that can support adaptive management of Gulf hypoxia and within-region water quality are those that can best inform targeting of the most effective actions at the lowest cost. Three forms of targeting are especially important:

- targeting subregions or watersheds (of perhaps the 8 or 12 digit HUC size) that have a disproportionate effect on hypoxia and local water quality;
- targeting the type and placement of conservation practices within those watersheds to achieve the greatest gains at the lowest cost; and
- targeting the timing of nutrient flows to best attenuate the hypoxic zone.

Both SWAT and IBIS/THMB models directly address targeting of practices by simulating the effects of farm/plot scale best management practices (BMPs) directly, whereas the absence of BMP simulation capability is a weakness in SPARROW (acknowledged in Alexander et al., 2008). Simulation of these BMPs is vital for the evaluation of successful management of nitrogen and phosphorus runoff in MARB. SWAT and IBIS/THMB include these practices but would benefit from additional verification that their mechanistic characterizations represent the range of field and watershed-scale processes present in the MARB. SPARROW shortcomings in simulation of BMPs are being addressed by ongoing efforts to evaluate farm and plot scale BMPs by USGS and others using data in the same sense as SPARROW is fitted (e.g., identifiability). When coupled with SPARROW, the results should yield a useful predictive model for the impact of BMP and other practices within MARB on Gulf hypoxia and should include uncertainty analysis.

Practices should be evaluated both for their impacts on total annual or long-term average nutrient loads and loads on a seasonal or other short-term time frame. Within-year timing of pollutant loads is simulated by SWAT and IBIS/THMB (both models operate on a daily basis; see Table 3.6) but not the annual-based SPARROW. Timing issues are critical for addressing seasonal water quality concerns both locally and for the April-to-June loads that appear to govern hypoxia development.

All three models address spatial targeting of sources and implementation. SPARROW spatial resolution is tied to the resolution of available monitoring data; recent studies of the MARB have been conducted at the HUC8 level. SWAT has been applied at a range of watershed scales from collections of fields to Mississippi River basin, whereas IBIS/THMB tends to be most applicable at the larger watershed scales. Both SWAT and IMIB/THMB spatial resolutions currently are dictated by computing capacity for larger scale basins. Spatial-scale issues are critical in targeting implementation actions, funding, and resources to areas with the greatest potential for improvement. Model interpretation must consider the relationship of watershed spatial heterogeneity and model averaging of input variables, process algorithms, and outputs. All three models provide information to assist with spatial targeting of actions.

Finally, there is a need to integrate watershed models with economic models in making targeting recommendations. Due to the ability to assess the effectiveness of specific conservation practices in a subwatershed context, integrated economic-watershed models have largely relied on mechanistic models such as EPIC and SWAT. More discussion of these integrated models can be found in Section 4.3.

3.4.6 Model Uncertainty

Model predictions all have a degree of uncertainty, which should be addressed in presenting model results. Model uncertainty may be due to variability, inaccuracy, or inappropriateness of multiple factors: (1) model algorithms or methods, (2) inputs (known or measured values, such as climate data), (3) parameters (values estimated based on functional relationship with known inputs, such as soil hydraulic conductivity or runoff curve number), (4) calibration data (including measurement errors associated with streamflow estimation and sample collection, storage, and analysis), (5) boundary conditions (such as initial soil moisture), (6) temporal scale (such as rainfall intensity), and (7) spatial scale (such as topography) (Shirmohammadi et al., 2006). Model uncertainty can be assessed by describing the impact of uncertain inputs and parameters, treated as random variables, on output variability.

Uncertainty varies among models and differs among watersheds, depending on availability of data, appropriateness of model assumptions for the given watershed and climate, and skill of the modeler in applying the model and interpreting the results. One study evaluated the impact of input uncertainty on SWAT2000 model output variation and found that input uncertainty was transferred nonlinearly through the model (Shirmohammadi et al., 2006). Coefficients of variation (CVs) of 34 input parameters ranging from 10 to 76% resulted in a single-year output CV of only 28% for streamflow; lower CVs for ammonium and organic N (6–7%) and mineral P (12%); and higher CVs for nitrate (101%), organic P (58%), and sediment (36%). Measured streamflow, nitrate, and ammonium values were within one standard deviation (SD) of the mean modeled output, whereas sediment was within 1.5 SD.

An advantage of SPARROW (and other models that are parameterized using optimization criteria, such as least squares or maximum likelihood) is that the model provides error terms for prediction with respect to parameter uncertainty. This information has been used to identify 150 high priority watersheds for management (Robertson et al., 2009). Process (mechanistic) models, such as SWAT and IBIS/THMB, are "overparameterized," which means that the observational data are insufficient to provide unique/optimal estimates of model parameters. Thus, these process models typically have been parameterized according to modeler's best judgment, with important ramifications on model uncertainty. New approaches acknowledge that many different parameter sets may fit equally well for these mechanistic and process-based models (e.g., Beven, 2001). This approach avoids the difficult question of how the modeler chooses an optimal, single set of parameters.

Key Findings and Recommendations

Interactions of climate, land, water body, and management factors on nutrient yields and loads are incredibly complex. As such, management decisions should always consider multiple models with different modeling approaches. The models discussed in this book are capable of nitrogen and phosphorus load estimation on the scale of the MARB, yet each has strengths and weaknesses. Other models and modeling approaches also exist, and each has inherent strengths, limitations, and value to improving understanding of and informing decision making related to the MARB and Gulf hypoxia. Thus, a diversity of models needs to be developed and applied for load estimation, BMP evaluation, implementation targeting, and forecasting. Models should provide information about the direction, magnitude, and uncertainty of the impact of current and planned actions on ecosystem services at the appropriate temporal and spatial scale and at a resolution and precision that is appropriate to guide these decisions. With an enhanced modeling toolbox at their disposal, decision makers will need to select the model or models best suited to answering their questions and guiding their decisions.

The uncertainty of results for each model reflects the uncertainty of the model structure and algorithms, as well as that propagated by the input data, user parameterization, calibration process, and other user-defined conditions. Other than the model itself, each of these factors is influenced by the skill of the model user, making it difficult to make blanket generalizations about reliability or applicability of the models discussed.

Adaptive management will be more informative, particularly in the initial years post implementation, if monitoring data are used to improve models for the next iterative prediction. This requires that the monitoring be designed, at least in part, for this task. These monitoring data will also enhance the modeling effort. Rigor of model validation can be assessed through statistical comparison of calibration data with validation data provided through monitoring. Greater availability of monitoring data will allow a greater difference between calibration and validation data sets and provide a more rigorous model validation. For example, applying a model to a different watershed with different climate will better test the robustness of the model than a validation using a different period of climate data within the same watershed.

Adaptive management will require modeling flexibility as well as consideration of the compatibility between watershed models, economic models, and Gulf of Mexico hypoxia models. The various models need to have the capability to translate across temporal and spatial scales and to communicate how factors affecting ecosystem services are simulated in order to have a smooth interface. For example, watershed model output of total N at annual scales should be able to interface with a Gulf model requiring daily conditions of inorganic N. In addition, models need to be developed and used to assess effects of policy decisions and management practices. Characterization of the degree of uncertainty would assist interpretation of results and application of these results within an adaptive management framework. Based on these findings, the Study Group offers the following recommendations.

- A diversity of watershed modeling approaches, ranging from simple forecasting to complex statistical and mechanistic approaches, will be useful for describing loading and timing of nutrients to the NGOM.
- Model selection should depend on the question(s) being asked and the associated strengths and weaknesses of the various models; for Gulf hypoxia, watershed models should address issues of management option selection, spatial targeting of actions, and temporal delivery of nutrient loads to the NGOM.
- Water quality monitoring and the documentation of critical ancillary information (i.e., inputs and management practices) should be designed, at least in part, to support model use and assessment, and adaptive management.
- Uncertainty of model results should be assessed and reported. As much as possible, all potential sources of error should be explicitly recognized and discussed in reporting results. Further, confidence bounds should be reported for all applicable sources (e.g., parameter uncertainty). For those sources for which formal confidence intervals cannot be computed, sensitivity analysis or another form of uncertainty analysis should be undertaken and reported.

Chapter 4
Scientific Basis for Goals and Management Options

4.1 Adaptive Management

Adaptive management offers a way to address the pressing need to take steps to manage for factors affecting hypoxia in the NGOM in the face of uncertainties. The authors of a recent study undertaken by the National Research Council of the National Academy of Sciences identified six elements of adaptive management that are directly relevant to goal setting and research needs (National Research Council, 2004): (1) resources of concern are clearly defined; (2) conceptual models are developed during planning and assessment; (3) management questions are formulated as testable hypotheses to guide inquiry; (4) management actions are treated like experiments that test hypotheses to answer questions and provide future management guidance; (5) ongoing monitoring and evaluation is necessary to improve accuracy and completeness of knowledge; and (6) management actions are revised with new cycles of learning.

Perhaps the most important "take-home" lesson from their work is contained in the following statement:

> Adaptive management does not postpone actions until "enough" is known about a managed ecosystem (Lee, 1999), but rather is designed to support action in the face of the limitations of scientific knowledge and the complexities and stochastic behavior of large ecosystems (Holling, 1978). Adaptive management aims to enhance scientific knowledge and thereby reduce uncertainties. Such uncertainties may stem from natural variability and stochastic behavior of ecosystems and the interpretation of incomplete data (Parma et al., 1998; Regan et al., 2002), as well as social and economic changes and events (e.g., demographic shifts, changes in prices and consumer demands) that affect natural resources systems.

Thus adaptive management provides an appropriate way for decision makers to deal with the uncertainties inherent in the environmental repercussions of prescribed actions and their influences on hypoxia.

Adaptive management can be conducted at the several management scales that occur in the NGOM and MARB. On the basin scale, adaptive management requires measurements of both nutrient loadings and hypoxia extent (area). Although it will not be possible to relate these changes to specific changes in the basin, these data will provide better understanding of the relationships between nutrients and hypoxia. On smaller scales, specific management actions can be treated as experiments that test hypotheses, answer questions, and thus provide future management guidance at that scale (for example, small watersheds).

The adaptive management approach requires that conceptual models are developed and used and that relevant data are collected and analyzed to improve understanding of the implications of alternative practices (e.g., Ogden et al., 2005). To help illustrate what is meant by a conceptual model, the Study Group has developed a diagram that shows major factors that affect hypoxia in the NGOM (Fig. 4.1). The corresponding conceptual model would estimate the relative contribution of each influence. Those estimates could serve as hypotheses of relative effects, and the diagram could illustrate hypothesized interactions and feedbacks. Such a conceptual model organizes how adaptive management research is conducted in a framework where the testing of hypotheses and the new knowledge gained is then used to drive management adaptations, new hypotheses, and the collection of new data on end points. Unlike the traditional model of hypothesis-driven research, adaptive management implies coordination with stakeholders and consideration of the economic and technological limitations on management. Unlike traditional demonstration projects, adaptive management implies an understanding that complex problems will require iterative solutions that will only be possible through generation of new knowledge as successive approximations to problem solving are attempted.

Successful implementation of the adaptive management process is occurring in the Grand Canyon (Meretsky et al., 2000) and the Everglades (Sklar et al., 2005). In addition, steps toward adaptive management are being examined in the Upper Mississippi River basin (O'Donnell and Galat, 2008). That work documents the need for greater collaboration between scientists and management agencies to plan, design, and monitor river enhancement programs. Problems exist in setting quantifiable success criteria, developing appropriate monitoring designs, and disseminating

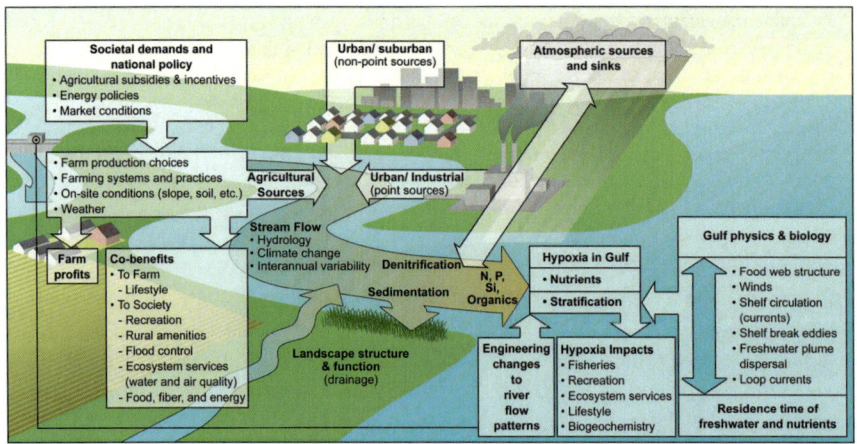

Fig. 4.1 A conceptual framework for hypoxia in the northern Gulf of Mexico

information. The Study Group expects similar difficulties in implementing adaptive management to occur in the MARB.

There needs to be a better understanding of the spatial and temporal aspects of basin-level responses to management practices and also a focus on other scales at which response can occur in a more timely fashion (Nassauer et al., 2007). Yet observations of a basin-level response to practices cannot be expected for some time, which calls for management and evaluation to be focused on a subbasin scale. Therefore it is important to obtain information at a scale where practices can be broadly and appropriately applied and where results are "meaningful and interpretable." The relevant scale would likely be at smaller subwatershed scales, where local water quality and quantity benefits may become evident more quickly. Furthermore, the demonstration of adaptive management within a small sub-watershed may enhance practice adoption at other locations. Thus conceptual models need to be developed for this scale of resolution as well. Focus at the small watershed scale will also provide local water quality and quantity benefits. The results from small watershed studies must be able to be extrapolated to other small watersheds in the subbasin and, preferably, the entire MARB, if they are to be useful in reducing hypoxia in the NGOM.

Experiments that could be applied at small watersheds to help to improve understanding of the effects of different practices have the following characteristics:

- Practices applied on the small watersheds should conform to accepted practice standards or make specific modifications of practices that can be implemented in new standards.
- Monitoring should be at appropriate intensities (time and space) to determine effects of practices on water quality and quantity.
- Monitoring should also measure co-benefits, including carbon sequestration, wildlife habitat, flood control, etc.
- Practices should be applied in suites or systems, and components should be monitored to determine effects of component practices.
- Changes in hydrology and crop productivity must be measured in addition to changes in water quality. Even at the small-scale, too many studies have focused just on nutrient concentrations in outflow water and neglected hydrologic or productivity changes.
- All components of the cost of adopting and maintaining these practices should be measured and monitored. Such costs include direct equipment and structural costs, yield effects, changes in management time, changes in risk, and other costs.
- These studies should be designed to improve our understanding at local, medium, and broad basin scale. Thus the experiments should be designed so that they can feed into conceptual models that operate at different scales.
- Within practical limits, studies should be part of an adaptive management research strategy for the MARB to optimize the efficiency of research investments and to assure that results are coordinated, complimentary, and consistent.

Integrated modeling and monitoring play an important role in adaptive management. The cornerstone of adaptive management is the concept of learning about the impacts of actions and using that new understanding to guide future actions. Models can assist that learning by being used to evaluate impacts and uncertainties of proposed actions, such as targeted practices and locations or proposed policies, on both MARB and NGOM responses. In addition, monitoring must also be part of an adaptive management strategy in order to verify that the actions are addressing the stated goals or to test hypotheses. Monitoring is needed to improve the next generation of models and model assessments and to eventually verify that projected changes occur.

Adaptive management is also important to building infrastructure and to strategic planning and policy development of mechanisms of conservation practice implementation. For example, adaptive management can be used to evaluate if incentive-based programs are effective at bringing about changes in conservation practice acceptance and adoption at a local or small watershed level. At a basin level, other programs might be needed to facilitate adaptation of strategies and policies, and there must be constant feedback among all vested parties. As the scale of system increases (i.e., from a small watershed to the entire MARB), the complexity of adaptive management increases dramatically.

Key Findings and Recommendations

Adaptive management can be used at several scales of resolution in the NGOM and MARB to provide a framework under which management activities can occur while monitoring and modeling the outcomes in order to provide information so that subsequent management can be improved. Therefore, the Study Group offers these recommendations.

- An adaptive management approach should be adopted to evaluate the success of reaching goals and for testing hypotheses (at the relevant scale).
- Conceptual models should be developed at appropriate scales of resolution to frame the adaptive management process in addressing factors affecting hypoxia in the NGOM.
- Both the use of quantitative models and the collection of data should be conducted within an adaptive management framework and at appropriate management scales so that the information gained from models and data are related to the critical questions about managing and understanding the system.
- Management actions should be designed as experiments within the context of evolving conceptual understanding of the system. The repercussions of management actions need to be monitored so the outcomes can be used to enhance learning and thus to improve future management actions.

4.2 Setting Targets for Nitrogen and Phosphorus Reduction

To reduce hypoxia in the bottom waters of the NGOM, the *Integrated Assessment* set a target that N loading should be reduced by 30% in order to shrink the 5-year running average size of the hypoxic zone to below 5,000 km² (1,930 mi²) by 2015. This reduction is significantly less than the three- to five-fold increase in N loading to the Gulf of Mexico due to human activity during the 20th century, and particularly in the past 30–50 years (Boyer and Howarth, 2008; Goolsby et al., 2001). Since the *Integrated Assessment*, a number of modeling efforts have provided a better depiction of how the area of hypoxia may respond to reduced N loading. The three available models were compared by Scavia et al. (2004), who concluded from these models that the 30% reduction in N is probably not sufficient to reach the goal of a hypoxia area of 5,000 km² or less (Scavia et al., 2004). The consensus from these models is that N loads probably need to be reduced by 40–45% to reach the hypoxia reduction goal. In addition, a number of studies suggest that the consequences of climate change need to be considered, and this may require an N load reduction on the order of 50–60% to meet the original *Integrated Assessment* goal for hypoxic area (Donner and Scavia, 2007; Justić et al., 2003b). However, predicting the consequences of climate change on nutrient fluxes and hypoxia remains a very uncertain business (Howarth et al., 2006). The Study Group finds that the consensus of models reported by Scavia et al. (2004) and the new model of Scavia and Donnelly (2007), which uses the latest available load estimates from the USGS, supports a target of reducing the 5-year running average of N loadings by at least 45%. This target should be reassessed as more monitoring data are obtained, current models are refined, and new models are developed.

Only recently has new evidence emerged for the need to control P inputs as well as N in the NGOM. Work by Sylvan et al. (2006) has shown P to be the limiting nutrient during periods of maximum primary production in the near-shore NGOM high-productivity zone. Because previous attention has focused on N, there has been limited effort to model the effects of P on hypoxic area. Scavia and Donnelly (2007) used the previously developed and calibrated model (Scavia et al., 2004) to evaluate both the effects of new USGS load estimates and to assess the potential for P to control hypoxia dynamics under current and historical conditions. Confirming the results of Sylvan et al. (2006), Scavia and Donnelly found that P could have become limiting in some areas and times because of the relative increase in N loads during the 1970s and 1980s. While they concluded that P did frequently control hypoxia in near-field zone of NGOM, they noted that a P-only strategy would likely reduce production in the near-field but possibly increase production in downfield N-controlled areas of NGOM. Their work, using the new USGS load estimates, reinforced the need for a dual nutrient strategy combining a 45% reduction in N with a 40–50% reduction in the 5-year running average of P loading. While the far-field effects could possibly be reduced through an N-only strategy, they suggested that a prudent approach would be to reduce both N and P, simultaneously. They also noted that an N-and P-reduction strategy would not only reduce hypoxia in the NGOM but would also help to remove P-induced Clean Water Act impairments in the MARB. Based on this recent modeling work, the Study Group finds that a comparable P reduction is needed, again based on 5-year running average fluxes. As with the N target, this P target should be reassessed over time as more monitoring information is gained and new models are developed.

The CENR report and Scavia et al. (2004) made recommendations on an N reduction target with reference to average fluxes for 1980–1996. These fluxes were calculated using different methods (see Section 3.1) than in this book, but the N reduction target proposed recently by Scavia and Donnelly (2007) used a combination of the newer USGS 5-year LOADEST and composite estimates since 1980. In this book we only use the 5-year LOADEST results, since the composite estimates are incomplete; however, they are very similar to each other (again, see Section 3.1).

During the past 5 years of record, annual water flux to the NGOM has declined by 5.8%, whereas nitrate-N and TKN have declined even more, leading to a total annual N reduction of about 21% (Table 4.1). Considering the original reduction target of a 30% reduction in total N, it would seem that substantial progress was made beyond the reduction that would occur from less flow alone. However, the largest reduction was in TKN, with a large part of this decrease from the Missouri River (discussed in Section 3.1). For the important spring flux of N, there was little reduction in nitrate-N beyond the reduced water flow (−11 and −12.4% declines in water and nitrate-N flux, respectively). Again, TKN was greatly reduced (−31.5%) during spring flows, leading to most of the decline in total N (−19.2%), beyond the reduction in water

Table 4.1 Annual and spring (sum of April, May, June) average flow and N and P fluxes for the MARB for the 1980–1996 reference period compared to the most recent 5-year period (2001–2005). Load reductions in mass of N or P also shown

	Flow (million m^3 (water) or million metric tons)			Flux (million metric tons)	
	1980–1996 flux	2001–2005 flux	Change (%)	45% reduction N target flux	45% reduction P target flux
Annual					
Water	692,500	652,500	−5.8		
Nitrate-N	0.96	0.81	−15.4	0.53	
TKN	0.61	0.43	−30.0	0.34	
Total N	1.58	1.24	−21.1	0.87	
Total P	0.137	0.154	+12.2		0.075
Spring					
Water	236,800	210,600	−11.0		
Nitrate-N	0.38	0.33	−12.4	0.21	
TKN	0.21	0.14	−31.5	0.12	
Total N	0.59	0.48	−19.2	0.32	
Total P	0.046	0.050	+9.5		0.025

4.2 Setting Targets for Nitrogen and Phosphorus Reduction

flux. This suggests that during the important high-flow spring period (April, May, June), reductions in nitrate-N flux to the NGOM have not occurred under management systems and programs now in place since the most recent report. However, the annual nitrate-N reduction indicates that the tile-drained corn and soybean systems in the Upper Mississippi and Ohio River subbasins seem responsive on an annual basis to the recent reductions in net N inputs, as discussed in Section 3.2. Whether spring nitrate-N loads will respond to these changes in NANI is uncertain at this time.

For total P flux, both annually and during the spring, there were increases of 12.2 and 9.5%, respectively. It is not clear why total P fluxes are increasing (with corresponding smaller water fluxes), and the result suggests that the reduction target of 45%, relative to the 1980–1996 period, is close to 50% for the 2001–2005 period. Likewise, the 45% N load reduction target, relative to the 1980–1996 period, is equivalent to a 30% reduction relative to the 2001–2005 period. Fertilizer P consumption in the MARB has been relatively constant since about 1984 and is similar to consumption during 1970–1975. Net P inputs to the MARB have declined since the 1970s and have been predominantly negative since the mid-1990s (see Section 3.2 and Fig. 3.25). Table 4.1 also indicates N and P reduction recommendations in units of mass with reduction targets of 45% N and 45% P, assuming that the reduction was spread across all forms of N and P, that occur both annually and during the spring.

While the Study Group finds that both N and P reductions are warranted, additional modeling and dose–response research are needed to refine the reduction targets, particularly for P loading. Scavia and Donnelly (2007) presented the only model results that relate P loads to hypoxia in the NGOM. Further, there are no experimental data relating phytoplankton responses there to different levels of P. Ideally, targets for reducing P based on water quality should have greater model support and should consider dose–response relationships for P responses by the in situ phytoplankton communities. In the meantime, the response of the Gulf system to a specific amount of P reduction remains uncertain and must await the formulation of new models and dose–response relationships for the receiving waters. Water quality models aimed at evaluating the effects of these reductions will also rely on this information. Dose–response relationships should be developed using in situ bioassays designed to "ask the phytoplankton" what the response relationships and bloom thresholds are. These bioassay experiments are a logical follow-up to the work of Sylvan et al. (2006), which has shown P to be the limiting nutrient during periods of maximum primary production in the near-shore NGOM high-productivity zone. Bioassays are needed on a seasonal basis, where the effects of hydrologic variability and changing N:P input (loading) ratios on primary production, phytoplankton community composition, and biogeochemical and trophic fate can be evaluated.

In Section 4.5.8 on Most Effective Actions for Industrial and Municipal Sources, the Study Group provides some ballpark estimates of possible N and P reductions from upgrading major municipal wastewater treatment plants. The Study Group's example calculations demonstrate that sewage treatment plant upgrades to achieve total N concentration limits of 3 mg/L and total P concentrations of 0.3 mg/L could create reductions in total annual N flux to the Gulf by about 10% and the total spring N flux by about 6%. Upgrading to achieve P concentrations of 0.3 mg/L would create reductions in P fluxes from sewage treatment plants from 41,000 metric tons P/year (45,000 ton P/year) to 10,500 metric tons P/year (11,600 ton P/year) or about a 75% reduction in annual flux from sewage treatment plants to the MARB. These reductions, in turn, would translate into reductions of total annual P flux to the Gulf by about 20% and the total spring P flux by about 15%. If further investigation and data collection confirms the Study Group's calculations, upgrades to major wastewater treatment plants in the MARB could accomplish nearly half of the Study Group's recommended P reduction targets. This would represent very significant progress for both improving water quality in the MARB and reducing hypoxia in the NGOM.

Despite the need for additional model and bioassay work, the proposed target of a 45% reduction in annual P load should be used in an adaptive management framework to allow development of strategies that optimize both N and P reductions while more knowledge is acquired on P reduction impacts on near-field hypoxia. Unlike N, the P reduction strategy will help address water quality impairments in the MARB. Given the evidence that both N and P should be reduced in the NGOM, setting a goal for P reduction should not await the development of new models and availability of new experimental data. Enough information exists now to set a goal in an adaptive management context beginning with the P reductions that are already feasible given existing technologies and options.

In 2000, USEPA recommended nutrient criteria to states and tribes for use in establishing their water quality standards consistent with Section 303(c) of the Clean Water Act (CWA) (USEPA, 2000c). USEPA's recommended criteria represent an estimated "reference condition," and it is assumed that the reference condition concentration would protect all designated uses (including the most protected uses, such as high-quality fisheries, sensitive aquatic life). The Study Group asked USEPA for a comparison of the Study Group's recommended 45% reductions for TN and TP flux to the reductions in nutrient levels that would correspond to USEPA's ecoregional nutrient criteria for reference conditions (USEPA, 2006b). This comparison is provided in Appendix E. Although a number of assumptions were required to make this comparison (see the caveats in Appendix E), USEPA's preliminary analysis suggests that the Study Group's recommended targets for reducing TN and TP are, for most regions, not likely to be as stringent as would be obtained if states adopted USEPA's recommended reference condition values into state water quality standards for all waters. This comparison should not be interpreted as the Study Group's endorsement of USEPA's recommended nutrient criteria but rather an emphasis on the need to consider both within-basin nutrient criteria and NGOM load reduction goals. Numeric nutrient standards being developed by the states of the MARB will almost certainly be concentration rather than load based and may be most stringent during warmer, lower flow periods when absolute loads can be relatively low but when local waters are most frequently impaired by excess nutrient levels. It will be important for USEPA and other agencies to evaluate and, if necessary, reconcile within-basin water quality standards with load reduction goals for

4.2 Setting Targets for Nitrogen and Phosphorus Reduction

the NGOM. Strategies are needed for integrating standards throughout the MARB to better manage hypoxia as well as local water quality.

A mechanism in the Clean Water Act for addressing water quality impairments is the development of Total Maximum Daily Loads (TMDLs), though it is important to note that the focus of TMDL development is identification of the source and causes of water quality impairment, rather than on implementation of change for improving water quality. Under Section 303(d) of the Clean Water Act, states, territories, and authorized tribes are required to develop lists of impaired waters (i.e., waters that have not met water quality standards). The law requires that the appropriate jurisdictions develop TMDLs for these impaired waters. The TMDLs specify the maximum amounts of pollutants that waterbodies can receive and still meet water quality standards. In addition, TMDLs allocate pollutant loadings among point and nonpoint sources.

The status of nutrient criteria and TMDL development along the Mississippi River has been reviewed by the National Academy of Sciences (National Academy of Sciences, 2007). The National Academy of Sciences notes that none of the 10 Mississippi River mainstem states currently have numeric criteria for nitrogen or phosphorus applicable to the River and, that without such standards, there is little prospect of significantly reducing or eliminating hypoxia in the Gulf of Mexico. The National Academy of Sciences also describes how the process of developing numeric nutrient criteria and TMDLs for the Mississippi River could lead to water quality improvements in the Gulf of Mexico. NAS suggests that through such a process, USEPA could adopt the necessary numerical nutrient criteria for the terminus of the Mississippi River and waters of the northern Gulf of Mexico. Maximum nutrient loads could be assigned to each state and the loads could be translated into water quality criteria. Each state would then be required to develop a TMDL for waters that failed to meet the applicable criteria, and a coordinated effort could be undertaken to reduce point and nonpoint source loads to meet allocations established by the TMDLs. Thus, the NAS report identifies an approach through existing legislation (the Clean Water Act) that could be used to redress Gulf Hypoxia, but the SAB stresses that a great many steps exist between calling for "a coordinated effort" and implementing the full set of actions that must be undertaken for water quality to actually improve in the Gulf.

Key Findings and Recommendations

Based on findings since the *Integrated Assessment*, a N reduction target of greater than 30% will be needed to reduce the hypoxic area to 5,000 km^2 (1,930 mi^2). Recent research indicates N reductions of at least 45% will be needed to achieve the target in most years and reductions may have to exceed 50% due to effects of climate change. Research by several investigators provides evidence that P may limit primary production in the river outflow, near-field areas of the Gulf. Based on new research with the same model used to establish the N target, reductions in P loads of 40–50% are needed to reduce P-controlled hypoxia in the near-field areas of NGOM. P reductions in the MARB will not only benefit the NGOM but will also help to address P impairments in the MARB. Based on these findings, the Study Group offers the following recommendations.

- To reduce the size of the hypoxic zone, the total N flux to the NGOM from the combined Mississippi and Atchafalaya Rivers must be reduced by at least 45% from 1980 to 1996 average fluxes, to no more than 790,000 metric ton N/year (870,000 ton/year), and 290,000 metric ton N (320,000 ton) during the spring (April, May, June), both on a 5-year running average.
- To reduce the size of the hypoxic zone, commensurate reductions in P are needed. The total P flux to the NGOM from the combined Mississippi and Atchafalaya Rivers should be reduced by at least 45% from 1980 to 1996 average fluxes, to no more than 68,000 metric tonne P/year (75,000 ton P/year) on a 5-year running average.

4.3 Protecting Water Quality and Social Welfare in the Basin

The Study Group has been asked whether social welfare can be protected while reducing hypoxia and improving water quality in the Basin. To thoroughly answer this question would require quantification of the full costs of all activities undertaken to reduce the necessary nutrient loading into the Gulf (from agricultural sources, point sources, air deposition, etc.) and the full benefits accruing from those activities. The benefits would include the direct benefits of reducing the size of the hypoxic zone (commercial fishery effects, recreational fishery gains, the value placed on preserving intact ecosystems, biodiversity, etc.) and the "co-benefits" (such as improved local water quality, increased wildlife habitat, flood control, aesthetic values).

Since the costs, benefits, and co-benefits will depend on the extent of coverage and specific locations of control options, a complete answer to the question would require knowing the details of how such nutrient reductions would occur. For example, if these reductions are to be achieved entirely through restoration of wetlands and tighter municipal source controls, it would be necessary to know where the wetlands would be located and where the point-source reductions would occur in order to estimate their costs and their co-benefits. In contrast, an entirely different set of co-benefits and costs would likely result from relying on a broader array of control options that also included nutrient management, increased perennials, riparian buffers, drainage management, and reductions in air deposition. Further, the exact policy approach (e.g., expanded EQIP funding, mandates, or taxes) would need to be specified if estimates of the incidence of the costs are to be estimated (i.e., whether the costs would ultimately be borne by taxpayers, by consumers, or by farmers and landowners).

To date, no set of models and/or studies have been undertaken that address all of the necessary components on a basin-wide scale to estimate the effects on social welfare. However, a number of studies, beginning with the research in the *Integrated Assessment*, have been done that address substantial components of this question. More complete efforts at quantifying the control costs than the benefits have been undertaken, though there remains a need for much more work on both sides of the equation. Integrated models at multiple levels and scales are needed to support this effort. The existing research focuses largely on agricultural nonpoint source control. This section summarizes findings from the limited set of large-scale economic-watershed models of agricultural nonpoint sources that have been applied to date.

4.3.1 Assessment and Review of the Cost Estimates from the CENR Integrated Assessment

Doering et al. (1999) in the *Integrated Assessment* undertook an ambitious cost-effectiveness analysis of several policy approaches to reach the N loss reduction goal of 20% established as part of the *Integrated Assessment*. The central modeling system they used was the US Mathematical Programming (USMP) model, which represents the agricultural sector in 45 production regions throughout the United States with 10 crops, 16 animal products, retail and processed products, and a range of domestic and international supply and demand relationships. Management practices include crop rotations, five tillage options, and varying fertilizer rates.

The environmental effects of various management practices and land uses in USMP are predicted by the EPIC model (the Environment Productivity Impact Calculator). USMP uses EPIC to predict changes in N loss, P loss, and sediment loss at the edge of the field from changes in land use and conservation practices. Donner et al. (2002) chose a 20% N loss reduction goal as "the best combination of sizable nitrogen loss reductions and acceptable economic costs" (Doering et al., 1999 p. 37). The remainder of their analyses focused on the evaluation of several policies that might achieve this environmental goal. Some key predictions from the modeling system include

- A 20% reduction in fertilizer N application rates would result in the reduction of edge-of-field N loss by about 11%. In contrast, a 45% reduction mandate and fertilizer tax set to achieve a 45% reduction is predicted to result in the target goal of N loss reduction of about 20%. The less than proportional reduction in N loss coming from reduced fertilization in this modeling system is a result of predicted changes in acreage resulting from the feedback effect of price changes. Specifically, higher crop prices due to lower yields from the reduced fertilization rates induce more acreage planted to the fertilized crop, thereby partially

offsetting the reduction in N. Whether the magnitude of the yield effects embedded in these models is accurate is an important question. For further discussion of this issue, see Section 4.5.6.
- Some 7.29 million hectares (18 million acres) of wetland restoration would achieve the 20% reduction in N loss goal at a cost of over $30 billion.
- Restoration of 10.9 million hectares (27 million acres) of riparian buffers was estimated to cost over $40 billion and generated relatively small reductions in N losses, suggesting that this strategy is not cost-effective for hypoxic zone control. In light of current evidence that phosphorous is also of concern, this result should be reconsidered as there is significant evidence that buffers can be quite effective in reducing sediment and phosphorous loss.
- A "mixed policy" with a 2.02 million hectares (5 million acres) wetland restoration program in conjunction with a 20% fertilizer reduction is more cost-effective than most of the previous approaches, but the 45% reduction in fertilizer is more cost-effective yet.
- The introduction of point–nonpoint source trading across the basin where the cap applies only to point sources will not achieve the 20% N loss reduction due to the relatively small magnitude of N contribution from point sources. Even with a stringent standard on point sources, only about 5% of the needed reductions occur.
- These policies are likely to produce large "co-benefits" (i.e., other environmental benefits occurring within the basin and on-farm productivity benefits not immediately captured in the current profitability resulting from the policies). For example, the authors estimate that restoration of 405,000 hectares (1,000,000 acres) of wetlands would yield total benefits in the basin that exceed the costs, even without considering any benefits of hypoxia reduction.

Cost estimates used for the *Integrated Assessment* for a 20% reduction in N discharge coming from agricultural nonpoint sources range from $15 billion to $30 billion; however, these estimates suffer from a number of shortcomings including consideration of only a few options for reducing nutrient discharge and limited targeting. More inclusive assessments with better targeting of options to locations where they are most appropriate may reduce these costs.

In follow-up research, some of the same study coauthors (Ribaudo et al., 2001) compare nitrogen reduction methods with wetland restoration and low and high levels of N loss reduction. They find that nutrient management is more cost-effective at low levels of N loss reduction while wetlands restoration is more cost-effective at high levels. Tables 4.2 and 4.3 (listed at the end of this discussion) briefly summarize the key components of these studies and the other large-scale studies that are reviewed in the following discussion.

Due to limits on the understanding of the economics and natural science at the time, the work in the *Integrated Assessment* and its follow-up is based on assumptions that, in light of more recent research and availability of data, assessments could be improved upon in future work. The USMP model represents a wide variety of agricultural raw inputs and intermediate products at a relatively aggregate scale. However, it does not contain detailed description of land use, soil characteristics,

4.3 Protecting Water Quality and Social Welfare in the Basin 123

Table 4.2 Summary of study features of basin-wide integrated economic-biophysical models

Authors	Study region	Models used	Environmental measures	Comments
Doering et al. (1999)	Entire United States with policies simulated in Mississippi River basin	USMP and EPIC	N, P, and sediment in the MARB (not delivered to the NGOM)	Original CENR study
Ribaudo et al. (2001)	Entire United States with policies simulated in Mississippi River basin	USMP and EPIC	N, P, and sediment in the MARB (not delivered to the NGOM)	Extension of CENR study
Greenhalgh and Sauer (2003)	Entire United States with policies simulated in Mississippi River basin	USMP and EPIC with SPARROW derived transport coefficients	N delivered to the Gulf, greenhouse gas emissions, P and N, soil erosion in the MARB	Study focuses on co-benefits of policies
Wu et al. (2004)	Upper Mississippi River basin	Econometric model and EPIC based metamodels	N leaching, N runoff, wind erosion, and water erosion in UMRB	Finer spatial detail than USMP but no price feedbacks
Ribaudo et al. (2005)	Entire United States with policies simulated in Mississippi River basin	USMP and EPIC	N in MARB	Follow-up to original CENR study
Wu and Tanaka (2005)	Upper Mississippi River basin	Econometric model and SWAT	N delivered to the NGOM	Finer spatial detail than USMP but no price feedbacks
Kling et al. (2006)	Upper Mississippi River basin	Econometric model and SWAT	N, P, and sediment in UMRB and N delivery to the NGOM	Finer spatial, but no price feedbacks

Table 4.3 Summary of policies and findings from integrated economic-biophysical models

Study	Policies/actions evaluated	Key findings[a]
Doering et al. (1999)	1. Fertilizer reduction mandates/fertilizer taxes 2. Wetland restoration 3. Riparian buffer 4. Mixed policy (wetlands and fertilizer reduction) 5. Water quality trading	1. Cost-effective approaches exist to reducing nitrogen losses in the 20% range 2. Wetland-based strategies are more expensive than fertilizer reduction 3. Buffers are not cost-effective for reducing N losses 4. A combination of 5 million acres wetland restoration with 20% fertilizer reduction is most cost-effective 5. These cost-effectiveness measures do not take into account the transport of nitrogen to the Gulf and the rankings of preferred alternatives could change
Ribaudo et al. (2001)	1. Reduce fertilizer rates 2. Wetland restoration	1. Below 26% reduction in N losses, fertilizer reduction/management is most cost-effective 2. Above this rate, wetland restoration is most cost-effective
Greenhalgh and Sauer (2003)	1. N trading between point and nonpoint sources 2. Greenhouse gas trading 3. N trading with additional payments for GHG reduction 4. N fertilizer tax 5. Conservation tillage payment 6. Expansion of CRP to 40 million acres nationwide	1. Nutrient trading (point/nonpoint) with tighter discharge limits could reduce nitrogen reach the NGOM by 11% annually 2. Nutrient and greenhouse gas trading were the lowest cost policies, but nutrient trading was the most cost-effective 3. The co-benefits of these policies in terms of greenhouse gas reductions, phosphorous, and sediment can be significant
Wu et al. (2004)	1. Conservation payments for conservation tillage 2. Crop rotations	Crop rotations not a cost-effective strategy for N reduction
Wu and Tanaka (2005)	1. Fertilizer tax 2. Payments for conservation tillage 3. Payments for land retirement 4. Payments for crop rotations	Fertilizer tax is the most cost-effective of policies considered
Booth and Campbell (2007)	Targeting CRP to watersheds with the greater proportion of fertilizer used. Hence, CRP rises in direct proportion to fertilizer/cropping intensity	Targeting CRP and enrolling an additional 2.7 million hectares in those areas with the greatest fertilizer intensity would increase annual agricultural subsidies to the MARB by 6.2% (over the combined commodity support and conservation funding in 2003)

4.3 Protecting Water Quality and Social Welfare in the Basin

Table 4.3 (continued)

Study	Policies/actions evaluated	Key findings[a]
Ribaudo et al. (2005)	N trading between point and nonpoint sources	Trading between waste water treatment plants and nonpoint/agricultural sources to meet the reductions achievable by installing advance nutrient removal technology at treatment plants would have large welfare gains
Kling et al. (2006)	Implementation of a set of targeted conservation practices including conservation tillage, land retirement, terraces, contouring, grassed waterways, and reduce fertilization rate on corn	1. Annual costs of $800 million per is predicted to achieve 22% reduction in N loading to the NGOM 2. Within the UMRB, sediment loads were reduced by 40–66%, total P was reduced by 6–47%, and N by 9–29%

[a]Doering et al. (1999) also conclude that fertilizer restrictions are more cost-effective than a fertilizer tax, but they apparently incorrectly count tax revenues as a cost rather than a transfer. The restrictions and tax have the same welfare effects, though different distributional implications.

yields, etc., at the individual field and/or subbasin scale. This inability to target finer scales could result in overstating the costs of meeting a particular reduction goal because significant cost savings can accrue from targeting land-management strategies.

The *Integrated Assessment* assumed a one-to-one relationship between the reduction in edge-of-field nitrogen loss and reduced loadings to waterways without incorporating the geographic differences in movement of N from the field of origination to the Gulf. Whether this shortcoming over- or understates the costs is an empirical question, but the results coming from a model that explicitly incorporates the fate and transport of nutrients and sediment might suggest very different results concerning the cost-effectiveness.

4.3.2 Other Large-Scale Integrated Economic and Biophysical Models for Agricultural Nonpoint Sources

Since completion of the *Integrated Assessment*, several basin-wide studies have evaluated policies that might reduce Gulf hypoxia and/or have effects on other environmental amenities that could be considered co-benefits (including carbon sequestration and upstream, local water quality indicators). The models can be divided into those that use the USMP modeling framework and those based on econometric estimates of behavioral response to economic drivers.

Booth and Campbell (2007) used a regression model to estimate the cost of reducing N losses when targeting conservation dollars to those areas with the highest proportion of fertilizer use. They modeled a hypothetical case in which conservation enrollment rises in direct proportion to the nonlinear rise in nitrate flux that occurs as fertilization intensity increases. The result was an increase in the amount of land in the high fertilizer watersheds enrolled in the Conservation Reserve Program by 2.7 million hectares (6.67 million acres) (a 29% increase over 2003 CRP levels) at a cost of $448 million. Booth and Campbell (2007) describe this as a 6.2% increase over the combined cost of commodity support and conservation programs. They account for the drop in commodity support spending that would accompany the enrollment of commodity-farmed land in the CRP. Booth and Campbell (2007) do not specify the percentage reduction in nitrate loading that would result from this scenario.

Wu et al. (2004) and Wu and Tanaka (2005) developed an econometric model of crop choice and tillage choice using the National Resources Inventory for the upper Mississippi River basin. They estimated the probability of adopting conservation tillage and crop choice based on a variety of physical and economic variables including land quality, slope, climate conditions, and profits. They used over 40,000 crop land points observed for 16 years, although only a subset of the observations were used for model fitting. These adoption models then simulate adoption profiles under alternative policies. Finally, the environmental effects of the policies are predicted with a biophysical model. Wu et al. (2004) used a set of environmental production functions estimated via a meta-modeling approach (Wu and Babcock, 1999), based on data generated from the EPIC model. They found that crop rotations are not a cost-effective strategy to N reduction.

Wu and Tanaka (2005) used the SWAT model to predict water quality changes from the policies. They considered the same two policies as Wu et al. (2004), as well as a policy that would increase the amount of land set-aside in a Conservation Reserve-type program and a fertilizer tax at various rates. They found a fertilizer tax to be the most cost-effective of policies they considered.

Kling et al. (2006) employed a similar econometric modeling approach. Like Wu et al. (2004), they used the National Resource Inventory data to link the cost data with the SWAT model. They estimated the costs and water quality benefits of implementing a set of conservation practices associated with implementation rules based on distances to a waterway, slope, and erodibility indices. The conservation practices assessed include grassed waterways, nitrogen management, terraces, buffers, land retirement, and conservation tillage. They estimated that this placement of conservation practices on the landscape would cost over $800 million annually (or roughly $16 billion if viewed as a lump sum cost assuming a 5% rate of discount) and would achieve a 22% reduction in N loadings into the upper Mississippi River basin at Grafton, IL. Within the UMRB, they estimated a 40–66% reduction in sediment loads, a 6–47% reduction in P loads, and a 9–29% reduction in N loads. These estimates (like those from all of the studies reviewed here) are likely to be very sensitive to the set of conservation practices included and the specific scenarios studied.

Greenhalgh and Sauer (2003) used the USMP augmented in two important ways: (1) they configured the model by watersheds and added information on municipal waste water treatment plants and (2) they included "attenuation" coefficients derived from the SPARROW model to reflect the transport component of N flows between watersheds. The focus of their work was on policy options for hypoxia that also contribute to greenhouse gas reductions. The policies they considered include N trading between point and nonpoint sources, GHG trading assuming external carbon prices of $5/ton and $14/ton, N trading with additional payments for GHG emission reductions, an N fertilizer tax, a subsidy to farmers willing to shift from conventional to conservation tillage, and an expansion of the CRP program to 16 million hectares (40 million acres) nationwide. Of the policies evaluated, none achieved the 20% reduction goal of the Doering et al. (1999) analysis. The largest reductions were achieved in their simulation of point/nonpoint source trading with a stringent N standard. The most cost-effective policies were also the trading programs.

Ribaudo et al. (2005) also considered the possibility of N trading between point and nonpoint sources using the USMP model. They found that trading has significant potential to reduce costs relative to a requirement that wastewater treatment plants be required to install stringent nutrient removal technology.[1]

These studies shed light on the costs of addressing the hypoxia problem from conservation practices in the agricultural sector and the way these costs may vary depending on the policy instrument chosen (trading program, conservation payment, tax, etc.). These studies also directly bear on the question of how much it will cost to address local water quality in the MARB. However, as noted above, shortcomings of the integrated models have prevented assessment of many policies as well as conservation practices and sinks. None of the models include point source and nonpoint source control options. With the exception of Booth and Campbell (2007), most models have not adequately addressed the cost savings associated with targeting. Nonetheless, results to date suggest that there is large variability in the costs of alternative policies. The issue of who pays these costs may also be important to consider since the incidence (who must pay the costs) may differ dramatically across policies. A notable example is a fertilizer tax, which has the same social costs as a restriction but which may have a much higher incidence on farmers.

Improved estimates of the costs of installing and maintaining conservation practices could be generated with the current suite of models by considering alternative sets of conservation practices. This can be accomplished using the following steps: (1) identifying conservation practices that are most likely to be effective in reducing nutrients important for hypoxia and (2) identifying scenarios that place these conservation practices on the landscape. These scenarios could be based on rules of thumb (identifying, for example, a particular conservation practice to be used on cropland with specific climate and soil characteristics); algorithms for optimal placement to minimize costs; multiple goals, such as maximizing in basin co-benefits or income support; or policy-relevant methods, such as the use of an environmental benefits index; or computing cost estimates from economic models and water quality changes from watershed models.

---

[1] It is important to recognize that these studies assume a perfectly efficient water quality trading program with no trading restrictions; current water quality trading programs do not match the modeled system

4.3.3 Research Assessing the Basin-Wide Co-benefits

As noted above, many of the same practices that could contribute to reductions in the hypoxic zone could also have significant effects on local water quality, carbon sequestration, wildlife habitat, flood control, and other ecosystem services. The physical co-benefits of many conservation practices and sinks are described in Section 4.5.10. On the basin-wide scale, there are a few studies that provide physical measures of one or more co-benefits that are associated with implementation of conservation practices that would address hypoxia, particularly related to carbon sequestration and water quality (see, for example, Feng et al. (2005), Lewandrowski et al. (2004), Greenhalgh and Sauer (2003)). These studies consistently indicated that significant co-benefits are present, but these estimates are not monetized and are reported in physical units. Further, the policies analyzed are not focused on hypoxia reduction.

Thus, the work reported in the *Integrated Assessment* remains the most complete coverage to date of the potential value to MARB residents of the water quality and other co-benefits. The estimates provided there suggested that the monetized value of the benefits to the basin were larger than the costs based primarily on benefit estimates of the value of erosion control and wetlands restoration. A more complete accounting of these benefits could be developed using benefits transfer techniques, although there are many ecosystem services for which currently accepted methods are not likely to adequately fully capture the value of the benefits. But, in any case, because the *Integrated Assessment* was not able to quantify all co-benefits, total co-benefits within the basin would almost certainly be larger than those estimated.

Due to the incredible complexity in this system, as well as limits in data, modeling, and research, definitive statements on social welfare are not possible. For example, there is incomplete information on the costs of farm-level actions to reduce edge-of-field nutrient losses. There is even greater uncertainty in quantifying the effectiveness of farm-level nutrient control actions in reducing watershed-level nutrient flux and about the relationship between watershed-level nutrient flux and the spatial and temporal dimensions of the hypoxic zone. These uncertainties are further exacerbated by the possibility of regime shift in the Gulf of Mexico, whereby the system could become more susceptible to hypoxia following the initial occurrences. If regime shift is a factor, then historic data on the relationship between nutrient flux and the size of the hypoxic zone does not provide guidance on the decrease in nutrients required to achieve a given reduction in the size of the hypoxic zone. Hence, a return to historic lower levels of nutrient fluxes might not be adequate to return to a corresponding size of the hypoxic zone.

There are many sources of uncertainty in the economic, hydrologic, and Gulf systems that make it difficult to render definitive conclusions about social welfare. Indeed, it is precisely because of these many uncertainties and need for additional research that we recommend an approach based on an adaptive management strategy that aims to move in a "directionally correct" fashion, rather focusing on achieving a precise outcome.

While we cannot definitely say that we can achieve the 5,000 km^2 (1,930 mi^2) goal while maintaining social welfare, there is evidence that suggests it is feasible to do so. First, and perhaps most importantly, welfare losses in the Basin will be at least partially or even totally offset by co-benefits of nutrient reduction actions. For example, if wetlands restoration is used to control nutrient flux, it will result in improvements in wildlife habitat and local water quality, both of which will improve welfare in the Basin. Findings from the Doering et al. (1999) assessment point out that the benefits accruing locally from wetlands restoration might well exceed the costs, even without any Gulf hypoxia reductions. Similar estimates are reported in Hey et al. (2004) for substantial restoration of wetlands in flood plains (see Section 4.4.2). Management actions that reduce farm-level nutrient losses may lead to better local water quality, thereby improving welfare for affected residents within the Basin. If management actions are undertaken to control air emissions, thereby reducing atmospheric deposition of nitrogen, it will result in improvements in air quality, reduction in acid precipitation, lower emissions of greenhouse gasses, etc. Thus, co-benefits within the Basin will at least partially and perhaps fully offset welfare losses associated with the costs of implementing management actions. And in the longer term, a transition from corn to perennial crops could benefit farmers and other Basin residents. Thus, there may be larger scale transitions in the agronomic system that provides opportunities to reduce nutrient flux while maintaining welfare in the Basin.

A second reason for optimism is that cost-effective approaches, such as targeting low cost sources and using emissions trading, have not yet been applied. These approaches have the potential to reduce the costs of nutrient control, possibly considerably, thereby reducing the burden of complying with the goal. Thus, there may be opportunities to control the cost of nutrient reduction.

4.3.4 Principles of Landscape Design

Another perspective for protecting social welfare can be drawn from the principles of landscape design. A landscape perspective involves broad-scale consideration of how decisions affect resources, particularly in the long run. Guidelines have been proposed as a way to facilitate land managers considering the ecological ramifications of land-use decisions (Dale et al., 2000). These guidelines are meant to be flexible and to apply to diverse land-use situations, yet require that decisions be made within an appropriate spatial and temporal context. These landscape design guidelines can serve as a checklist of factors to be considered in making decisions that relate to implications for hypoxia in the Gulf.

- *Examine the impacts of local decisions in a regional context.* The spatial array of habitats and ecosystems shapes local conditions and responses (e.g., Patterson, 1987; Risser, 1985) and local changes can have broad-scale impacts over the landscape. Hypoxia is a classic example of such impacts (Russo et al., 2008), for fertilizer applications in the Midwestern states can affect oxygen conditions in the Gulf of Mexico. This guideline notes that it is critical to examine both the constraints placed on a location by the regional conditions and the implications of decisions for the larger area. Therefore, it is critical to identify the surrounding region that is likely to affect and be affected by the decision and examine how adjoining jurisdictions are using and managing their lands. Forman (1995) suggests that land-use planning should first determine nature's arrangement of landscape elements and land cover and then consider optimal spatial arrangements and existing human uses. Following this initial step, he suggests that the desired landscape mosaic be planned first for water and biodiversity; then for cultivation, grazing, and wood products; then for sewage and other wastes; and finally for homes and industry. Of course, planning under pristine conditions is typically not possible. Rather, the extant state of development of the region generally constrains opportunities for land management.

- *Plan for long-term change and unexpected events.* Impacts of decisions can, and often do, vary over time as a result of delayed and cumulative effects. Future options are often constrained by the decisions made today as well as by those made in the past. For example, areas that are urbanized are unlikely to be available for any other land uses because urbanization locks in a pattern on the landscape that is hard to reverse. Thus, management actions should be implemented with some consideration as to the physical, biological, esthetic, or economic constraints that are placed on future uses of resources. External effects can extend beyond the boundaries of individual ownership and thus have the potential to affect surrounding owners. Planning for the long term also requires consideration of the potential for unexpected events, such as variations in temperature or precipitation patterns or disturbances. Long-term planning must also recognize that one cannot simply extrapolate historical land-use impacts forward to predict future consequences of land use. The transitions of land from one use or cover type to another often are not stable over time because of changes in demographics, public policy, market economies, and technological and ecological factors.

- *Preserve rare landscape elements, critical habitats, and associated species.* This guideline implies a hierarchy of flexibility, and it implicitly recognizes ecological constraints as the primary determinants in this hierarchy. For example, a viable housing site is much more flexible in placement than an agricultural area or a wetland dedicated to improving water quality and sustaining wildlife. Optimizing concurrently for several objectives requires that planners recognize lower site flexibility of some uses than others. However, given that most situations involve existing land uses and built structures, this guideline calls for examining local decisions within the regional context of ecological concerns as well as in relation to the social, economic, and political perspectives that are typically considered.

- *Avoid land uses that deplete natural resources over a broad area.* Depletion of natural resources disrupts natural processes in ways that often are irreversible over long periods of time. The loss of soil via erosion that can occur during agriculture and the loss of wetlands and their associated ecological processes and species are two examples. This guideline requires the determination of resources at risk, which is an ongoing process as the abundance and distribution of resources change. This guideline also calls for the deliberation of ways to

avoid actions that would jeopardize natural resources and recognition that some land actions are inappropriate in a particular setting or time, and they should be avoided.
- *Avoid or compensate for effects of land use on ecological processes.* Negative impacts of land-use practices might be avoided or mitigated by some forethought. To do so, potential impacts need to be examined at the appropriate scale. At a fine scale, farm practices may interrupt ecoregional processes. At a broad scale, patterns of watershed processes may be altered, for example, by changing drainage patterns as part of the land use. Therefore, how proposed actions might affect other systems (or lands) should be examined. For example, human uses of the land should avoid uses that might have a negative impact on other systems; at the very least, ways to compensate for those anticipated effects should be determined. It is useful to look for opportunities to design land use to benefit or enhance the ecological attributes of a region.
- *Implement land-use and -management practices that are compatible with the natural potential of the area.* Local physical and biotic conditions affect ecological processes. Therefore, the natural potential for productivity and for nutrient and water cycling partially determine the appropriate land-use and management practices for a site. Land-use practices that fall within these limits are usually cost-effective in terms of human resources and future costs caused by unwarranted changes on the land. Nevertheless, supplementing the natural resources of an area by adding nutrients through fertilization or water via irrigation is common. Even with such supplements, however, cost-effective management recognizes natural limitations of a site. Implementing land-use and -management practices that are compatible with the natural potential of the area requires that land managers understand a site's potential. For example, land-management practices such as no-till farming reduce soil erosion or mitigate other resource losses. Often, however, land uses ignore site limitations or externalize site potential. For example, building shopping malls on prime agriculture land does not make the best use of the site potential. Nevertheless, land products are limited by the natural potential of the site.

Together these guidelines form the basis of a landscape design perspective that should improve the ability to understand and manage the complex system that is affecting hypoxia in the Gulf of Mexico.

Key Findings and Recommendations

The large-scale policy models that have been developed to date each have strengths and weaknesses. None of the models adequately address the full range of management options (wetlands, buffers, nutrient management, etc.) or the full range of policy instruments in a geographically explicit manner. In fact, no single model is likely to be adequate for the full range of decision making that adaptive management of this complex system requires. Moreover, the focus of prior analyses was on cost-effective strategies to reduce N loss, which was the concern at the time. Given that the best current science suggests P is also a limiting nutrient in the Gulf, it is important to seek cost-effective practices that affect both N and P while considering possible trade-offs between them.

The CENR study remains the only research effort to consider the overall costs and benefits of controlling hypoxia in the Gulf of Mexico. The study suffers from a number of shortcomings (many control options and sources of nutrients were not considered, the hydrology of fate and transport was ignored, and no sensitivity analysis concerning key assumptions was undertaken to name a few). The evidence from this work and other studies suggests that it is probable that social welfare in the basin can be maintained while achieving the goal of a 5-year running average of 5,000 km^2 for the hypoxic zone. Most importantly, welfare losses from costs incurred to control hypoxia in the Basin will be offset, at least in part, by co-benefits of nutrient reductions. For example, research on wetlands in the MARB suggests that the benefits of large-scale restoration efforts would exceed the costs. Second, only limited targeting of control options that focus on hypoxia reduction and its co-benefits have been undertaken. Given the significant gains in cost savings that targeting can achieve, this suggests that it may be possible to achieve hypoxia reduction at lower cost than predicted in models that do not consider complete targeting. Based on these findings, the Study Group offers the following recommendations.

- The management of factors affecting hypoxia within the MARB should be viewed as components of a designed landscape so that costs and benefits at various spatial and temporal scales are explicitly considered.
- Integrated economic and watershed models are needed to support an adaptive management framework. Models are needed that represent land use and costs of conservation at both the fine scale, such as the 8 or 12-digit HUC size, as well as a larger scale that encompasses the entire MARB.
- Research that assesses the optimal suites of conservation practices to maximize both local water quality and other co-benefits and Gulf hypoxia reduction is needed. This will require improved understanding of the watershed-scale benefits of these control measures and their costs.
- To reduce hypoxia and protect social welfare in the MARB, control measures that both reduce hypoxia cost-effectively and provide co-benefits in the MARB should be targeted whenever possible. Targeting control measures can reduce the costs and increase co-benefits associated with measures to control hypoxia in the Gulf of Mexico.

4.4 Cost-Effective Approaches for Nonpoint Source Control

While the *Action Plan* and this Study Group urge the reliance on adaptive management principles, a variety of tools can be used as the vehicle for implementation within adaptive management. The current *Action Plan* indicates a principle of encouraging "actions that are voluntary, practical, and cost-effective" (page 9). Additionally, the plan will "utilize existing programs, including existing State and Federal regulatory mechanisms," as well as identify needs for additional funding. These statements include a variety of tools ranging from purely voluntary programs (those with no associated financial incentives) to current conservation programs funded by state and federal agencies (such as the Conservation Reserve Program [CRP] and the Environmental Quality Incentive Program [EQIP]) to water quality trading. Research assessing the costs and effectiveness of these approaches is addressed in this section.

Complicating the design of cost-effective approaches is the geographic distance between the sources of nutrients and the receiving waters downstream. Two identical farm fields in different locations (with resulting differences in the hydrology of the local watershed) will send differing amounts of nutrients to the Gulf. Hence, the effectiveness of a practice or sink in a particular location depends on what sources and sinks are present elsewhere in the watershed. Whether it is cost-effective to install a buffer at a particular location may depend upon whether there is a wetland at the base of the watershed, whether conservation tillage is being practiced elsewhere, etc. Thus, rather than focus on individual practices, policy options that can simultaneously encourage the adoption of practices and sinks that are jointly cost-effective will best protect social welfare in the Basin.

It is important to clarify the concept of "costs." Here, "costs" refers to the least amount of compensation needed to effect change, e.g., the compensation that would be necessary for a landowner or farmer to adopt a conservation practice. This is the standard concept of economic cost, relevant to any good or service. This cost includes "direct" costs, such as the cost of new equipment, building of structures, and labor to manage a practice, as well as a myriad of potential "indirect" costs, such as lost profits from adopting the practice and compensation for added risk from the practice. Components of these costs can be negative; that is, it may actually increase profitability to adopt some practices (conservation tillage in certain circumstances is a notable example).

Second, the focus of most economic studies is on total costs with little or no consideration paid to what subset of society actually bears the costs (incidence) of the policy. This focus on efficiency (seeking the lowest cost approach) is based on the premise that compensation could always be paid to those bearing the cost in some form so that society will be best off if the lowest cost option is pursued. However, since such compensations are rarely paid, the issue of who pays is likely to enter the policy decision. Complete information on the incidence of alternative tools in this context is not available, but where appropriate, we note the likely incidence considerations.

4.4.1 Voluntary Programs – Without Economic Incentives

There is a small and growing literature concerning the effectiveness and optimal design of voluntary agreements that do not have positive or negative financial incentives associated with them (Morgenstern and Pizer, 2007; National Research Council, 2002). Key insights were presented in a game-theoretic model by Segerson and Miceli (1998), who identified the conditions under which voluntary agreements are likely to yield efficient pollution levels without significant economic incentives. They studied voluntary agreements that are based on threats of harsher outcomes if the goals are not met, using the example of mandatory abatement requirements if the voluntary agreement does not succeed in meeting the pollution goal. The premise is that firms will voluntarily agree to reduce pollution if they can avoid the costs that future mandatory controls would otherwise bring. In the absence of financial compensation, the presence of a positive probability of a penalty (or cost in the form of mandatory control) is required to support Segerson and Miceli's findings that there are situations in which efficient levels of pollution control can be achieved with voluntary agreements (without economic incentives). They found that pollution reduction is likely to be small when the background threat is weak.

Empirical work also sheds light on the efficacy of voluntary agreements that do not have financial incentives. Mazurek (2002) identified 42 voluntary environmental initiatives sponsored by the federal government since 1988. Although the programs she identified are largely outside the realm of agriculture, her conclusions are relevant. Mazurek concluded that a variety of implementation problems have led to "lower-than-expected" environmental results for voluntary (without financial incentive) agreements, a result consistent with findings of a 1997 USGAO (1997) report concerning four voluntary agreements related to climate change.

In the same National Research Council report (2002), Randall identified three essential functions for government if voluntary agreements (without financial incentives) are to be effective. These key functions are meaningful monitoring to back up a threat of government inspection, "credible threat of regulation" if the goals are not met, and a clear liability system to punish "blatant polluters and repeat offenders." Randall concluded that "voluntary (or negotiated) agreements, industry codes, and green marketing should be viewed as promising additions to the environmental toolkit, but they should supplement, not supplant, the regulatory framework. They make a nice frosting on the regulatory cake. But the cake itself must be there."

Finally, Morgenstern and Pizer (2007) presented seven case studies on voluntary agreements (without economic incentives) in the United States and elsewhere. Point estimates of environmental improvements attributable to the voluntary programs ranged from negative values (actual declines in environmental performance) to a maximum of 28% improvement in environmental performance. Morgenstern and Pizer concluded "that voluntary programs have a real but limited quantitative effect...."

Given the historical aversion to imposing mandatory requirements in agriculture, the collective weight of these studies suggest that voluntary agreements that do not have incentives associated with them are not likely to be adequate on their own to achieve significant reductions in nutrient runoff. In short, voluntary programs without incentives can have small effects but cannot be relied upon to induce major environmental improvements.

4.4.2 Existing Agricultural Conservation Programs

Currently, the largest incentive-based conservation programs related to agriculture are the EQIP and CRP. A potentially significant program introduced in the 2002 Farm Bill was the Conservation Security Program (CSP), which has been funded only partially and implemented incrementally. The CRP pays farmers to retire land, and the other two pay farmers to implement conservation practices on their farms (EQIP is a cost-share program; CSP was intended to cover the full costs of adoption). Numerous studies undertaken by USDA's Economic Research Service and others have estimated the magnitude of environmental benefits from these programs in physical terms (e.g., tons of erosion reduction, acres of habitat preserved, acres of wetlands restored) and some efforts have been made to monetize these benefits [see Claassen et al. (2004) for a summary of CRP studies as well as Haufler (2005)]. The Conservation Effects Assessment Program (CEAP) was initiated in an attempt to provide nationwide estimates of the benefits provided by the full suite of conservation programs; a national assessment of the water quality benefits is being developed currently (Bob Kellogg, presentation to SAB Hypoxia Advisory Panel, December 6, 2006).

The CRP pays landowners to take their land out of crop production and place it in perennial vegetation or trees, depending on the region of the country, with a goal of creating wildlife habitat and reducing erosion (and originally to reduce crop production). The CRP enrolls about 10% of total US cropland, nearly all in 10-year contracts although there is significant concern that high corn prices due to ethanol expansion may rapidly reduce this amount. A number of studies have identified large environmental benefits associated with the CRP [Smith and Alexander (2000), Feather et al. (1999)]. The program has used an Environmental Benefits Index (EBI) since 1990 to prioritize parcels for inclusion in the program that gives points to land based on particular environmental attributes and cost. The movement from targeting erodible lands (prior to 1990) to the use of the EBI for targeting has been estimated to have doubled the benefits from the program (Feather et al., 1999). Ribaudo (1989) estimated that a CRP enrollment that targets lands based on environmental damages (benefits) would have significantly greater benefits still. By redesigning the weights in this index, the program could target land that is predicted to contribute high nutrient loadings to the Gulf.

Many other studies have addressed the cost-effectiveness of land retirement to achieve environmental benefits within the context of the CRP. In a series of papers assessing the efficiency of the Conservation Reserve Enhancement Program (CREP) in Illinois, Khanna et al. (2003) linked the AGNPS model with site-specific characteristics of parcels to examine the relative efficiency of alternative targeting mechanisms (Yang et al., 2003, 2004, 2005). Extremely large gains from targeting were reported; for example, Yang et al. (2004) estimated that with targeting, 30% less cropland could have been retired (at almost 40% less total cost), while achieving 20% reductions in erosion instead of the actual 12% reduction.

The EQIP program is a cost-share program for conservation practices in livestock facilities and on land that remains in agricultural production. A prospective benefit cost analysis (as required by Executive Order 12866) predicted over $5 billion in net benefits from the EQIP program as implemented under the 2002 Farm Bill, even though not all of the benefits could be monetized (US Department of Agriculture, 2003).

The Wetland Reserve Program (WRP), Grassland Reserve Program (GRP), and Wildlife Habitat Incentive Program (WHIP) are all smaller land retirement programs that also could potentially benefit efforts to reduce Gulf hypoxia. Additional information on the large-scale potential for wetlands is provided by Hey et al. (2004), who addressed the question of whether the social benefits from restoring up to 2.83 million hectares (7 million acres) of cropland in the 100 years floodplain of the upper Mississippi River basin to wetlands exceed the costs. The benefits include reduced flood-related crop damages; reduced crop subsidies; and non-flood-related recreation benefits of wetland conversion, including fishing, hunting, and general recreation usage. These benefits were compared to estimates of the costs of cropland conversion comprised of farm rental rates (representing the present value of farmland income) and the costs of wetland construction and maintenance. Hey et al. (2004) estimated that the benefits exceed the costs in all locations considered except one county in Missouri. In the context of NGOM hypoxia, this difference is especially striking because the benefits exceed the costs for this conversion even without considering any benefits from reduction of the hypoxic zone. As the authors carefully pointed out, the social efficiency of converting 2.83 million hectares (7 million acres) does not mean that private benefits will exceed the private costs for all parties. Individual landowners would stand to lose while recreationists accrue benefits.

These findings represent an important addition to the assessment of wetlands in the *Integrated Assessment*. While Doering et al. (1999) concluded that wetland restoration was less cost-effective than fertilizer reductions, their analysis did not include cost savings from crop subsidy reductions nor flood-related crop damages. In addition, the Hey et al. (2004) work focused on wetlands targeted in flood plains. The study suggests two points of key importance for NGOM hypoxia: (1) there is a large amount of acreage that is situated in locations that potentially could serve as nutrient sinks in the upper Mississippi River basin, and (2) the co-benefits of this action are large enough, in and of themselves, to justify the social efficiency of converting this land to nutrient sinks even without considering the benefits associated with reducing Gulf hypoxia.

The programs mentioned above can be categorized into one of two groups: land retirement programs and "working" land programs. Both the CRP and WRP are examples of land retirement programs, since landowners receive payments in exchange for taking land out of active agricultural production and putting the land into perennial grasses, trees, or wetlands restoration. In contrast, EQIP and the CSP are examples of working land programs whereby landowners or producers receive payments to cover part or all of the costs of making changes in conservation practices or management decisions on their land that remains in agricultural production. Some research has addressed the cost-effectiveness of working land programs versus land retirement programs. For example, Feng et al. (2006) found that a cost-effective allocation of resources to sequester carbon in agricultural soils favors working land (via conservation tillage subsidies) over land retirement (via payments to retire land and plant it in perennial grasses). It is important to note, however, that this study focused on stylized working land and land retirement programs rather than attempting to address the cost-effectiveness of existing conservation programs as actually implemented.

The existing working land and land retirement programs are implemented with features that likely affect the cost-effectiveness of the programs for achieving environmental gains in different ways. For example, the CRP uses an EBI that favors admitting land into the program that achieves environmental benefits at relatively low costs. All else equal, this component of the program will improve its cost-effectiveness. In contrast, the CSP provides payments for ongoing stewardship of farmers so that program expenditures are used to reward past behavior rather than to change existing behavior. This, all else equal, will reduce the program cost-effectiveness for achieving environmental gains. The lack of competitive bidding and clear targeting also reduces the cost-effectiveness of this program. Finally, it is worth noting that targeting and competitive bidding were explicitly disallowed in the EQIP program during its most recent reauthorization. Again, this will reduce its cost-effectiveness.

4.4.3 Emissions and Water Quality Trading Programs

Emission trading is a regulatory approach that sets a maximum allowable level of overall emissions and then allows sources to exchange pollution allowances. A properly structured trading program can reduce the costs of achieving emission standards by allowing the flexibility necessary to focus pollution reductions on sources that are less expensive to control. In theory, a broad-based emissions trading program could help to reduce the air and water contributions of nutrients to the NGOM. Water quality trading is simply the name given to the extension of emissions trading to achieving water quality objectives.

In a recent survey of the programs to support water quality trading in the United States, Breetz et al. (2004) identified 40 water trading initiatives and an additional six state policies with specific programs related to water quality trading. USEPA has supported these programs (USEPA, 2004a) and has produced explicit policies related to their implementation. Many states and regions also have explicit policy guidance. However, the effectiveness of these programs appears to have been quite limited as very few trades are actually occurring. Further, little evidence of environmental improvement associated with these programs exists (Breetz et al., 2004).

A key problem with these programs is the lack of a required water quality improvement necessary to generate adequate demand for credits (King, 2005). To achieve "cap and trade," an effective cap is necessary. A cap could come from a tight enough cap on point sources such that they would find it cost-effective to purchase credits from agricultural nonpoint sources. Alternatively, the cap could be extended to agricultural sources. While some have conjectured that the Total Maximum Daily Load (TMDL) program may eventually play this role, there is no current mandate for agricultural sources to restrict nutrient runoff. Also problematic are a range of restrictions on allowable trading, such as requirements that a particular baseline set of conservation practices be in place with credits accruing only for additional conservation activity.

While trading could be a significant contributor to cost-effective nutrient control, the necessary institutions for water and/or air emissions trading to be an effective policy instrument are not broadly in place. In addition to clear and enforceable limits on emissions or water quality contributions (from point and/or nonpoint sources), enforceable rules concerning trading ratios, liability when standards are not met, monitoring, etc. must be established before these markets can flourish. Ideally, a trading program to address NGOM hypoxia would be broad based and include highly diverse sources (such as air deposition and many agricultural nonpoint sources) to maximize the potential for cost savings.

4.4.4 Agricultural Subsidies and Conservation Compliance Provisions

US farmers have been the recipients of farm payments for decades. These payments support prices and/or income, especially of farmers growing bulk commodities such as corn and soybeans. Economic theory suggests that, all else equal, such payments will increase the intensity and acreage of farming, possibly resulting in increased water quality problems. Research by Reichelderfer (1985) provided empirical evidence that these payments encourage crop production on highly erosive land. Likewise, a recent study from USDA's Economic Research Service (Lubowski et al., 2006) quantified the effect of one major program, subsidized crop insurance, on the location and acreage of cropland and its environmental effects. Lubowski et al. (2006) estimated that about a million hectares (2.5 million acres) were brought into production as a result of the program and that these lands are more vulnerable to erosion, are more likely to include wetlands, and have higher levels of nutrient losses than average.

To some extent, USDA's conservation programs (see Section 4.4.2) exist to counteract the "perverse effects" or unintended consequences of its crop subsidies inasmuch as government financial support has encouraged farmers to choose commodity crops that require more fertilizer, maximize yield without regard to soil and water quality consequences, and cultivate marginal land. Restructuring or eliminating existing subsidies could serve to mitigate some of these perverse effects (e.g., by shifting subsidies to reward less fertilizer-intensive crops as well as by requiring, as a condition of receiving subsidies, certain conservation practices).

Taheripour et al. (2007) provided additional evidence on this point. First, their model suggests that removal of all crop subsidies would reduce nitrogen pollution by 8.5% and that the reduced need for distortionary income taxes to support these subsidies could increase social welfare by $1.2 billion. Further, they found that tax-neutral policies to achieve nitrogen reduction can generate significant double dividends (a double dividend refers to a situation where a policy not only internalizes an externality but also reduces the deadweight losses associated with distortionary taxation, such as an income tax). They provide an estimate of the magnitude of the double dividend for a range of nitrogen-reduction goals and policy approaches including a nitrogen tax, a nitrogen reduction subsidy, a tax on output, and a combined output tax and nitrogen reduction subsidy and find that a double dividend from these instruments can be significant.

While environmental improvements associated with agriculture have largely been pursued via cost-share or subsidy programs, one significant regulatory approach has been the implementation of environmental compliance provisions that require farmers who receive farm program payments (including price support and income support) to undertake some environmental performance practices. Specifically, in the 1985 Food Security Act, conservation compliance provisions required owners of highly erodible land (a categorization of land based on its slope and soil type) to implement soil conservation plans, and a "swampbuster" provision disallowed payments to go to farmers who converted wetlands to crop land. Claassen et al. (2004) estimated that up to 25% of the reduction in soil erosion that occurred between 1982 and 1997 was attributable to conservation compliance. Many believe these gains could have been higher if there had been stronger enforcement of the mechanism. While no direct estimates are available of the increased benefits that could come from more enforcement, there is evidence of very limited reporting and penalizing of violations (Claassen, 2000).

Claassen et al. (2004) assessed the prospect for reducing nutrient losses from the Mississippi River basin by extending compliance requirements to nutrient management. They used "nutrient management" to refer to the range of activities related to the timing and level of fertilization decisions that best minimizes soil nutrients in excess of crop needs at any point in time. They noted that the ideal set of nutrient-management practices will vary considerably across farms and regions and that the costs of these activities will also vary notably across this space. Using data from the EQIP program, they summarized the distribution of incentive payments needed to induce willing adoption of nutrient management practices as defined under EQIP. For the Heartland region (ERS Farm Resource Region), the average annual incentive payment is about $7/ac, and 95% of the payments are $12/ac or less.

While these data provide an excellent starting point for assessing the cost-effectiveness of nutrient management methods addressing local water quality and NGOM hypoxia, several additional pieces of information would be needed for a full assessment. First, these costs represent the compensation needed for those farmers who have already adopted practices under the EQIP program; those who have not adopted are likely to have at least as high costs, possibly substantially higher. In this regard, these costs could be viewed as a lower bound. Second, these costs are specific to the EQIP requirements for nutrient management. Whether these requirements are effective enough to yield substantial off-site benefits is not addressed. Nonetheless, based on this cost assessment and a comparison with the annual commodity program payments farmers typically receive, Claassen et al. (2004) concluded that substantial nutrient management could occur with extension of conservation compliance provisions to nutrients.

Claassen et al. (2004) also considered whether buffer practices could be induced under conservation compliance provisions. They included riparian buffers, filter strips, grassed waterways, and contour grass strips in their discussion of buffer practices. To assess the costs of these practices and how they vary across locations, they looked at information on producers' willingness to accept compensation for adoption of the practices observed for continuous CRP priority areas. Owners of these lands received an average payment of about $90/year in addition to 50% cost share for installation of the buffer practice. Based on this analysis, as an example, Claassen et al. (2004) computed the annual costs per area for a filter strip and concluded that, in many cases, this payment would be below the average subsidy received by producers, thereby suggesting that buffer practices might also be successfully adopted under nutrient compliance provisions.

Finally, Claassen et al. (2004) noted that conservation compliance provisions are likely to have few transaction costs relative to other policies (although enforcement costs would need to be considered) and require very low budgetary outlays beyond the payments that are already provided for commodity or insurance programs. Claassen et al. (2004) also argued that conservation compliance requirements have been relatively cost-effective due to the flexibility with which they can be implemented. Producers in different regions of the country, with differing soil and weather conditions, can meet their compliance obligations with different practices. This flexibility means that the most appropriate technologies can be used for the location of the practice.

4.4.5 Taxes

The use of a per unit tax to internalize the costs of externalities of production is well known to be highly cost-effective when the tax is placed directly on the externality generating activity; these "Pigouvian" taxes are the equivalent of placing the appropriate price on the pollutant (Baumol and Oates, 1988). Taxes can be a powerful market signal, communicating the need to change behavior, Baumol and Oates (1988) demonstrated that subsidies (essentially just negative taxes) can also be designed that provide the equivalent market signals for changes in behavior. This argument is often used to support the design of environmental programs that pay participants for the provision of environmentally friendly practices rather than using taxes to change behavior. A potentially important exception to this equivalence can occur when the provision of a positive payment induces entry into the farming sector generating production on otherwise unprofitable lands. This possibility was addressed in Section 4.4.4 in the context of general agricultural subsidies and conservation compliance.

A tax directly on an input into production that is highly correlated with the pollutant can be an efficient second-best policy. The possible use of a nitrogen fertilizer tax was considered in Doering et al. (1999) and found to be as cost-effective as any of the policies they considered (they note that the initial incidence falls on farmers). Fertilizer taxes already exist in some states but are set at much smaller levels than those studied by Doering et al. (1999). The inelastic demand for fertilizer (Denbaly and Vrooman, 1993) means that the magnitude of taxes needed to induce behavioral change would likely be large.

The incidence of a tax (and thus determination of who pays the costs) is likely to fall on farmers and consumers of food products made from crops that use fertilizer. In contrast, the incidence of conservation program payments is largely on taxpayers. Finally, it is important to note that tax instruments will be more efficient the more broadly they are applied to the various nutrient sources identified as pollutant contributors; so ideally a tax would be applied to all nutrient sources rather than singly to fertilizer.

4.4.6 Eco-labeling and Consumer Driven Demand

The idea that environmentally friendly producer behavior can be induced by consumer demand is one basis for eco-labeling and certification programs. Dolphin safe tuna (Teisl et al., 2002) and organic fruits and vegetables (Loureiro et al., 2001) are two successful examples. Research analyzing the effectiveness of eco-labeling suggests some promise.

Thogersen (2002) summarized three schemes, all implemented in Europe, that have been credited with significant reductions in emissions from heating appliances and paint solvents (the German "Blue Angel" brand) and reductions in pollutants from paper production and household chemical and laundry emissions (the Swedish "Good Environmental Choice" label and the Nordic "Swan" label). Although not specific to a particular product, Clark and Russell (2005) noted that several studies of the Toxic Release Inventory have shown that information can affect firms' choices.

Could consumer-driven demand affect the changes in land-use and agricultural management necessary to contribute notably to nutrient flows into the Gulf? This approach would require the labeling of food and fiber products made from agricultural outputs in the MARB to indicate that they were produced in such a way as to reduce or eliminate nutrient contributions to hypoxia. Consumers would then need to respond to this labeling by purchasing products, presumably at a higher cost, in adequate quantity to change the market behavior. Given that much of the grain produced in the Corn Belt is used for livestock feed and not directly traceable to its field of origin, it will be difficult to distinguish products that were produced with "hypoxia-friendly" production practices from those that were not. It is not clear that labeling can credibly be produced without significant government involvement and expense (Crespi and Marette, 2005). Nor is it clear that consumer response would be adequate to drive changes in production practices, even if the labeling challenges could be overcome. One area in which labeling may prove effective is in animal agriculture, where the tracking of an individual unit from producer to final consumer is more straightforward.

Key Findings and Recommendations

Voluntary agreements with no accompanying economic incentives are not likely to be adequate to obtain significant reductions in N and P. While there may still be some low-cost conservation practices that can be implemented in some locations (better "crediting" for manure spreading for example), nutrient reductions that face agricultural producers with costly trade-offs cannot be expected without strong economic signals. These economic incentives can take many forms: conservation payments such as those in many current agricultural conservation programs, taxes, restructuring or removal of subsidies (such as conservation compliance provisions).

Water quality trading programs have not yet demonstrated the ability to improve environmental performance and/or reduce costs of meeting environmental targets primarily due to an absence of effective emissions restrictions. However, with clearer water quality improvement mandates and more flexible rules for trading, these programs could develop into cost-effective instruments.

Numerous studies have demonstrated that existing incentive-based conservation programs, specifically the CRP, WRP and EQIP, have provided significant environmental benefits. However, these programs can be much more cost-effective with additional targeting and competitive bidding mechanisms. Given the menu of existing programs, it is possible to reduce hypoxia and protect water quality in the MARB without significant new government funding, although the distributional consequences of the various approaches will differ. Based on these findings, the Study Group offers the following recommendations.

- To achieve N and P reductions from agricultural sources of the magnitude needed to affect hypoxia, economic incentives are needed to induce adequate adoption of conservation practices. These incentives can take many forms: conservation payments, taxes, and/or restructuring of existing farm subsidy and compliance requirements.

- To maximize the N and P reductions achieved with federal and state conservation dollars (e.g., CRP, WRP and EQIP), targeting and competitive bidding mechanisms are needed so that lands enrolled in these programs achieve maximum environmental benefits at lowest cost. Strategically placed wetlands in the upper Mississippi River basin could serve as effective nutrient sinks. Research has demonstrated that the local co-benefits are large enough, in and of themselves, to justify restoring these wetlands. The additional benefits associated with reduction in Gulf hypoxia reinforce the conclusion of the desirability of wetlands restoration.
- Water quality trading programs hold promise, but, without enforceable caps (water quality standards), these programs cannot be expected to achieve much nutrient reduction.
- To minimize the adverse effects of existing agricultural subsidy programs, conservation compliance requirements that target reductions in nutrients could be very cost-effective, but only with adequate enforcement.
- To select policies and programs with maximum economic efficiency, all co-benefits should be considered regardless of which policy tools are used. For example, since wetlands provide valuable habitat and flood control in addition to water quality benefits, there may be instances in which it is desirable to control nutrients by restoring wetlands, even if it is less costly to reduce nutrients by managing croplands.

4.5 Options for Managing Nutrients, Co-benefits, and Consequences

4.5.1 Agricultural Drainage

The *Integrated Assessment* reports identified several research needs related to agricultural drainage. Brezonik et al. (1999) emphasized the importance of agricultural drainage in nutrient transport from cropland and identified increased spacing of subsurface drainage tile and controlling water table levels (controlled drainage) among those practices that could potentially reduce nitrate losses from cropland. Mitsch et al. (1999) noted that controlled drainage was not widely practiced in US Corn Belt and that most of the research on controlled drainage had been conducted in more southern climates.

4.5.1.1 Alternative Drainage System Design and Management

Relatively few field studies have addressed the effects of subsurface drain depth and spacing on N losses from cropland. Overall, results suggest a trend of decreased subsurface flow and decreased N loss at wider tile spacing or decreased tile depth. Reported reductions in nitrate export are primarily due to reductions in the volume of flow rather than reductions in nitrate concentration. Drain flows and N loss can be affected by both drain spacing and depth (Hoffman et al., 2004; Kladivko et al., 2004; Skaggs et al., 2003, 2005), and use of drainage intensity (Skaggs et al., 2005) normalizes some of the variability in results of drainage spacing studies. Drainage intensity increases with deeper tile depths and closer tile spacing. Research suggests that reducing drainage intensity by either shallower tile depth or wider tile spacing will reduce subsurface flow and nitrate loss. However, adjustments in tile spacing and depth are only possible when drainage systems are being installed, and the Corn Belt is already extensively drained. As these systems are replaced, repaired, and upgraded over the next few decades, there will be opportunities to consider alternative drainage designs to minimize nutrient losses. In the meantime, there may be opportunities to achieve similar benefits by retrofitting existing drainage systems with control structures that allow some management of subsurface drainage.

Drainage management (controlled drainage) is currently an area of active research and development (http://extension.osu.edu/~usdasdru/ADMS/ADMS index.htm). Research suggests that drainage management could reduce nitrate transport from drained fields by 30% for regions where appreciable drainage occurs in the fall and winter (Cooke et. al., 2008). Although water table management could potentially alter nitrification and denitrification reactions, reported reductions in nitrate export with controlled drainage are primarily due to reductions in the volume of flow rather than reductions in nitrate concentration. Some uncertainty arises from difficulties in closing water balances (and therefore N balances) in field studies, and an unknown amount of subsurface flow reduction could be due to lateral seepage and/or increased surface runoff (Cooke et al., 2008). Simulation studies predict increased surface runoff when higher water tables are maintained using controlled drainage (Singh and Helmers, 2006; Skaggs et al., 1995), suggesting a potential tradeoff between reduced subsurface drainage and increased surface runoff. Although raising the water table can decrease the volume of infiltrating water entering drainage tile, higher water tables can also increase surface runoff resulting in increased erosion and loss of particulate contaminants such as soil bound phosphorous.

Controlled drainage requires relatively flat and uniform topography, and slopes of less than 0.5 or 1% are recommended (Cooke et al., 2008; Frankenberger et al., 2006). Concerns for erosion and surface runoff increase with increasing slope, and slopes greater than 0.5–1% can require an impractical number of control structures. There has been speculation that new technologies could make the practice economically feasible at slopes of 2% or more, but this would raise even greater concerns over surface runoff. Although tile drainage is widespread throughout the Corn Belt, it is not clear what portion of this tile drainage can be retrofitted with structures for controlled drainage. A first approximation might be an estimate of the fraction of tile-drained lands with slopes less than 0.5–1%, but this approach requires higher resolution topography than is generally available in the Corn Belt. These estimates are available for a few large drainage districts in north central Iowa for which very high resolution topography were developed. Although 50–75% of the cropland in these drainage districts is tile drained, only about 10% has a slope less than 1% and only about 3% has a slope less than 0.5% (Matt Helmers, Iowa State University, Ag Drainage Website, http://www3.abe.iastate.edu/agdrainage). These results suggest that controlled drainage may be applicable to a relatively small fraction of tile-drained land in Iowa, but this may not be representative of other regions of the Corn Belt. Based on STATSGO soils data, Illinois, Indiana, and Ohio may have twice as much cropland suitable for controlled drainage as Iowa (Dan Jaynes, National Soil Tilth Lab, Ames, IA). High-resolution topography could provide a much better basis for this assessment.

4.5.1.2 Bioreactors

Denitrification bioreactors have been installed in the field as treatment systems for tile drain effluent (Van Driel et al., 2006) and as denitrification walls (a trench filled with carbonaceous material to intercept subsurface flow) (Robertson et al., 2000; Schipper et al., 2004, 2005; Schipper and Vojvodic-Vukovic, 1998, 2001). Bioreactors on tile drains are typically bypassed during high flows and "are most usefully applied in the treatment of baseflows rather than peak flows." Current knowledge indicates that denitrification walls are effective for at least 5–7 years with little or no loss of nitrate removal capacity (Robertson et al., 2000; Schipper and Vojvodic-Vukovic, 2001). A variety of materials such as corn stalks, wood chips, and sawdust are potential organic amendments to enhance denitrification in bioreactors. Continued research is needed to determine whether denitrification bioreactors could be installed around lateral tile drain lines and whether this would be technically and economically feasible. Future redesign of tile drain systems may include integrated denitrification enhancements around tile lines and at the outlets of smaller tile lines.

Key Findings and Recommendations

Alternative drainage designs with reduced drainage intensity due to shallower tile depths and/or wider tile spacing could significantly reduce nitrate losses but can be expected to increase surface runoff and losses of particulate contaminants. Controlled drainage could significantly reduce nitrate losses where appreciable drainage occurs in the fall and winter but can be expected to increase surface runoff and losses of particulate contaminants. Controlled drainage is most appropriate for areas having slopes of less than 0.5–1%, and it is not clear what fraction of tile-drained lands are suitable for application of controlled drainage. In some areas, slope could seriously constrain applicability of the practice. Bioreactors can significantly reduce nitrate concentrations but typically must bypass peak flows during which much of the nitrate load is transported. Based on these findings, the Study Group offers these recommendations.

- Additional research is needed to evaluate topographic constraints on the applicability of controlled drainage including developing high-resolution topography for the Corn Belt.
- Additional research is needed to fully characterize water and nutrient balances for alternative drainage design and management, most critically using small watershed-scale studies (less than 2,500 hectares or about 10,000 acres) to document effects when scaled up.
- A strategy for implementation of alternative drainage design or management should be developed that includes consideration of potential trade-offs between reduced nitrate loss through tile drains and increased P loss through surface runoff.

4.5.2 Freshwater Wetlands

If wetlands are to serve as long-term "sinks" for nutrients, reductions in nutrient loads must reflect net storage in the system through accumulation and burial in sediments or net loss from the system, for example through denitrification or vegetation removal. The effectiveness of wetlands in reducing N export from agricultural fields will depend on the magnitude and timing of NO_3 loads and the capacity of the wetlands to remove NO_3 by denitrification. In contrast to NO_3, gaseous losses of P are insignificant, and sediment accretion of bound inorganic P and unmineralized organic P is the primary mechanism by which wetlands serve as long-term P sinks. With the exception of P associated with suspended solids, wetlands are generally less effective at retaining P than at removing NO_3 (Reddy et al., 1999).

4.5.2.1 Nitrogen

The effectiveness of wetlands in NO_3 reduction is a function of hydraulic loading rate, hydraulic efficiency, NO_3 concentration, temperature, and wetland condition. Of these, hydraulic loading rate and NO_3 concentration are especially important for wetlands intercepting nonpoint source loads. Hydrologic and NO_3 loading patterns vary considerably for different landscape positions and different geographic regions. The combined effect of variation in land use, precipitation, and runoff means that loading rates to wetlands receiving nonpoint source loads can be expected to vary by more than an order of magnitude and will, to a large extent, determine NO_3 loss rates for individual wetlands.

Mitsch et al. (2005a) examined NO_3 retention in Mississippi River basin wetlands receiving nonpoint source NO_3 loads either directly or through diversion of river water. Their study extended the earlier analysis of Mitsch et al. (1999) to include additional wetlands and to include wetlands outside the agricultural regions of the Corn Belt. They found that 51% of the NO_3 mass reduction by the wetlands examined could be explained by a nonlinear regression based on annual mass load of NO_3 per area of wetland. However, when the analysis is restricted to Corn Belt

4.5 Options for Managing Nutrients, Co-benefits, and Consequences

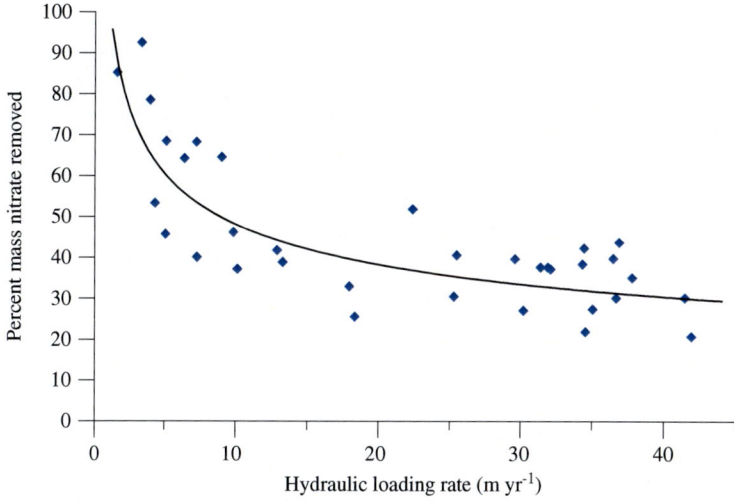

Fig. 4.2 Percent mass nitrate removal in wetlands as a function of hydraulic loading rate. Best fit for percent mass loss = 103 * (hydraulic loading rate)$^{-0.33}$ ($R^2 = 0.69$). Adapted from Crumpton et al. (2006, 2008)

wetlands that receive seasonally variable water and nutrient loads (i.e., subjected to nonpoint source loading regimes), the relationship is much weaker (Crumpton et al., 2006, 2008). Based on 34 "wetland years" of available data (12 wetlands with 1–9 years of data each) for sites in Ohio (Mitsch et al., 2005a; Zhang and Mitsch, 2000, 2001, 2002, 2004), Illinois (Hey et al., 1994; Kovacic et al., 2000; Phipps, 1997; Phipps and Crumpton, 1994), and Iowa (Crumpton et al., 2006; Davis et al., 1981), percent mass NO_3 removal is much more closely related to hydraulic loading rate (HLR) (Fig. 4.2, $R^2 = 0.69$) than to mass loading rate ($R^2 = 0.22$).

Hydraulic loading rate explains relatively little of the variability in NO_3 mass removal, which can vary considerably more than percent NO_3 removal among wetlands receiving similar hydraulic loading rates. However, much of the variability in mass NO_3 removal can be accounted for by explicitly considering the effect of HLR and flow-weighted average (FWA) NO_3 concentration (Crumpton et al., 2006, 2008). For the wetlands in Fig. 4.2, mass NO_3 removal rate can be predicted as the product of percent removal (estimated as 103 * HLR$^{-0.33}$) and mass load (estimated as HLR * FWA). This simplifies to the function [mass removal in kg N/ha-year = 10.3 * (HLR in m/year)$^{0.67}$ * FWA NO_3 concentration in g N/m^3] and explains 94% of the variability in mass NO_3 removal for the wetlands considered here (Fig. 4.3). The isopleths on the function surface in Fig. 4.3 represent the combinations of HLR and FWA that can be expected to achieve a particular mass loss rate and illustrate the benefit of targeting wetland restorations in areas with higher NO_3 concentrations. The wetlands examined by Mitsch et al. (2005a) had a median loading rate of 600 kg NO_3-N/ha-year, at which they predicted losses of 290 kg NO_3-N/ha-year. This mass loss rate is near the lower mass loss isopleth of Fig. 4.3 as would be expected for

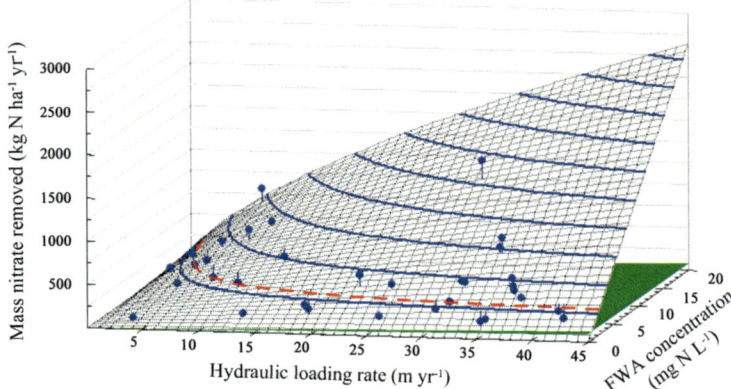

Fig. 4.3 Observed NO_3 mass removal (*blue points*) versus predicted NO_3 mass removal (*blue surface*) based on the function [mass NO_3 removed = $10.3*(HLR)^{0.67} * FWA$] for which $R^2 = 0.94$. *Blue lines* are isopleths of predicted mass removal at intervals of 250 kg/ha-year. The *dashed, red line* represents the isopleth for mass removal rate of 290 kg/ha-year suggested by Mitsch et al. (2005a). The *green plane* intersecting function surface represents organic N export. Adapted from Crumpton et al. (2006, 2008)

either low FWA concentrations at moderate to high HLRs or higher FWA concentrations at lower HLRs. Half of the wetlands considered by Mitsch et al. (2005a) had NO_3 concentrations below 3 mg N/l. NO_3 concentrations in tile drainage water commonly exceed 10–20 mg N/l (Baker et al., 1997, 2004, 2008; David et al., 1997; Sawyer and Randall, 2008). The greatest benefit of wetlands for mass NO_3 reduction will be found in those extensively row-cropped and tile-drained areas of the Corn Belt where NO_3 concentrations and loading rates are highest. For these areas, NO_3 mass removal rates could be several times higher than predicted by Mitsch et al. (1999, 2005a).

Total and organic N data were available for about half of the wetlands represented in Fig. 4.3. All of these wetlands were sinks for total N, but most were net producers of organic N, although in comparatively small amounts (and none were net producers of NH_4). On an average, FWA organic N discharged from the wetlands increased by approximately 0.2 g N/m^3 (range from <0 to 0.3 g N/m^3) relative to incoming concentrations, with no relation to HLR or NO_3 concentrations. The mass export of organic N was small compared to NO_3 removal and had relatively little impact on reductions in total N, especially at higher NO_3 concentrations. For comparison to mass NO_3 loss, mass organic N export can be estimated as the product of HLR and the increase in FWA organic N and is represented by the green plane intersecting the function surface in Fig. 4.3. At elevated NO_3 concentrations, wetlands are nearly as effective in reduction of total N as in reduction of NO_3. At very low NO_3 concentrations, organic N production could equal NO_3 removal, in which case wetlands would not function as total N sinks.

There is some concern over increased N_2O emissions in wetlands exposed to high nitrate loads, and N_2O emissions do increase in wetlands at elevated nitrate levels. However, N_2O accounts for a very small fraction of N removal in wetlands receiving non point source nitrate loads, and N_2O emission rates from these systems are very low (Hernandez and Mitsch 2006; Paludan and Blicher-Mathiesen 1996; Stadmark and Leonardson, 2005). N_2O emission accounted for only 0.3% of total N loss in wetlands receiving river flows with elevated nitrate levels (Hernandez and Mitsch, 2006), and less than 0.13% of total nitrate loss in a wetland recharged by groundwater with elevated nitrate levels (based on maximum flux rates reported by Paludan and Blicher-Mathiesen, 1996). N_2O emission rates in wetlands receiving nonpoint source nitrate loads average around 1 μmol N_2O m^{-2} h^{-1} (Hernandez and Mitsch, 2006; Paludan and Blicher-Mathiesen, 1996), which is very similar to rates reported for cultivated crops in the Midwest (1–2 μmol N_2O m^{-2} h^{-1}) (Grandy et al., 2006; Parkin and Kaspar, 2006). The available research suggests that wetlands restored on formerly cultivated cropland for the purpose of nitrate removal would have little or no net effect on N_2O emissions.

4.5.2.2 Phosphorus

P removal in wetlands is controlled by three sets of processes: (1) sorption or release of P by existing sediments, (2) accumulation of P in new biomass, and (3) accumulation of P associated with the formation and accretion of new sediments/soils (Reddy et al., 2005). Existing sediments will have a finite capacity for sorption of P, determined in part by Al and Fe content in acid soils and by Ca and Mg content in alkaline soils. There will also be a finite capacity for the accumulation of P in new biomass. Of the three sets of processes, only the last contributes to long-term, sustainable P retention by wetlands: the accumulation of bound inorganic P and unmineralized organic P associated with the formation and accretion of new sediments and soil.

P sorption on both antecedent and newly accreting wetland soil is largely controlled by Fe, Al, and Ca. Reducing conditions found in wetlands may decrease sorption of P as insoluble complexes formed with Fe^{+3} are released upon reduction to Fe^{+2}, solubilizing the P (Patrick et al., 1973). High S levels may enhance P flux from soils due to the binding of iron by sulfides (Bridgham et al., 2001; Caraco et al., 1989). Alkaline wetland soils are more conducive to P sorption than acidic wetland soils due to the presence of Ca in the alkaline wetland soils and the formation of insoluble Ca-bound P (Bruland and Richardson, 2006; Richardson, 1999). These two studies indicate that wetlands developed on soils rich in calcite and exchangeable Ca are likely to be more effective sinks for P under the reducing conditions necessary for denitrification. More research is needed to understand (1) the effects of wetland creation such as is being done in the upper Mississippi River basin and (2) whether wetlands created/restored on Mollisols will be effective P sinks due to formation on Ca–P complexes in addition to sedimentation and SOM formation. Bruland and Richardson (2006) determined that marshes with a higher soil P sorption index (amount of P sorbed by soil from a phosphate solution in 24 h incubation) would be the best P sinks and that specific marshes could be targeted based on this index. It is important to remember, however, that antecedent soils of restored wetlands have a finite P retention capacity. The long-term sustainable capacity of these systems to retain P is determined primarily by the accumulation of

P associated with the formation and accretion of new sediments and soils. Studies of wetlands constructed to intercept nonpoint source nutrient loads in the MARB confirm the importance of sediment accretion for P retention (Anderson et al., 2005; Mitsch et al. 2005b) but also demonstrate that wetlands can become a P source if sediments are remobilized (Mitsch et al., 2005b). Most of the MARB studies represent recently constructed wetlands, and the long-term sustainable capacity of these systems to reduce P loadings is unclear.

Wetlands created, enhanced, and restored for N removal could also function for P removal, but limits to sustainable P removal must be recognized. Both NO_3 and P removal in wetlands will be enhanced by longer retention times and accretion of organic rich sediments. Long-term solutions for P load reduction in the MARB will likely depend more on reduction in sources than will long-term N load reduction. It will be important to manage restored wetlands so they do not become long-term sources of P after nonpoint sources of P have been reduced.

Key Findings and Recommendations

As concluded in the *Integrated Assessment*, wetlands can be very effective in NO_3 removal. Recent data, though limited, support the *Integrated Assessment*'s conclusion that N_2O evolution from wetlands restored as NO_3 sinks would be a low percentage of total denitrification. Wetlands receiving significant nonpoint source NO_3 loads at moderate-to-high NO_3 concentrations export comparatively small amounts of organic N and are nearly as effective in reduction of total N as in reduction of NO_3. This situation is less true for wetlands receiving loads at low NO_3 concentrations. Hydraulic loading rate and NO_3 concentration are especially important determinants of NO_3 removal rates in Corn Belt wetlands. Additional information is needed on created, restored, and enhanced wetlands including long-term monitoring for total N and P retention. Based on these findings, the Study Group offers the following recommendations.

- Wetland restoration should be evaluated for its full range of benefits.
- For greatest basin-wide reduction in nitrate load, wetland restorations should be targeted in those extensively row-cropped and tile-drained areas of the Corn Belt where nitrate concentrations and loading rates are highest and sized based on expected hydraulic loading rates and load reduction goals. For these areas, nitrate mass removal rates could be several times higher than previously predicted.
- Although limits to sustainable P removal by wetlands must be recognized, wetlands restored for N removal should be managed for P retention as well.

4.5.3 Conservation Buffers

Conservation buffer practices include riparian buffers (forests and herbaceous cover), field borders, filter strips, contour buffer strips, grass waterways, windbreaks, hedgerows, and other practices. They are part of the suite of conservation practices that are applied by farmers to achieve productivity, stewardship, and environmental quality goals. Conservation buffers differ from other conservation practices in that they will require long-term set aside of critical lands from continued agricultural production. Although often installed under the Conservation Reserve Program (CRP), conservation buffers differ from other uses of CRP because conservation buffers allow most land to remain in production while using critical areas as buffers for the agricultural land.

Prior analysis of nutrient control in the MARB focused on riparian forest buffers, one prominent type of conservation buffer (Mitsch et al., 1999). Studies conducted over the past decade in the Corn Belt have shown conservation buffers, especially riparian forest buffers and riparian herbaceous buffers, to be effective sinks for nutrients and sediment in landscapes with a significant portion of water moving as either surface runoff or shallow subsurface flow. If nitrate is transported from cropland primarily in tile drain flow as in much of the Corn Belt, riparian buffers and vegetated filter strips will have little opportunity to intercept nitrate loads. It is likely that if drainage management is changed to limit subsurface discharge through tile drains with concomitant increases in surface runoff and shallow water table flow, riparian buffers will be critical to achieve water quality goals.

Reduction of nitrogen by riparian buffers is generally determined by soil type, watershed hydrology (artificial drainage, groundwater flow paths, saturation), and subsurface biogeochemistry (organic matter supply, redox conditions) (Mayer et al., 2006). Control of P depends more on infiltration, surface roughness, and runoff retention. Many riparian buffers have been restored or established, but few have been studied to quantify water quality benefits. Richard Schultz, Tom Isenhart, and others developed the Riparian Management System for application in areas of the Corn Belt dominated by tile-drained systems. Modifications to the original USDA Riparian Buffer specification included integration of wetlands to intercept and remove tile drainage nitrate. Lee et al. (2000; 2003) reported rates of nutrient and sediment removal in multi-species buffer strips intercepting surface runoff in these systems. They found that switchgrass and switchgrass/woody buffers retained 50–80% of total N, 41–92% of NO_3-N, 16–93% of total P, and 28–85% of dissolved reactive P from surface runoff produced in simulated rainfall events.

Riparian herbaceous cover helps reduce sediment and other pollutants in surface runoff through the combined processes of deposition, infiltration, and dilution. Those functions are due to the cascading influence of perennial vegetation on soil quality when compared to soils under annual row crops. A series of studies on Bear Creek compared soil quality and related processes within riparian soils in a corn–soybean rotation with those soils in which perennial herbaceous vegetation had been reestablished (Schultz et al., 2004). Six years after establishment of riparian switch grass, those soils contained more than eight times the belowground biomass as adjacent crop fields (Tufekcioglu et al., 2003). As a result, soils in riparian herbaceous cover amassed up to 66% more total organic carbon in the top 50 cm (20 in) than crop-field soils (Marquez et al., 1999). This resulted in a two-and-a-half-fold increase in microbial biomass and a four-fold increase in denitrification in the surface 50 cm (20 in) of soil when compared to crop-field soils of the same mapping unit. As a result of increased soil quality, infiltration was nearly five times faster in soils under perennial vegetation than in row-cropped fields (Bharati et al., 2002). Riparian Management Systems such as those on Bear Creek are well suited to intercept increased overland flow that might be associated with changes in drainage management.

Several researchers have investigated the combined effects of these processes within riparian herbaceous vegetation and reported that sediment and nutrients in surface runoff can be reduced in the range of 12–90% compared to unbuffered crop fields (Dosskey, 2001; Lee et al., 2003). Major differences in impacts on the soil ecosystem depend upon the photosynthetic pathway of the dominant vegetation (e.g., C3 [cool-season grasses] or C4 [warm-season grasses]) in a buffer. Riparian herbaceous cover can help improve the quality of shallow groundwater, much like filter strips or riparian forest buffers. Hydrogeologic setting, specifically the direction of groundwater flow and the position of the water table in thin sand aquifers underlying the buffers, generally, is the most important factor determining buffer efficiency (Dosskey, 2001).

When applied as part of a conservation management system, the effectiveness of conservation buffers can be enhanced. There are few data on the field or landscape level effectiveness of conservation buffers applied with or without other conservation measures. Most data are from plot studies. Plot studies are inadequate, especially for studies of grass waterways (GWW), which are designed to convey overland flow from fields and stream bank restoration designed to reduce loss of sediment and sediment bound chemical from unstable banks. Because GWW are installed in areas of known water flow, they avoid problems of runoff bypassing filter strips and field borders. The few studies of GWW conducted at the field scale show that they are very effective at both runoff reduction and sediment trapping. In Germany, unmanaged grass waterways reduced runoff and sediment delivery by 90 and 97%, respectively, compared to adjacent fields with no GWW (Fiener and Auerswald, 2003). A GWW that was mowed closely was less effective, with reductions of 10 and 27% for runoff and sediment delivery, respectively. In New Brunswick, Canada, Chow et al. (1999) compared up- and downslope cultivation of potatoes and grain to the same crops with a terrace and grass waterway system. The conservation system reduced runoff by 31% and sediment delivery by 78%. On three small watersheds in the claypan soils region of Missouri, sediment and TP loss increased as the extent of GWW decreased (Udawatta et al., 2004).

There are ongoing efforts by USDA to estimate the impacts of conservation buffers on water quality in all watersheds with significant amounts of agriculture. The Conservation Effects Assessment Project (CEAP) will eventually provide model-based estimates of the water quality impacts of conservation practices in the MARB (Kellogg and Bridgham, 2003). Conservation buffers are an important

4.5 Options for Managing Nutrients, Co-benefits, and Consequences

component of USDA conservation programs. Table 4.4 summarizes the extent of seven major conservation buffer practices installed in the six subbasins of the MARB in federal fiscal years 2000 through 2006 (October 1999–October 2006) (M. Sullivan, personal communication, based on USDA-NRCS-Performance Results System, http://ias.sc.egov.usda.gov/prshome). An estimated 0.94 million hectares (2.31 million acres) of conservation buffers were installed in the MARB in 1999–2006. As shown, each hectare of conservation buffer treats 1 or 3 hectares of adjacent agricultural land, giving an estimated 3.46 million hectares (8.55 million acres) of agricultural land that has been treated by these six conservation buffer practices (Table 4.4).

Information on the extent of other conservation practices established from FY 2000 through FY 2006 is also available from the NRCS Performance Results System. Practices that are applied each year such as conservation tillage, residue management, and nutrient management may be reported more than once during the record period if there is a change in owner/operator, a new conservation plan is developed and associated practices are reported. There may have also been some systematic annual reporting in the early years of the record period (2000–2003) (Personal communication, Mike Sullivan, USDA-NRCS). All conservation tillage and residue management practices combined were applied on as much as 8.42 million hectares (20.8 million acres), and nutrient management was applied on as much as 7.4 million hectares (18.3 million acres) in the MARB in FY 2000–FY 2006 (Mike Sullivan, USDA-NRCS, Personal Communication, based on NRCS-PRS). Wetland creation, enhancement, and restoration were applied on 0.57 million hectares (1.42 million acres), drainage water management was applied on 756 hectares (1,867 acres), and stream bank restoration was installed on 3,155 km (1,972 mi). The values for 2002–2005 were reported in the USEPA Management Action Review Team report (MART, 2006a) and are similar to the above numbers when put on the same year basis.

Currently, no national databases allow a more detailed estimation of the environmental benefits of these conservation practices, including conservation buffers. This is the goal of the CEAP project. Estimates can be made based on acreage values, but these cannot take into account either placement or efficacy of practices. Cumulatively, conservation buffers, residue management, nutrient management, and wetlands have impacted up to 21 million hectares (51.9 million acres) of agricultural land in the MARB based on the FY 2000–FY 2006 areas of conservation practices. This area is the sum of residue management, nutrient management, conservation buffer acreage, wetland acreage and the potential land treated by conservation buffers (Table 4.4) and wetlands (assuming 3 hectares treated for 1 hectare of wetlands). In reality, conservation practices are applied as a system of practices, and it is likely that the total area treated through these practices is less than 21 million hectares (51.9 million acres). Additionally, the databases used are likely to include some duplicate reporting for the annual practices. The nutrient load reductions for these practices could be estimated based on amounts of N and P load retained. Although these would be crude estimates, they would provide numbers for comparison to the nutrient load reduction goals and provide a rough idea of where conservation programs stand relative to those goals.

Table 4.4 Areas (ha) of conservation buffers installed in the six subbasins of the MARB for FY 2000–FY2006

Subbasin	Contour buffer strips (ha)	Field border (ha)	Filter strip (ha)	Grassed waterway (ha)	Riparian forest buffer (ha)	Stream bank protection (km)	Windbreaks and shelterbelts (ha)	Conservation buffers applied (ha)
Ohio	3,362	5,441	50,617	21,346	32,497	755	794	114,832
Tennessee	196	1,914	10,724	817	10,752	418	2	26,025
Upper Mississippi	22,217	7,357	159,604	43,421	75,139	722	8,448	317,422
Lower Mississippi	165	7,541	10,274	661	56,106	503	391	75,486
Missouri	7,374	16,413	116,755	31,067	31,492	470	39,377	256,693
Arkansas White-Red	1,883	15,631	79,658	8,197	29,745	287	2,173	145,290
Sum	35,196	54,298	427,631	105,507	235,731	3,155	51,185	935,748
Area treated (ratio)	1:1	1:1	3:1	3:1	3:1	NA	3:1	
Area treated	70,393	108,595	1,710,525	422,030	942,926	NA	204,739	3,459,207

* Kilometers are shown for stream bank protection. Conservation buffers applied include areas in other practices not shown here that are cumulatively small areas compared to the practices shown. The areas treated are based on the ratios shown and assumes that each hectare of buffer treats either 1 or 3 hectares of adjacent agricultural land. Areas of practices are from Mike Sullivan, USDA-NRCS, Personal Communication, and are derived from NRCS-PRS, http://ias.sc.egov.usda.gov/prshome.

> **Key Findings and Recommendations**
>
> Conservation buffers and other conservation practices have affected a significant acreage of MARB cropland through existing federal, state, and private programs. The Study Group offers the following recommendations.
>
> - Continued, new, and enhanced small watershed-based studies of suites of conservation practices as applied on farms and in agricultural watersheds are necessary. Analysis of effects of conservation buffers and other conservation practices in the MARB should be coordinated with the ongoing USDA Conservation Effects Assessment Project.
> - Conservation buffers and other conservation practices in the MARB should be refocused on N and P retention with special attention given to the interactions of buffers with other practices. Environmental benefits indices should be calculated in a way as to provide extra weight for N and P retention.

4.5.4 Cropping Systems

Current cropping systems within the MARB are well established, but advances in N fertilizer production technology, innovative crop rotations, inter-seeding with cover crops, and alternative mulches or crop residues provide opportunities to improve water and nutrient use efficiency as well as to decrease leaching and runoff of nutrients and sediments. For example, inter-seeding of a leguminous cover crop within existing crop rotations could enhance N and P use efficiencies, as long as the cover crop is carefully managed. Also, greater adoption of perennial systems, which could include cellulosic production, has the potential to influence nutrient export via reduced N and P applications as well as altered water budgets. Evapotranspiration and infiltration will likely be greater with perennial than annual cropping systems, contributing to a decrease in potential runoff. Hydrologic and water quality issues related to perennials and cellulosic production are discussed in more detail in Section 4.5.9.

A continuous corn rotation typically results in annual N fertilizer applications between 150 and 250 kg N/ha (134 and 223 lb N/ac). This is a large amount of N fertilizer relative to amounts applied to other crops. Including other crops (particularly legumes) in a crop rotation usually reduces annual N fertilizer applications needed. In addition to applying less N, perennial crops, such as alfalfa or other grass mixtures, have longer effective growing seasons and are more efficient N users than annual crops, which translate to greater water use and less nitrate leaching.

Randall et al. (1997) compared tile drainage and nitrate loss for corn–soybean and corn–corn rotations to alfalfa and Conservation Reserve Program (CRP) grassland. From 770 to 905 mm (30–36 in) of tile water was recorded for the corn–corn and corn–soybean rotations from 1988 to 1993, whereas 416 to 640 mm (16–25 in) of tile water was recorded for alfalfa and CRP. Flow-weighted nitrate-N concentrations were less than 5 mg/L for alfalfa and CRP but ranged between 13 and 40 mg/L for the rotations including corn and soybean. The 4-year nitrate-N loss from continuous corn or corn–soybean rotations was 202–217 kg N/ha (180–194 lb N/ac), while for alfalfa and CRP the loss was less than 7 kg N/ha (6 lb N/ac). Similarly, Jaynes et al. (2001) showed for a corn–soybean rotation in central Iowa that even at economically optimum N fertilizer rates for corn (67–172 kg N/ha or 60–154 lb N/ac), NO_3 loss in tile drainage water increased from 29 to 43 kg N/ha (26–38 lb N/ac) with application rate. Also, a net N mass balance indicated that N was being mined from the soil at economically optimum N fertilizer rates and the system would not be sustainable (Jaynes et al., 2001).

Besides crop selection to enhance N and P removal, crop rotation also can be managed to maximize nutrient removal and minimize leaching. Together, crop selection and rotation can influence the amount of N and P in a soil profile as well as water available for nutrient leaching. As mentioned, legumes, such as alfalfa and soybean, that do not require supplemental N can effectively use or "scavenge" residual inorganic N remaining in the soil from previous crops. Some crops take up more P, and deep-rooted crops can remove N and P from subsoil horizons. For example, root development of a typical 3-year continuous corn system (maximum depths in May through September) does not always coincide with time of high NO_3 leaching potential (generally February–April). An alternative cropping system comprised of corn–winter wheat–alfalfa provided a much different root development pattern, one that should more efficiently retain N because it has deeper roots that are present most of the year (Sharpley et al., 2006b). Olson et al. (1970) found that NO_3 concentrations at a depth of 1.2–1.5 m (3.9–4.9 ft) in a silt loam soil were lower for an oat–meadow–alfalfa–corn rotation than for continuous corn when ammonium nitrate was applied to both systems. The reduction in NO_3 leaching was directly proportional to the number of years that oats, meadow, or alfalfa was grown in rotation with corn. The reduction was attributed to the combined recovery of NO_3 by shallow-rooted oats, followed by deep-rooted alfalfa (Olson et al., 1970). The potential for NO_3 leaching in such rotations is, therefore, less when compared with continuous annual monocropping systems.

Clearly, including perennial crops in a rotation, as well as conversion to perennial systems, can reduce NO_3 leaching, partly due to the fact that perennials are generally more efficient users of N than annuals. As a result, Randall and Vetsch (2005) raises a key question of whether significant reductions in nutrient (especially NO_3) loadings to surface waters are possible without changing from the predominant annual cropping system of corn–soybean rotation to a mixed system that includes perennials. While annual grain crop production is an essential component of agricultural systems in several areas of the MARB, the development of economically viable continuous cropping systems will help improve in-field nutrient use efficiency and decrease off-site loads. Additional co-benefits of perennials, such as switchgrass, are that they have the potential to accumulate large amounts of below-ground biomass and are effective in sequestering C (McLaughlin and Lszos, 2005; McLaughlin and Walsh, 1998). Costello et al., (2009) estimate that switching from corn to cellulosics for ethanol production could reduce NO_3 output from the MARB by 20%. On the other hand, corn-based ethanol production increases export of dissolved inorganic nitrogen (Donner and Kucharik, 2008).

Retirement of land through the Conservation Reserve Program has demonstrated different results for various cropping systems. For lands previously in corn, the reduction in N delivered to the Mississippi River may have been as much as 25–30 kg N/ha-year (22–27 lb N/ac-year). For soybean it would have been somewhat less, and for small grains, particularly wheat in the High Plains, smaller reductions in the range of 10 kg N/ha-year (8.9 lb N/ac-year) may have been realized (see Section 3.1.2). Where CRP has been used to establish buffers, not only are reductions from the retired lands realized, but the buffers can also be effective in reducing inputs of N and P from upslope cropland entering water courses via surface runoff and shallow subsurface flow. It should be noted, however, that most land enrolled in CRP is primarily sloping, erosive land that is not tile drained. For instance, McIsaac and Hu (2004) studying N flux in several Illinois rivers between 1977 and 1997 found that riverine N flux was about 100% of net N input for the tiled-drained region (27 kg N/ha-year or 24 lb N/ac-year). In the nontile-drained region, riverine N flux was between 25 and 37% of net N input (23 kg N/ha-year or 20 lb N/ac-year).

Key Findings and Recommendations

Cover crops and other living mulches can improve water and nutrient use efficiencies and reduce nitrate leaching. Further research and demonstration is needed in the MARB in several areas: examining the benefits of intercropping cover crops with annuals such as corn, determining if leguminous cover crops reduce fertilizer N requirements, and assessing how changes in cropping patterns can impact nutrient loss at both local and basin-wide scales. If farmers could be encouraged to switch to a rotation of perennial crops as compared to the predominant corn–soybean rotation system, significant N and P reductions would result. Based on these findings, the Study Group offers these recommendations.

- Cover, relay, and perennial crops should be considered in alternative cropping systems that will reduce nutrient loss. Cropping systems that efficiently include cover crops in grain and row cropping should also be encouraged in the Corn Belt region of MARB. This should focus on the use of fall-planted, small-grain cover crops more suited to the short growing season after harvest and cold winters of the upper Midwest.
- Where corn–soybean production systems exist and/or where it is not feasible to plant cover crops, it is even more important to encourage off-field conservation practices.

4.5.5 Animal Production Systems

4.5.5.1 System Development and Nutrient Flows

While overall production livestock numbers in the MARB have declined (see Section 3.2), there has been an intensification of operations in certain areas (see Figs. 4.4 and 4.5, and Appendix C). Farmers adopted the animal-feeding-operation (AFO) paradigm because of competitive pressures, changing marketing practices, a need to be responsive to consumer demand for quality meat products at a low cost, and declines in income from traditional grain crops in certain areas of the MARB with inherently infertile soils (Lanyon, 2005). This critical socioeconomic shift must be considered when proposing changes within the MARB that decrease the impact of AFO and manure management on nutrient export.

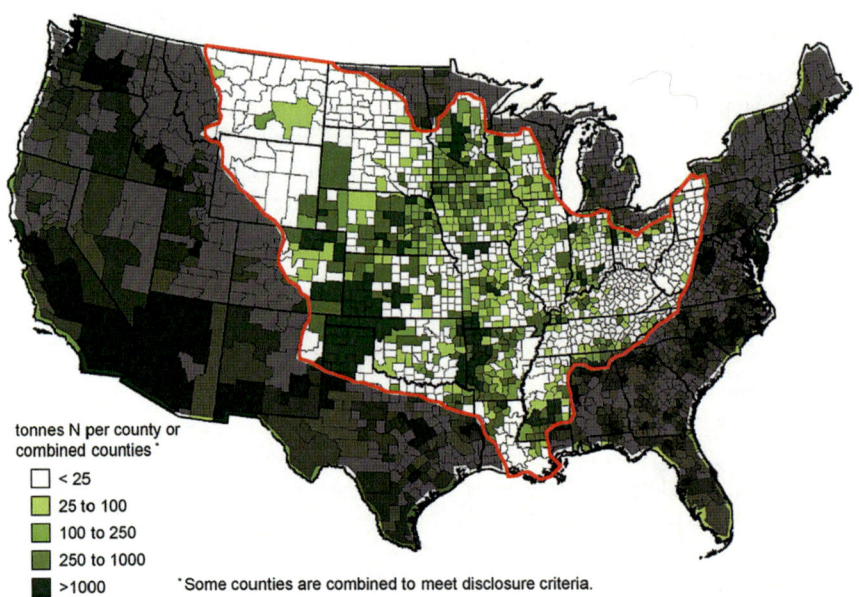

Fig. 4.4 Recoverable manure N, assuming no export of manure from the farm, using 1997 census data. Adapted from USDA (2003) with the author's permission

As a consequence of the spatial separation of crop and animal production systems, fertilizer N and P is imported to areas of grain production. The grain (harvested N and P) is then transported to areas of animal production, where inefficient animal utilization of nutrients in feed (less than 30% is utilized) are excreted as manure. This system has led to a large-scale, one-way transfer of nutrients from grain- to animal-producing areas within the MARB and dramatically broadened the emphasis of nutrient and manure management strategies from field to watershed to basin scales. For the MARB, farm-level nutrient excesses are estimated at 337 million kg N (743 million lb N) and 242 million kg P (534 million lb P) (Gollehon et al., 2001).

4.5 Options for Managing Nutrients, Co-benefits, and Consequences

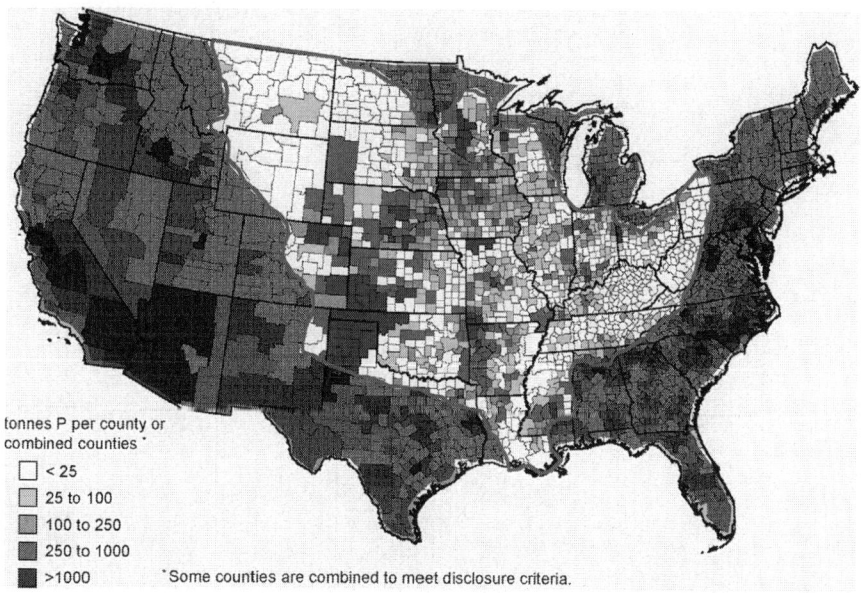

Fig. 4.5 Recoverable manure P, assuming no export of manure from the farm, using 1997 census data. Adapted from USDA (2003) with the author's permission

The land application and discharge of nutrients in manure from AFOs are regulated under the National Pollutant Discharge Elimination System (NPDES), which generally defines an AFO as an operation where livestock are confined for an extended period of time (at least 45 days in a 12-month period) and there is no grass or other vegetation in the confinement area during the normal growing season (USEPA, 2000a). This definition is intended to differentiate confinement-based operations from pasture-based operations, which are excluded from the Confined Animal Feeding Operations (CAFO) regulations. The NPDES permit is required to control pollutants at an AFO and keep them from entering surface waters. More explicitly, the USEPA (2000a) defines CAFOs as livestock operations that meet one of the following characteristics:

- Confine more than 1,000 animal units (AU), where 1,000 AUs are defined as 1,000 slaughter and feeder cattle, 700 mature dairy cows, 2,500 swine (other than feeder pigs), 30,000 laying hens or broilers if the facility uses a liquid system, and 100,000 laying hens or broilers if the facility uses continuous overflow watering.
- Confine between 300 and 1,000 AU (as defined above), and either a man-made ditch or pipe carries manure or wastewater from the operation to surface water or animals come into contact with surface water running through the area where they are confined.

These regulations are enacted at a national level, and thus, recommendations and controls on the land application or utilization of manures and their component

Table 4.5 Status of implementation of permits under the 2003 CAFO rule for states within the MARB. Data provided by USEPA Office of Wastewater Management, 2007

State	Number of CAFOs	Number of CAFOs with permits to date	Permit coverage for CAFOs under 2003 rule
Alabama	558	440	79
Arkansas	2,110	70	3
Colorado	225	33	15
Illinois	500	8	2
Indiana	584	413	71
Iowa	1,859	113	6
Kansas	476	462	97
Kentucky	150	67	45
Louisiana	150	2	1
Michigan	198	56	28
Minnesota	1,007	1,000	99
Mississippi	433	190	44
Missouri	492	492	100
Montana	TBD	75	TBD
Nebraska	1,000	303	30
New Mexico	151	47	31
North Carolina	1,222	1,200	98
North Dakota	47	0	0
Ohio	162	64	40
Oklahoma	625	163	26
Pennsylvania	462	165	36
South Dakota	369	303	82
Tennessee	129	130	101
Texas	1,204	639	53
Virginia	150	0	0
West Virginia	30	0	0
Wisconsin	161	161	100
Wyoming	51	47	92
Total	**14,505**	**6,643**	**46**

nutrients are in place at a state level in the MARB. Based on a USEPA summary of CAFO permit implementation completed in the first quarter of 2007, less than half of the CAFOs in the MARB were permitted (46%; Table 4.5). States included are in Table 4.5, if part of the state drains into the MARB. The approximate number of permitted CAFOs in the MARB is similar to the national average (44%; USEPA, 2007); but clearly, rule implementation varies among states.

4.5.5.2 Manure as a Component of N and P Mass Balances

Within the MARB, counties with the greatest excess of recoverable manure N and P (if applied on the farm where it is generated) tend to be in the western and drier areas of the basin, Arkansas, and central Minnesota (Figs. 4.4 and 4.5). Recoverable manure is defined as the portion of manure *as excreted* that could be collected from buildings and lots where livestock are held and, thus, would be available for land application. Recoverable manure nutrients are the amounts of manure N and P that would be expected to be available for land application (USDA, 2003). They are estimated by adjusting the quantity of recoverable manure for nutrient loss during collection, transfer, storage, and treatment. Recoverable manure nutrients are not adjusted for losses of nutrients at the time of land application. Where riverine N export is the greatest (upper Mississippi and Ohio River basins with tile drainage), manure N excess tends to be less. Lower Mississippi River basin states, particularly Arkansas and northern Missouri, clearly have more manure P on some farms than land area to apply it (Fig. 4.4). Although N from manure can be important in specific areas, basin-wide N loss is a result of the dominant inputs of fertilizer and N_2 fixation on tile-drained corn and soybean fields. For P, manure is a more important source, particularly on the western side of the basin (Fig. 4.5).

Large-scale consolidation has created much larger AFOs, which makes economical utilization and redistribution of manure to croplands difficult, and has profound consequences for regional nutrient transfer and management within the MARB. Furthermore, the potential for co-locating AFOs with areas of the corn production for ethanol generation may exacerbate the accumulation of manure-based nutrients in these areas. This co-location stems from the use of by-products from ethanol production (distiller's grain) as animal feed (for more information see Section 4.5.9).

4.5.5.3 Remedial Strategies

Manure is a valuable resource for improving soil structure and increasing vegetative cover, thereby improving water quality via reduced runoff and erosion potential. Manures have been historically applied at rates designed to meet crop N requirements. This has resulted in the accumulation of soil P above levels required for crop production, and a concomitant increase in the potential for N and P loss via runoff, leaching and N_2O emission within the MARB (Table 4.6; Aillery et al., 2005; Sharpley et al., 1998). In the past, separate strategies for either N or P have been developed and implemented at farm or watershed scales. The Study Group recognizes that this approach needs to change; N and P need to be managed jointly in order to improve water quality. Because of different critical sources, pathways, and sinks controlling N and P export, remedial strategies directed at only N or only P control can negatively impact the other nutrient. For example, basing manure application on crop N requirements to minimize nitrate leaching can increase soil P and enhance P losses (Sharpley et al., 1998; Sims, 1997). In contrast, reducing surface runoff losses of P via conservation tillage can enhance nitrate leaching in some cases (Sharpley and Smith, 1994).

Long-term sustainable management of nutrients in manure begins with sound feed decisions, which generally lie with the integrator in the CAFO industry rather than the individual farmer. Nutrient inputs to a farm should be matched as closely as possible with export as animal or crop products. If a farm's N and P budget is rich in imports, regardless of any other management decisions, there will be an ongoing accumulation of N and P on the farm, which in the long

Table 4.6 Estimates of manure production and N and P loss to water and air from Animal feeding operations within the Mississippi River basin. Total manure in millions of milligrams; other materials in millions of kilograms. Based on information from the 2002 US Census of Agriculture (adapted from Aillery et al., 2005)

Region of MARB	Number of operations	Total manure	N runoff	N leached	N emissions	Total N loss	P runoff
Lake States (MI, MN, WI)	52,498	62.52	32.89	0.36	164.45	198	5.58
Corn Belt (IA, IL, IN, MO, OH)	71,252	85.09	39.73	0.47	234.89	275	11.78
Northern Plains (KS, ND, NE, SD)	26,087	72.27	36.31	0.37	168.44	205	6.99
Appalachia (KY, NC, TN, VA, WV)	22,776	79.57	54.65	0.91	259.16	315	15.79
Delta States (AR, LA, MS)	12,252	19.97	8.92	0.15	62.57	72	4.47
Southern Plains (OK, TX)	10,500	49.19	21.96	0.20	119.74	142	7.72
Total	195,365	368.63	194.46	2.46	1009.26	1206	52.34

term will ultimately increase the potential for nutrient loss to water or air when manure is land applied. Nevertheless, the short-term impacts of land-applying manure or litter on nutrient loss can be reduced by the adoption of conservation practices detailed by USDA-NRCS (ftp://ftp-fc.sc.egov.usda.gov/NHQ/practice-standards/standards/590.pdf). However, conservation measures at both farm and watershed scales involves a complex suite of options, which must be customized to meet site-specific needs (for more information see Section 4.5.10 and Appendix C).

4.5.5.4 Alternative Manure Management Technologies

Reducing farm-gate inputs of N and P in animal feed presents one of the best nutrient management opportunities to effect a lasting reduction in N and P loss (Appendix C). Other measures, generally aimed at reducing the potential for N and P losses, are seen as short- rather than long-term solutions to environmental concerns. For instance, long-term monitoring of P budgets in Ohio showed that after nearly 20 years of BMP adoption and despite continually increasing soil test P levels, manure applications and timing have been managed better, resulting in more efficient use of P and reduced P loss to surface waters (Baker and Richards, 2002). Manure-related conservation practices include the following:

- manure amendments, such as alum, to reduce ammonia volatilization and sequester P in less soluble forms;
- coagulant and flocculent techniques to separate and concentrate nutrients in liquid manure systems; and
- combining manure with biosolids and woodchips to reclaim soils that have been disturbed (e.g., by mining or urban development).

As the cost of N fertilizer increases, it is clear that new markets for alternative uses or products for manure will open up. For example, on-farm and regional energy production via burning of manure is of increasing cost-effectiveness. Ash production via burning, while rich in P, will be appreciably less bulky and, thus, enable cost-effective transportation further from the source of generation. The bulky nature of manures, and resulting high cost of transportation, has always been a major limitation to more effective redistribution of N and P to nutrient deficient areas of the MARB.

Recent efforts to exclude cattle from streams as part of the Conservation Reserve Enhancement Program (CREP) were estimated to have resulted in a 32% decrease in P loadings to streams within the Cannonsville watersheds in south central New York (James et al., 2007). Thus, exclusionary programs like CREP and stream bank fencing are working to reduce nutrient loading by fencing cattle out of the stream and adjacent riparian zones. Clearly, grazing management and placement of stream bank fencing are important to minimizing watershed export of P. For instance, herd size, pasturing time, and cattle type could all be used to prioritize sites for stream bank fencing installation. In addition, field observations [such as those by James et al. (2007)] show installation of alternative watering sources do not necessarily preclude continued use of streams as a preferred water source.

The wider adoption of manure hauling that links producers with buyers will greatly enhance the sustainability of AFOs. At a state level, the Discovery Farms program is conducting research on privately owned Wisconsin farms in different geographic areas, facing different environmental challenges (see http://www.uwdiscoveryfarms.org/new/index.htm). The Discovery Farms program has been very successful at gaining farmer support in at-risk catchments in efforts to find the most economical solutions to overcoming the challenges environmental regulations place on farmers. At a watershed level, the Illinois River Watershed Partnership (see http://www.irwp.org/index.html) was established in 2005 to improve and protect water quality in the Illinois River in Arkansas and Oklahoma by working at a grassroots level with watershed citizens and other organizations.

Key Findings and Recommendations

The impacts of animal production systems are mainly expressed at a local rather than MARB scale. Overall, numbers of animals in the MARB have decreased, but localized increases have occurred in several regions, which have had an impact on local water resources. The economic and environmental sustainability of AFOs hinges on reducing the nutrient imbalance at farm and watershed scales through carefully managed feeding strategies. The wider adoption of manure transportation that links producers with buyers will greatly enhance the sustainability of AFOs. The large-scale consolidation of AFOs, co-siting with biofuel production facilities (by-product grains used as animal feed), and increases in N fertilizer prices will likely create the economies of scale and alternative technologies for on-farm or localized manure use and management more feasible.

The success of nonprofit programs supported by watershed agricultural councils, industry, and state agencies should provide valuable demonstration models. If energy prices remain at current levels, bioenergy production from manures could provide an off-farm market for manures and reduce localized nutrient surpluses. Continuing educational efforts with farmers and the public regarding the importance and impact of conservation practices will be essential to reach environmental goals. Based on these findings, the Study Group offers the following recommendations.

- Strategies need to be implemented to encourage further development of alternative uses for manures, such as in composting, pelletizing, and granulation, and as a soil amendment in nutrient deficient areas of the MARB.
- Land-management planning and implementation of conservation practices should be designed to identify and avoid applications in critical loss areas, to use buffers or riparian zones, to manage grazing, to exclude stream banks, and to use subsurface injection with innovative applicators.
- Incentives to encourage on-farm and local bioenergy production from manure sources should be provided.

4.5.6 In-Field Nutrient Management

4.5.6.1 Fertilizer Sources

The principal fertilizer N sources (>90% of fertilizer N) used in the MARB are anhydrous ammonia, urea–ammonium nitrate solutions, and urea. Anhydrous ammonia use in several leading corn-producing states (IL, IN, IA, MN, NE, OH) has tended to decline in recent years, perhaps with the exception of consumption in Illinois and Indiana (Fig. 4.6) (Vroomen, H., Vice President, Economic Services, The Fertilizer Institute, 820 First St., NE, Washington, D.C., 2002, personal communication, 2007). The largest decline has been in Nebraska, where use of anhydrous ammonia N has declined about 40% since the mid-1980s.

4.5 Options for Managing Nutrients, Co-benefits, and Consequences

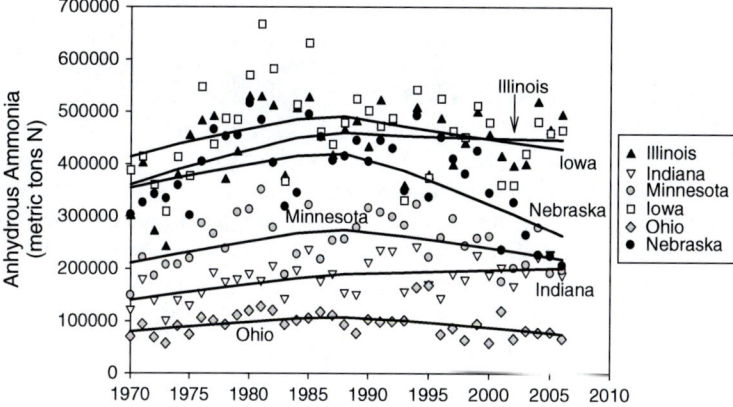

Fig. 4.6 Fertilizer N consumption as anhydrous ammonia in leading corn-producing states for years ending June 30

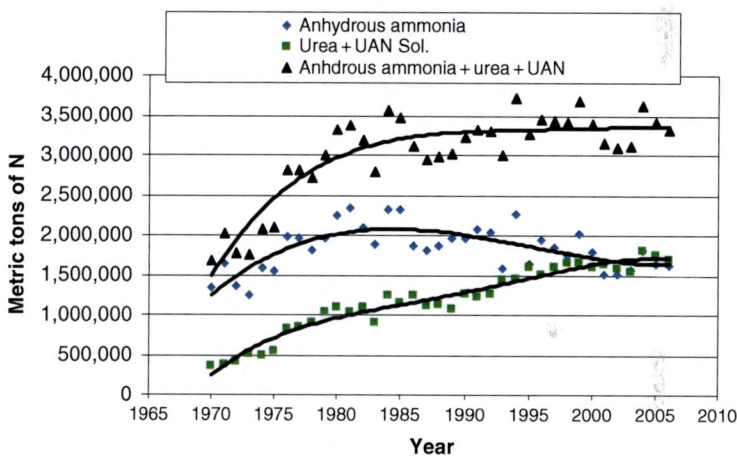

Fig. 4.7 Changes in the consumption of principal fertilizer N sources used in the six leading corn-producing states (IA, IL, IN, MN, NE, and OH) for years ending June 30

The combined N consumption of urea and urea–ammonium nitrate solution has increased and recently surpassed anhydrous ammonia tonnage in these six leading corn-producing states (Fig. 4.7). Although these data illustrate shifts in fertilizer N sources used, they do not allow conclusions about the portion of the annual anhydrous ammonia consumption that may be applied in the fall.

4.5.6.2 Fertilizer Use and Application Technology

The *Integrated Assessment* (CENR, 2000) concluded that "discharges of nitrogen from farms to streams and rivers could be reduced by implementing a wide variety of changes in management practices." These practices include switching from fall fertilizer N to spring N applications and applying nitrogen fertilizer and manure at not more than agronomically recommended rates. Application rate and timing are linked for N because the closer application is to the time of crop need, less N is lost to the atmosphere and water, and less N is needed. Research at five Management System Evaluation Area (MSEA) sites in the MARB (OH, IA, MN, MS, NE) reaffirmed BMPs for water quality, including soil nitrate tests, improved water management, and improved N timing and placement relative to crop needs (Power et al., 2000). Determining N sufficiency by monitoring for plant greenness and use of field or remote-sensing technologies followed by site-specific N applications hold promise to manage N more precisely.

Application timing. The risk of N loss with corn is greatest when fertilizer is applied some time before the period of rapid plant growth. Data on fall application are not directly available for the MARB and even seasonal data on fertilizer sales are not kept by all states in the MARB (Terry, 2006). Fertilizer sales records for Iowa (from July 2002 to June 2006) showed that 48% of N fertilizer was sold in the period from July to December and 52% from January to June. For anhydrous ammonia, the most common N form used and the primary form applied in the fall, 54% was sold in the period from July to December and 46% was sold from January to June. July to December sales of anhydrous ammonia accounted for 273,000 tons of actual N (data from http://www.agriculture.state.ia.us/fertilizerDistributionReport.htm). For Illinois, there has been an increase in fall N sales from the 1970s and 1980s to present, from about 25% to 40–50% (Fig. 4.8).

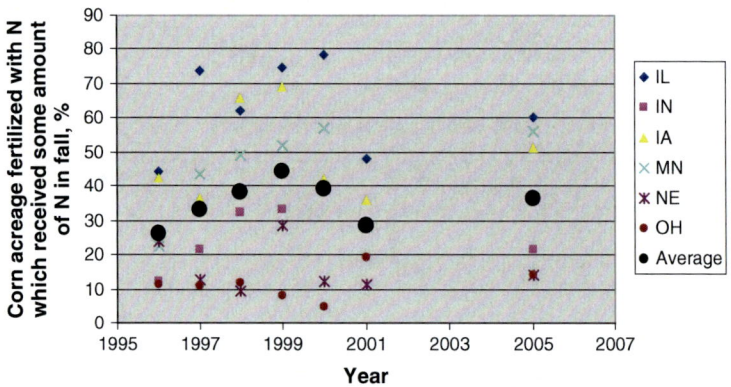

Fig. 4.8 Percentage of N-fertilized corn acreage that received some amount of N in the fall

Although it is not possible to correlate fall N application directly with fall fertilizer N sales, it is likely that a large fraction of the fall N sales represents N applied in the fall (Czapar et al., 2007). A portion of this fall N tonnage sold may also be stored at dealerships or in on-farm storage vessels for application the following spring. The USDA Agricultural Management Resource Survey (ARMS http://www.ers.usda.gov/Data/ARMS/app/Crop.aspx) data provide some

4.5 Options for Managing Nutrients, Co-benefits, and Consequences

insight into fall N applications, yet they are not sufficiently complete (i.e., key years are missing) to determine if the percentage of the acreage that receives some amount of fall N is increasing, decreasing, or remaining static (Fig. 4.8). The data do indicate that Minnesota, Iowa, and Illinois tend to fall apply some N on a larger fraction of their corn acres, compared to the other three states shown in Fig. 4.8. For three states, USDA ARMS data (Fig. 4.9) were used to calculate the total fraction of N applied

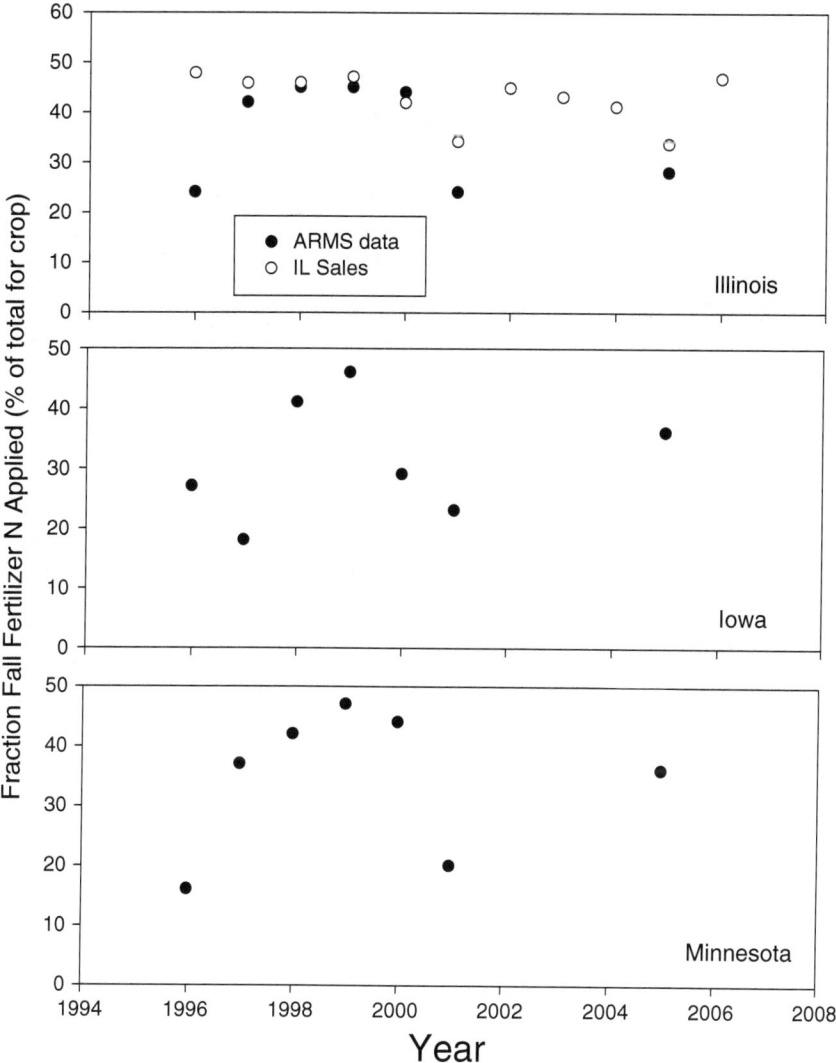

Fig. 4.9 USDA ARMS data for the three states with highest fall N application, showing total amount of fall-applied N for that crop. Also shown are Illinois sales data for the same period

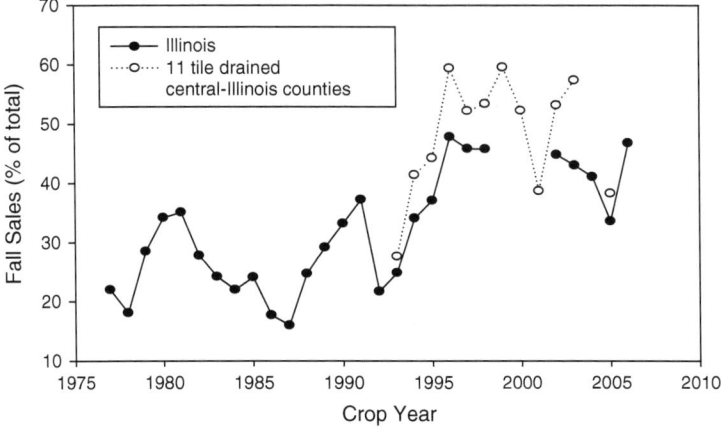

Fig. 4.10 Fraction of annual fertilizer N tonnage in Illinois sold in the fall

to corn in the fall, and IL sales were also compared (Fig. 4.10). As fewer producers in the Corn Belt farm the existing acreage, there has been greater pressure to complete fertilization in the fall, because of the numerous logistical challenges (labor demands, transportation and application equipment availability, weather uncertainty, and fertilizer supply and cost uncertainty) in the spring.

Randall and Sawyer (2005) contacted State Extension soil fertility specialists and State Fertilizer Associations to determine the fertilizer N amount that is applied in the fall. Based on these data, they estimated 25% (5.1 million hectares or 12.9 million acres) of the 20.5 million hectares (50.6 million acres) of corn in an 8-state area (IA, IL, IN, MI, MO, MN, OH, WI) received N in the fall. States with the largest amount of fall-applied N were Minnesota (1.85 million hectares or 4.56 million acres), Iowa (1.42 million hectares or 3.52 million acres), and Illinois (1.33 million hectares or 3.28 million acres). It is likely that tile-drained portions of these eight states have higher proportions of N applied in the fall either because of a greater dominance of corn/soybean agriculture or because regional soil temperatures are also cold enough to help minimize the conversion of ammonium-N to nitrate-N (nitrification) in the fall. Fall N application for corn as anhydrous ammonia is currently a recommended practice by virtually all Land Grant universities in the Corn Belt, where soil temperatures are consistently below 50°F at the 1.2–1.8 cm depth (0.47–0.72 in. or about $\frac{1}{2}$–$\frac{3}{4}$ in.), and the risk of environmental loss is not considered high or a pragmatic concern (Snyder et al., 2001). Additional guidance is usually provided in publications by Land Grant universities to maximize the benefits of fall N application and to help to minimize the risk of economical and environmental N losses (e.g., Bundy, 1998; Shapiro et al., 2003).

In a 2003 phone survey of Champaign Co., IL (a dominantly tile-drained area), 61% of the 352 respondents reported applying some N in the fall, and 49% of respondents applied all of their N in the fall (von Holle, 2005). Overall, the farmers who fall fertilized applied an average of 79% of their annual N needs before January 1, 2003. Data from 11 tile-drained central Illinois counties showed generally greater fall N fertilizer sales than the state as a whole (Illinois Department of Agriculture fertilizer tonnage reports). This difference was primarily due to the southern and nontile-drained portion of the state having winter soil temperatures that are too warm for fall application, where it is not recommended.

The effects of fall N application versus spring N application on nitrate transport in tile drainage depend on many factors, including soil temperatures, soil texture, precipitation, and drainage intensity. Randall and Sawyer (2005) reviewed the timing of N application and determined that spring application in Minnesota will typically result in 15% less nitrate-N loss than with fall application. In areas with warmer nongrowing season temperatures (such as central Illinois) that are tile-drained, losses of fall-applied N may be greater. Watershed-scale studies of changing from fall to spring application (sidedressed) and changing the rate to account for more efficient use of spring-applied N showed at least a 30% reduction of nitrate concentrations in tile drain water (Jaynes et al., 2004). These studies indicate there is a great potential in some years for substantial reductions in N loss by applying N closer to when the crop can utilize it efficiently.

If these various estimates of N use and nitrate-N loss are combined, changing from fall to spring application may affect at least 25% of the corn acreage and reduce nitrate-N losses to streams from those acres by perhaps 10–30%. Split applications of N do not always result in increased N efficiency and reduced nitrate-N losses just because of improved N synchrony with crop uptake demands. The literature to support this practice indicates mixed results (Randall and Sawyer, 2005).

Nitrification inhibitors delay the conversion of ammonium to nitrate in soil. In Illinois, it is estimated that a nitrification inhibitor is added to about 50% of the fall-applied anhydrous ammonia (Czapar et al., 2007). Application of a nitrification inhibitor with anhydrous ammonia in the fall increased apparent recovery of N fertilizer in the corn grain from 38% without a nitrification inhibitor to 46% with an inhibitor, compared to 47% with spring application with no nitrification inhibitor in long-term research results in Minnesota (Randall et al., 2003; Randall and Sawyer, 2005). Ferguson et al. (2003) found that in Nebraska the benefits of nitrification inhibitors (either increased yield or reduced NO_3-N leaching) are strongly dependent on specific conditions and are most likely to be observed at suboptimal N rates (i.e., less than the economically optimum N rate [EONR; the point where the last increment of N returns a yield increase large enough to pay for the additional N]). They also reported that nitrification inhibitors can reduce crop yields with late sidedress N applications. It is well known that time of N application will largely govern any benefits from the use of nitrification inhibitors. Assuming increased N recovery by the crop translates to less nitrate leaching, nitrification inhibitors can potentially provide an economic benefit to farmers while reducing leaching.

Although the fertilizer N use trends indicate increased urea and urea–ammonium nitrate (UAN) solution use in the Corn Belt and lower anhydrous ammonia use (Fig. 4.7), there is a need for more research to document the benefits of split N applications of these two sources versus the more traditional fall anhydrous N applications. Use of urea and UAN solutions may provide greater flexibility in N management than has been experienced with anhydrous ammonia. Studies are underway to evaluate the crop and water quality effects associated with different N sources and time of application (e.g., see reports of work by Gyles Randall and others at the University of Minnesota: http://sroc.cfans.umn.edu/research/soils/index.html). In years when corn growth proceeds rapidly, timely sidedressing can be difficult, and delayed application can severely reduce yields (Randall, G., University of Minnesota, 2007, personal communication).

Application rate. Current N recommendations are usually applied across large geographic regions and may provide erroneous results for field-specific soil–crop–climate conditions (Gehl et al., 2005; Sawyer and Nafziger, 2005). Grouping soil types with similar drainage characteristics, rooting depth, and organic matter content is a feasible approach for determining more localized N recommendations and may result in more environmentally friendly N management (Oberle and Keeney, 1990). Remote sensing, geographic information systems, and variable application technologies offer an opportunity to develop and implement site-specific N recommendations, but the agronomic understanding of yield response to N on a site- and season-specific basis lags behind the technological innovations. There are instances, however, where considerable progress has been made in developing site-specific N recommendations (Raun et al., 2005).

Application of N near rates that provide the EONR usually results in drainage tile flow having nitrate-N concentrations in the range of 10–20 mg/L NO_3-N for soybean–corn rotations and 15–30 m/L NO_3-N for continuous corn (Sawyer and Randall, in press). Application of N above the EONR further increases NO_3-N losses and reduces net economic return. To the extent that N is being applied above the EONR, reductions in N loss through tile drains can be achieved with concurrent positive effects on net return (Sawyer and Randall, in press).

A review of the effects on N rates on corn–soybean systems in the upper MARB was conducted by Sawyer and Randall (2008) who found that in order to achieve a 30% reduction in tile drainage nitrate-N load, based on a study in Illinois, the N rate had to be reduced by 78 kg N/ha (70 lb N/ac) below the EONR, resulting in a large net economic loss ($67/ha or >$27/ac). These results illustrate an example of the risk of potentially large economic losses to farmers (and their communities) if they are asked to reduce N rates below their maximum net return or EONR (Sawyer and Randall, 2008). The potential environmental benefits of any N rate reductions are highly site specific and will also depend on how farmer's past N rates match their site-specific EONRs.

Economically optimum N rates are not the same across the Corn Belt states, and the same is true for other crops because of differences among soils, adapted crop varieties, climate, management, and many other factors that influence production and crop N requirements (Hong et al., 2006; Sawyer and Nafziger, 2005). Corn N needs vary widely both among and within fields (Lory and Scharf, 2003; Scharf et al., 2005). In some fields, in some areas of the MARB, where farmer's N rates have exceeded the EONR (especially where elevated N concentrations have been observed in water resources) there may be opportunities to reduce N rates for corn (Mamo et al., 2003) and other crops. Nitrogen application rate reductions must be economical for the farmer while also protecting water resources. Prior history of many management inputs including fertilizer N, manure, and tillage can affect crop N response and EONR interpretations. Farmers should carefully consider N rates and evaluate results over several years, in the same fields or plot areas. Rate reduction results obtained in 1 year can be highly affected by environmental conditions. For example, it is not uncommon to observe year-to-year variations in rain-fed corn yields ranging above 3.1–4.5 Mg/ha (50–90 bu/ac), and economic N rates associated with those yields to vary by more than 60–84 kg N/ha-year (54–75 lb N/ac-year) (Jaynes et al., 2001; Mamo et al., 2003; Sawyer and Randall, 2008).

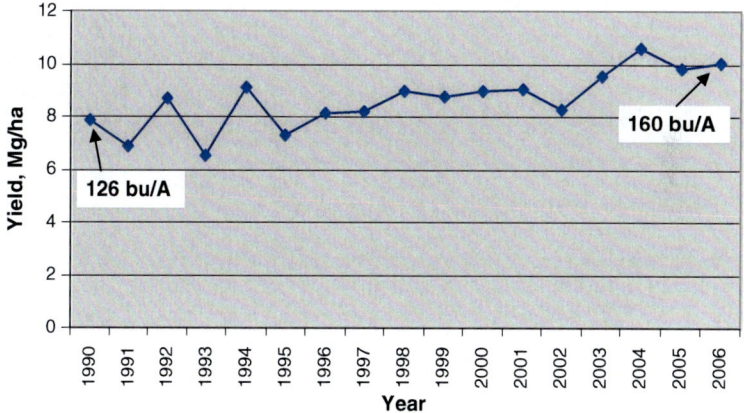

Fig. 4.11 Average corn yields in six leading corn-producing states (IA, IL, IN, MN, NE, and OH), 1990–2006 (Source: USDA National Agricultural Statistics Service)

As discussed in Section 3.2, higher crop yields (Fig. 4.11) have resulted in increased N removal in harvested grain, without increased N fertilization. Greater crop harvest N removal may have helped contribute to slight reductions in net N inputs in the entire MARB since about 2000, particularly in the Ohio and Upper Mississippi River subbasins (see Section 3.2), the two subbasins that also contribute the greatest annual and spring N flux to the NGOM. Increased crop yield trends, improved plant genetic selection, and pest control may also be contributing to the reduced nitrate-N transported to the NGOM since the mid-1990s, and the steady decline in total N delivered to the NGOM since the 1980s (see Section 3.1.1 and Fig. 3.8). Any reductions in N application rates could threaten attainment of high crop yields, which are vital to profitable production and which have contributed in some measure to the reductions in net N inputs and riverine N discharge mentioned above.

Challenges and complexities of determining the EONR in individual fields and farms prevent the ability to make any general conclusions regarding N rate reductions across the MARB that will achieve specific N load reductions to the NGOM.

Because of the complexity and dynamic nature of the N cycle, soil tests for N (nitrate, mineralizable N) have not met with much success in practical field applications (e.g., Scharf et al., 2006a). Some, like the Pre-Sidedress Nitrate Test (PSNT), have resulted in modest successes in N rate adjustments, particularly where there is a long history of manure applications and there has been a buildup of residual soil N (organic and inorganic). A new soil N test (ISNT) developed in Illinois offered promise of more reliably predicting mineralizeable soil N pools (Khan et al., 2001; Mulvaney et al., 2001); however, a recent report indicates the ISNT does not work well elsewhere (Barker et al., 2006a, 2006b; Laboski et al., 2006).

One of the key challenges in managing N in farm fields is to minimize unnecessary N applications in low-yielding years and to provide adequate N in high-yielding years to meet crop demands. Historically, it has been very difficult for even experts to predict residual soil N, recently applied fertilizer N, and mineralized N accessible by plants during a given growing season (e.g., Schlegel et al., 2005; Shehandeh et al., 2005). Furthermore, the inability to accurately predict the amount, intensity, or duration of rainfall in a given year makes it difficult to adjust N rates each year for a specific soil, crop variety/hybrid, tillage system, or cropping system.

4.5.6.3 Watershed-Scale Fertilizer Management

The first watershed-scale study of changing from fall to spring N application involves changes in both rate and timing (Jaynes et al., 2004). The Late Spring Nitrate Test (LSNT) is designed to help farmers add appropriate amounts of N in the spring instead of fall. Use of the LSNT for corn grown within a 400 hectares tile-drained watershed in Iowa resulted in at least a 30% reduction of nitrate-N concentrations in tile drain water. The LSNT involved changing timing, rate, and source of N fertilizer. Another Iowa study concluded that although watershed-scale implementation of LSNT had the potential to reduce nitrate loss through drainage water, it could also increase grower risk, especially when above-normal rainfall occurs shortly after the sidedress N is applied and N is lost to tile drainage or denitrification (Karlen et al., 2005). Development of affordable risk insurance or some other financial incentive by federal, state, or private agencies may be needed to stimulate adoption of the LSNT.

4.5.6.4 Controlled-Release Fertilizers

Controlled- and slow-release N fertilizers (CRN) are fairly commonly used in high-value applications, such as horticultural crop and turf production. Products include urea formaldehyde, isobutylidene diurea, sulfur-coated, and polymer-coated products. Use of CRN fertilizer is limited because of the high cost, with worldwide consumption less than 1% of all fertilizer N products. However, recent advances have brought some CRN products to an economical level for many agricultural crops. Controlled-release N fertilizers have the potential to significantly improve N use efficiency, maintain crop productivity, and minimize the potential for nitrate loss from fields (Blaylock, 2006).

4.5.6.5 Effects of N Management on Soil Resource Sustainability

It is well known that soil organic carbon (SOC) storage in Corn Belt Mollisols has been decreased by long-term cropping. For instance, in an Iowa study to determine the effects of cropping systems on SOC, there was 22–49% lower SOC than native prairie sampled in fencerows for all cropping systems that had been in place for 12–36 years (including continuous corn [CC]; corn soybean rotation [CS]; corn, corn oats, alfalfa; and corn oats alfalfa, alfalfa) (Russell et al., 2005). Current efforts to sequester carbon by restoring SOC and to obtain benefits of fertility and tilth associated with higher SOC in Mollisols should be considered in achieving nutrient load reductions from these crop production systems.

Nutrient management practices need to be assessed for their ability to enhance or maintain SOC content in addition to their impact on profit, yield, and water quality (Jaynes and Karlen, 2005). A careful review of the literature on this subject is warranted because of the potential that fertilizer management to achieve water quality improvements may lead to further soil quality degradation. Jaynes and Karlen (2005), based on Jaynes et al. (2001), find a partial N mass balance for three fertilizer N levels in a corn–soybean rotation on Mollisols in the Des Moines lobe region of Iowa. Tillage consisted of either moldboard or chisel plowing in the fall and use of a field cultivator for seedbed preparation and for weed control several times during the early growing season. The partial N mass balance shows that the 1X and 2X fertilizer N rates have a negative N mass balance, and the 3X rate has a positive mass balance. Although the 2X rate (134 kg N/ha or 120 lb N/ac on corn, no N applied to soybeans) was the economic optimum, the negative N mass balances may indicate a long-term decline in soil fertility. According to the authors, "the lower two N rates were thus effectively mining N from the SOM, which would result in a measurable decrease in SOM and a degradation of the soil resource over the long term." Although all treatments had average nitrate-N concentrations above 10 mg/L nitrate-N, there were large and consistent differences among N loads in drain tile (Table 4.7). The 1X and 2X treatments achieved drain tile nitrate-N load reductions of 39 and 27%, respectively, compared to the 3X fertilizer N rate (201 kg N/ha or 179 lb N/ac).

The N mass balance approach to determining long-term changes in SOC or SOM presents numerous problems. First, there is no mechanism for lower fertilizer N applications to directly stimulate increased SOM mineralization. Any effect on SOC would be due to lower residue, particularly during the corn phase of the rotation and during soil tillage. Second, although a very high-quality study, the partial N mass balances shown are subject to different interpretations if only small errors exist. For instance, the total mass balance residual is less than 5% of the total fluxes measured and is 6–14% of the estimated N fixation. Therefore, small imprecision in estimated or measured values could lead to different interpretations.

A number of studies have made direct measurements of SOC over long-term studies of fertilizer rates. At least six relevant studies (three in IA and one each in KS, MN, NE) have been conducted on Mollisols in the Corn Belt. The general conclusion from these studies is that high fertilizer N rates on continuous corn will

Table 4.7 Partial N balance for 4-year rate study by Jaynes et al. (2001). The last two columns were added here and were not part of original table

Fertilizer rate	N inputs			N outputs						
	Total fertilizer applied kg N/ha	Total wet and dry deposition kg N/ha	Total fixed kg N/ha	Total grain removed kg N/ha	Total drainage loss kg N/ha	Total runoff kg N/ha	Change of residual mineral N kg N/ha	N balance residual %	(Residual/ fixed)*100 %	(Residual/ total flux)*100 %
1×	144	43	395	522	119	0	6	−55	−14	−4.4
2×	289	43	397	590	142	0	13	−26	−6.5	−1.8
3×	414	43	394	606	195	0	−7	47	12	2.8

4.5 Options for Managing Nutrients, Co-benefits, and Consequences 175

lead to SOC increases and that suboptimal N rates lead to SOC depletion. There is no direct evidence for an effect of lower nonzero fertilizer rates near the economic optimum, leading to decreases in SOC from these studies.

Russell et al. (2005) analyzed studies of two Iowa sites (Kanawha and Nashua) for the impact on SOC of four N fertilization rates (0, 90, 180, and 270 kg N/ha-year or 0, 80, 161, and 241 lb N/ac-year) and four cropping systems (continuous corn [CC], corn soybean [CS]; corn–corn–oat–alfalfa [CCOA], and corn–oat–alfalfa–alfalfa [COAA]). One study had been ongoing for 23 years and the other for 48 years at the time of sampling of SOC in 2002. The only difference related to fertilizer rate was for the 23-year experiment (the Nashua site). In this experiment, the 270 kg N/ha-year (241 lb N/ac-year) for CC had higher SOC for only the 0–15 cm (0–5.9 in) depth. There were no differences among the 0, 90, and 180 kg N/ha-year rates for CC at the Nashua site for any depths. There were also no differences for the 0–100 cm (0–39 in) soil for any N rates used for CC, including the highest rate of 270 kg N/ha-year (241 lb N/ac-year). There were no other significant fertilizer N rate effects found in the study (Russell et al., 2005).

An earlier Iowa study that included the Nashua and Kanawha sites and a third site (Sutherland) reached similar conclusions as those of Russell et al. (2005). In that study, Robinson et al. (1996) found that N fertilizer rate on corn (0–180 kg N/ha-year or 0–161 lb N/ac-year) was not significant in determining SOC but only whether fertilization occurred. In both studies (Robinson et al., 1996; Russell et al., 2005), the cropping systems with alfalfa [termed meadow in Robinson et al. (1996)] had the highest SOC. Corn silage treatments and no fertilizer treatments had the lowest SOC (Robinson et al., 1996). A third Iowa study did not compare SOC under different fertilizer rates but did show that high fertilizer N (206 kg N/ha-year or 184 lb N/ac-year) resulted in increases in SOC over 15 years with continuous corn (Karlen et al., 1998a). The general conclusion from the Iowa studies is that, for either CC or CS systems, fertilizer rate has little or no effect in the 90–180 kg N/ha-year (80–161 lb N/ac-year) range. Given that the average N fertilizer application to corn in Iowa was 158 kg N /ha (141 lb N/ac) in 2005 (USDA ERS: http://www.ers.usda.gov/Data/ARMS/app/CropResponse.aspx) and the economic optimum rate ranged between 67 and 172 kg N/ha or about 60 and 154 lb N/ac (approximate mean of 137 kg N/ha or 122 lb N/ac) during 1996 and 1998 in the Iowa study by Jaynes et al. (2001), it seems unlikely that these rates would lead to a depletion of SOC due to a N rate effect. Corn yields with the moderate N rates in the Jaynes et al. (2001) study ranged around 10 Mg/ha (159 bu/ac), and the Iowa state average corn yield in 2005 was about 10.9 Mg/ha (173 bu/ac).

Results from other studies in the Corn Belt are mixed and have found no consistent effect of N rate on SOC. In Kansas, Omay et al. (1997) found no effect of either 224 or 252 kg N/ha (200 or 225 lb N/ac) versus no N for over 10 years of CC or CS. A small significant difference in SON (less than 5% decrease) was found on one soil for the zero N treatment. Increased residue inputs were attributed to N fertilization and inclusion of soybean in the rotation reduced SOC and soil organic N. In contrast, CC receiving 200 kg N/ha-year (179 lb N/ac-year) for 13 years had higher SOC than in the zero N treatment on a Minnesota Mollisol (Clapp et al., 2000). In an 18-year experiment in Nebraska, N rate (0, 90, 180 kg N/ha-year or 0, 80, 161 lb N/ac-year) had an effect on SOC in the 0–7.5 cm (0–2.9 in) soil after 8 years but had no effect after 18 years, presumably due to tillage differences (Varvel, 2006).

Recent work in Nebraska on an irrigated Mollisol compared long-term (initiated in 1999) continuous corn and corn–soybean rotations under recommended and intensive management and found that SOC was increased under recommended and intensive management of CC but not in the CS systems (Adviento-Borbe et al., 2007; Dobermann et al., 2007). These scientists also reported that greenhouse gas (GHG) emissions from agricultural systems can be kept low when management is optimized toward better exploitation of the yield potential. To accomplish SOC increases while keeping GHG emissions low, Dobermann et al. (2007) reported the following required factors: (1) choosing the right combination of adapted varieties or hybrids, planting date, and plant population to maximize crop biomass production; (2) tactical water and N management, including frequent N applications to achieve high N use efficiency and minimized N_2O emissions; and (3) a deep tillage (noninverting) and residue management approach that favors a buildup of SOC as a result of large amounts of crop residues returned to the soil.

If a fertilizer effect on SOC exists, it is more likely to occur under CC than CS because increased fertilizer generally leads to increased corn production. It is logical to assume that increased corn production (including grain, stover, and roots) should lead to increased SOC. In general in the published studies, this relationship does not hold, although applying zero N fertilizer generally leads to less SOC over time than high fertilizer N rates. In summary, although it is beyond the scope of the Study Group to review all the research relevant to changing SOC in Corn Belt soils, it is clear that inclusion of alfalfa in a rotation is very effective at building SOC. The effects of tillage are not clear. Based on the existing literature, there is evidence that changes in fertilizer rates within the range of those optimum for corn production are unlikely to lead to long-term SOC and SON declines. Although it is possible to build SOC under CC with relatively high fertilizer additions [e.g., 201–299 kg N/ha-year or 179–267 lb N/ac-year (Adviento-Borbe et al., 2007; Dobermann et al., 2007) and 206 kg N/ha-year (184 lb N/ac-year) Karlen et al., 1998b], care must be taken to ensure that these fertilizer additions are sustainable economically and that they do not harm water quality. From a global C balance perspective, it is also worth noting that there is a C emissions cost of producing N fertilizer that would need to be taken into account when doing C mass balances for higher fertilizer N rates on corn. However, if high-yield production is achieved, with good N use efficiency, these fertilizer C emissions may be offset (Adviento-Borbe et al., 2007). More research on the net effects of N fertilizer rates on SOC and GHG emissions is needed.

4.5.6.6 Precision Agriculture Management Tools for Nitrogen

Global positioning system (GPS) and geographic information system (GIS) technologies are becoming more widely adopted by farmers and show promise for developing management zones in fields that could target application rates for

4.5 Options for Managing Nutrients, Co-benefits, and Consequences

low- versus high-yielding areas (Schlegel et al., 2005) and reduce N applications in areas of the field most prone to N losses (Chua et al., 2003). Field-transect apparent electrical conductivity (ECa) or electromagnetic induction measurements can help define management zones, based on surrogate detection of soil texture differences (Davis et al., 1997; Kitchen et al., 1999). Reductions in N application rates for corn range from 6 to 46% when using site-specific management zone approaches as opposed to a uniform rate of N application (Koch et al., 2004). Dividing fields into a few management zones might reduce N loss, but because of within-field variability, more spatially intensive N management might provide greater economic and environmental benefits (Hong et al., 2006; Scharf et al., 2005).

Basing N applications on past yields has not proven to be an effective approach to variable-rate fertilization of N (Murdock et al., 2002; Scharf et al., 2006b). In-season crop N-sensing research (chlorophyll meter, remotely sensed multispectral color images, on-the-go and handheld optical reflectance sensors) (Scharf et al., 2006a), using reference "N-rich" or calibration strips or plots in targeted areas within fields (Raun et al., 2005), has shown the potential benefits of these newer technologies in providing in-season guidance to farmers and crop advisers for improved N nutrition management. This "N-rich" calibration approach appears to have been more successful with winter wheat than for corn, to date. Chlorophyll meters and remotely sensed crop reflectance have been used as an index for plant N status, and N-fertilizer use efficiency improved when these techniques were used (Osborne et al., 2002; Varvel et al., 1997). Crop N-sensing technologies present opportunities to reduce and better time fertilizer N applications; however, there have been few direct assessments of impacts of these approaches on residual soil N and nitrate losses. Further verification of the performance of these techniques is needed in order for implementation by farmers to be more widespread.

When technology costs are considered, economic returns to farmers are often inadequate to justify adoption of variable-rate N management. Frequently, the costs of spatial N management technologies exceed the cost of the fertilizer N saved, which are dependent on fertilizer prices. As a consequence, adoption of these technologies has proceeded at a slower rate than anticipated, partly because of high technology and equipment costs and spatially variable economic returns. Economics research suggests a number of reasons for this low slow adoption, including high fixed costs of adoption and uncertainty in returns. These factors suggest that incentives to encourage adoption may need to cover option values and that revenue insurance programs to address the risk may be appropriate instruments (Khanna, 2001; Khanna et al., 2000; Isik and Khanna, 2002, 2003).

Incentives have been used in Missouri in cost-sharing some of the expenses of precision technologies within the USDA EQIP program (Agronomy Technical Note MO-35, September 2006). Cost share in this Missouri USDA-NRCS Code 590 nutrient management program provides a farmer $49/ha ($20/ac) per year for a 3-year contract, with the full $148/ha ($60/ac) provided at the end of the first year. Farmers in this Missouri EQIP precision N-sensing program are advised to follow guidance for N-sensing interpretation based on work by Scharf et al. (2006a, 2006b).

4.5.6.7 Precision Agriculture Management Tools for Phosphorus

Spatial variability in soil test phosphorus (P) levels can be large, with levels often ranging from very low to very high (agronomic interpretation) in the same field (Bermudez and Mallarino, 2007; McGraw and Hemb, 1995; Reetz et al., 2001; Wittry and Mallarino, 2004). This variability can also be large in fertilized, manured, and grazed pastures (Mallarino and Schepers, 2005; Snyder and Leep, 2007). With the advent of commercially available GIS and GPS technologies in the early 1990s, crop advisers and farmers began to more precisely define the spatial variability of soil fertility levels, including soil test P (Fig. 4.12). In recent years, zone or grid (e.g., 0.25–1 hectare or 6–2.5 acres) sampling has been used to better define management units to receive different P application rates (Reetz et al., 2001), as opposed to the formerly recommended practice of whole-field composite sampling (e.g., Thom and Sabbe, 1994). In spite of considerable research effort, no widely accepted standard for soil sampling fields for precision or site-specific management has been established (Mallarino and Schepers, 2005), because soils are naturally heterogeneous and their spatial variability occurs at many scales. Recent soil sampling summary results for more than 3.3 million soil samples in North America from both public and private soil testing laboratories also showed wide variability in soil test P levels within and among states in the United States (PPI/PPIC/FAR, 2005). Snyder (2006) summarized the soil test results for the 20 major MARB states (over 2.1 million samples) and reported (1) 40% of the states have experienced a decline in soil

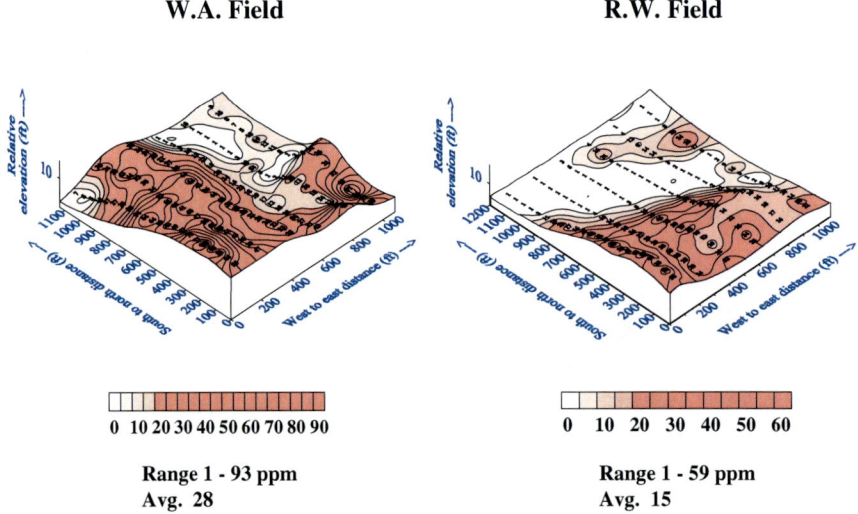

Fig. 4.12 Variability in soil test P levels in typical farmer fields in Minnesota (2007 personal communication with Dr. Gary Malzer, University of Minnesota)

4.5 Options for Managing Nutrients, Co-benefits, and Consequences

test P since 2001, and (2) 78% of the samples tested below 50 mg/kg (ppm) Bray 1 equivalent-extractable P and 94% tested 100 ppm or below. In fact, crop harvest removal of P exceeds fertilizer plus recoverable manure P in 11 of the 20 states (PPI/PPIC/FAR, 2002). These data are in agreement with the trends in net anthropic P input in the MARB, discussed in Section 3.2 of this book.

Early season detection of corn P deficiency may be possible with remote sensing, but detection of deficiencies later in the season, which correlate better with crop yield, has not been successful (Osborne et al., 2002). At this time, remote sensing or on-the-go sensing of plant P status does not appear to be as commercially viable as plant N sensing.

Variable-rate fertilization can result in better P fertilizer management. For example, Bermudez and Mallarino (2007) found that variable-rate technology applied 12–41% less fertilizer and reduced soil test variability on farmers' fields in Iowa, compared with the traditional uniform rate fertilization method. Perhaps one of the most important aspects of intensive soil sampling and variable-rate P application technologies is the capability to apply P fertilizer where it is needed while minimizing or reducing P applications in field areas which have elevated soil test P. In Iowa, variable-rate P application helped decrease soil test P in field areas with high soil test P, when applying manure (Fig. 4.13) or fertilizer (Fig. 4.14). As of yet, however, variable-rate or precision P fertilization has been shown to have little economic benefit in the major corn and soybean producing states compared to uniform applications (Lambert et al., 2006; Mallarino and Schepers, 2005). Further, there are ongoing efforts to update soil test P crop response calibrations and fertilizer recommendations to optimize P fertilization (Beegle, 2005).

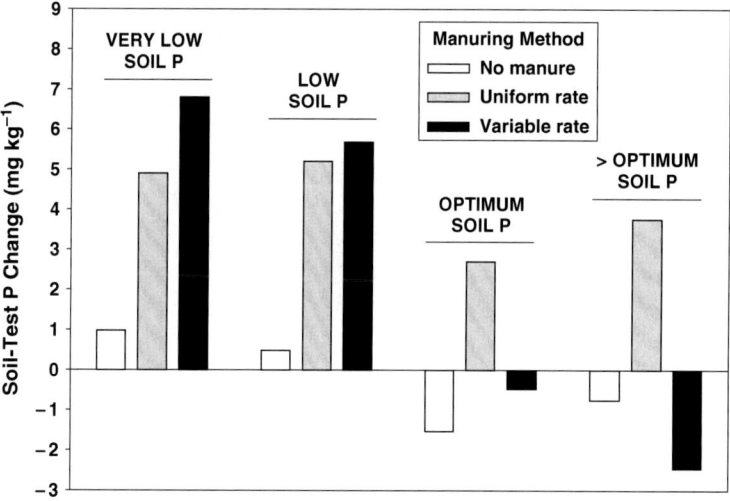

Fig. 4.13 Effect of variable-rate versus uniform-rate application of liquid swine manure on changes in soil test phosphorus in Iowa fields [2007 personal communication with Dr. Antonio Mallarino, Iowa State University and Wittry and Mallarino (2002)]

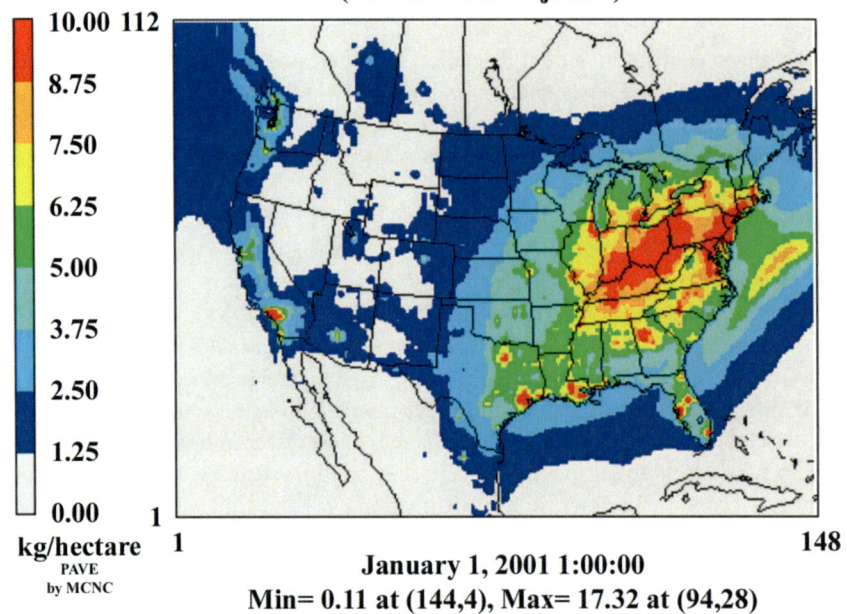

Fig. 4.14 Effect of variable-rate versus uniform-rate application of fertilizer P on soil test P in multiple Iowa fields across multiple years

Numerous studies have shown a strong relationship between soil test P levels and the concentration of dissolved P in runoff (Andraski and Bundy, 2003; Pote et al., 1999; Sharpley et al., 2006a, 2006ba) and tile drainage (Heckrath et al., 1995). Recent work by Gentry et al. (2007) showed that tile drainage P losses in Illinois can exceed 1 kg P/ha-year (0.9 lb P/ac-year), with much of the loss occurring during a few peak storm events in the spring. However, annual manure or fertilizer P applications can control the concentration of total and dissolved P in surface runoff (Pierson et al., 2001; Sharpley et al., 2001).

Soil test P thresholds alone cannot define the potential or risk of P losses from agricultural fields. Slope, hydrologic characteristics, tillage, P rate, and time after P application before a runoff producing rainfall, and other factors also affect the risk of P loss (Sharpley et al., 2006a). To address all factors influencing P loss from agricultural fields, an environmental risk assessment tool (the P Index) was proposed by Lemunyon and Gilbert (1993), which has been regionally modified and adopted by 49 of 50 states in the United States to identify and delineate the risk for agricultural P loss for use in the development of Comprehensive Nutrient Management Plans

(Sharpley et al., 2003). Use of P Indices has also been encouraged by industry, in recognition of the spatial variability in soil test P levels within fields, and the spatial variation in source and transport factors (Snyder et al., 1999). "Variable rate P application can be practically implemented on the basis of P index ratings for field zones, not just based on soil test P" (Wortmann et al., 2005). Variable-rate fertilizer P application is becoming more common in Nebraska, Iowa, Missouri, Kansas, and other states, and some custom applicators are beginning to apply manure at variable rates.

4.5.6.8 Nutrient Management Planning Strategies

A survey of 127 farms (90% of all farms) in two northeastern Wisconsin watersheds offers some insight into how successful nutrient management has been in reducing nutrient applications (Shepard, 2005). Farmers with a nutrient management plan (53% of farms) applied less N and P (139 kg N/ha and 31 kg P/ha or 124 lb N/ac and 28 lb P/ac) than farms without a plan (188 kg N/ha and 44 kg P/ha or 168 lb N/ac and 39 lb P/ac), but only half the farmers credited on-farm manure N, and only 75% fully implemented their plans on most of their acres.

For nutrient management planning to decrease nutrient loss, technical and financial assistance programs need to focus on plan implementation and maintenance in the MARB rather than on targeting the number of plans written in a given period. Despite programs subsidizing plan writing, a critical limitation is the lack of certified plan writers to meet the demand and deadlines. Further, there needs to be an effective mechanism to ensure plan adoption and regular updating of plans. Efforts are underway in the Heartland states of the MARB (IA, KS, MO, NE) to develop nutrient management plan assessment protocols. This aims to identify key factors that limit plan implementation so that practical solutions can be developed. One option is preparation of a simplified plan that farmers can quickly refer to. Also, documenting nutrient management plan implementation is being rewarded with financial credits in New York drinking water supply watersheds (Watershed Agricultural Council, 2004). These credits can be used to purchase or upgrade equipment that would need to be used to implement the plan, such as manure spreaders and injectors.

An assessment is needed of the socioeconomic barriers to successful adoption of nutrient management planning strategies in the MARB as well as the N and P loss reductions achievable. Such an assessment has been done in a drinking water supply watershed for New York City that claims a 93% participation in volunteer conservation programs (Watershed Agricultural Council, 2004). A survey of CREP participants showed that they were generally older and more likely to obtain information from extension agents, consultants, and watershed council personnel than nonparticipants, but there was no difference in educational level or farming status (full or part time) (James, 2005). Overall, negative attitudes toward voluntary adoption of BMPs were a result of the loss of productive land and loss of being able to decide independently what to do on their own land. These survey results illustrate the difficulties in gaining adoption of nutrient management BMPs by farmers in any watershed, transferring new BMP technology, and addressing the socioeconomic pressures faced.

> **Key Findings and Recommendations**
>
> Reductions in N losses and residual soil NO_3-N are possible with attention to improved infield N management. It may be possible to reduce N rates and alter N timing in some portions of the MARB. Such rate reductions may be accomplished through implementation of refined management, but they must be economical for farmers and care must be taken to protect soil resource sustainability. Crop N sensing and variable-rate N management implementation, using management zone approaches may prove useful in attainment of economic optimum N rates in individual fields, which may also help reduce N losses. Higher fertilizer, fuel, and machinery costs have stimulated increased interests in some newer N management technologies, as well as other means to improve fertilizer N effectiveness and efficiency; however, use of site-specific or precision technologies has not yet proven financially rewarding to many farmers, due to the high cost of sampling, ground-truthing, and application technology. Based on these findings, the Study Group offers the following recommendations.
>
> - Because of the importance of both N and P to Gulf hypoxia and as various cropping systems can have different positive and negative effects on N and P export reduction, remedial strategies must be directed at system-wide nutrient management rather than either N or P applications alone. Future research to evaluate the effects of different nutrient management impacts on crop production should include measures of water and air quality effects.
> - There is a lack of consistent year-to-year USDA nutrient management survey data, which hinders any broad nutrient use and management evaluation and interpretations. These data will become more important in monitoring and understanding changes in nutrient management practices as biofuel markets expand. Consistent year-to-year data collection on nutrient management of major crops and emerging energy crops is recommended.
> - Cost-share incentives like the USDA payment support for crop N-sensing and precision N management in Missouri, intensive educational programs (e.g., on-farm demonstrations), and/or other means should be explored to encourage the agricultural community to improve nutrient use efficiency and effectiveness with all nutrient sources (i.e., fertilizer, manure, biosolids, composts, by-products, etc.). Such programs may be especially helpful in corn systems in the upper Mississippi and Ohio River

subbasins, which have been identified as major contributors of spring nitrate-N flux to the NGOM.
- Although the economic and water quality impacts of controlled-release fertilizers in commercial field crop systems have not been fully proven, their beneficial use should be explored through additional research and demonstrations at field and watershed scales. Programs to stimulate greater adoption of locally proven technologies like urease and nitrification inhibitors (and controlled-release fertilizers, once proven economically and environmentally effective) to enhance crop nitrogen recovery and use efficiency should be considered as the shift toward greater urea and urea–ammonium nitrate N use continues.
- Watershed-scale evaluations of split applications of N in the spring for corn should be conducted to determine watershed-scale benefits of this N management approach compared to the more traditional application of anhydrous ammonia in the fall, especially in the upper Mississippi and Ohio River subbasins.
- More research on the net effects of N fertilizer rates on soil organic carbon (SOC) and greenhouse gas (GHG) emissions is needed.
- Crop and animal production systems are essential to the economic viability of agriculture in the MARB. Thus, an infrastructural assessment of how animal production can co-exist with grain and forage production is needed. Long-term strategies should be explored whereby more effective crop and animal production systems remedy or avoid excessive N and P loading to water and air resources.
- Cost–benefit ratios vary among farmers with, for example, labor availability, farm organization, and financial situation. However, past experience shows that adoption of conservation practices is not solely dependent on cost-effectiveness. Thus, there needs to be consideration of the socioeconomic barriers to, and impacts of, adoptions of nutrient management planning strategies in the MARB. New approaches should be investigated to overcome socioeconomic barriers, including incentive programs.

4.5.7 Effective Actions for Other Nonpoint Sources

4.5.7.1 Atmospheric Deposition

This section reviews actions for reducing NOx emissions that contribute to atmospheric deposition of nitrogen. Atmospheric deposition of oxidized nitrogen compounds released during fossil fuel combustion contributes an estimated 30% of the entire inputs of new nitrogen for the United States as a whole, (Howarth et al.,

2002). As discussed earlier in Section 3.2, atmosphere deposition of oxidized nitrogen is less important in the MARB but still accounts for an estimated 8% of nitrogen contributions to the upper MARB and 16% of the nitrogen inputs to the Ohio River basin. NOx emissions to the atmosphere in the United States could be virtually eliminated at reasonable cost using currently available technologies (Moomaw, 2002; Howarth et al., 2005). In addition to potential benefits concerning Gulf hypoxia, reducing NOx emissions in the MARB can contribute to improved local air and water quality and can reduce atmospheric transport of nitrogen to the northeastern states, where atmospheric deposition is an even more significant problem.

In addition to deposition of oxidized nitrogen, there is significant deposition of ammonia and ammonium (NHx) in some regions of the MARB. These are not considered in the mass balance approach for nitrogen in Section 3.2 because the NHx originates largely from volatilization from animal wastes and other agricultural sources and so does not represent new nitrogen inputs to the basin, but rather a recycling of nitrogen within the basin (Howarth et al., 1996). Nonetheless, high rates of volatilization followed by conversion to ammonium nitrate or sulfate can lead to significant long-distance transport and contribute to reactive N distribution in other sensitive areas. Furthermore, high rates of NHx deposition in the basin can result in increased leakage of nitrogen to downstream aquatic ecosystems. In Iowa, Minnesota, and Wisconsin, NHx deposition exceeds NOy deposition, and averages over 7.5 kg N/ha-year (6.7 lb N/ac-year) in Iowa (results from CMAQ model, Robin Dennis, NOAA, unpubl.)

Mobile sources account for approximately 55% of NOx emissions to the atmosphere on a national level (Melillo and Cowling, 2002). While automobiles have been subject to fairly strict NOx standards in recent years, emissions from light trucks have not historically been as strict. Tightening regulations on light trucks represents an opportunity for significant reduction in NOx emissions, as approximately half of new vehicle sales in recent years have been light duty trucks (Moomaw, 2002). Heavy diesel trucks, buses, and trains have accounted for a growing fraction of NOx emissions because of strict NOx standards on automobiles and the absence of similarly strict controls on heavy diesel vehicles.

Stationary sources account for approximately 45% of NOx emissions, with electric generating facilities accounting for roughly half of all stationary source emissions, and industrial fuel combustion account for slightly less than one-third. The remainder of stationary source NOx emissions is from nonfuel industrial processes (12%) and from commercial, institutional, or residential fuel combustion (8%) (USEPA, 2006a).

Stringent new source performance standards have greatly reduced emissions from new electric generating facilities. Low-emission, combined-cycle gas turbines account for most new electric generating capacity in recent years (Bradley and Jones, 2002). Unfortunately, some existing policies provide incentives that discourage more widespread adoption of new, cleaner technologies. For example, under the Clean Air Act, high NOx emissions by older, coal-fired power plants are "grandfathered," and therefore not subject to the stringent emission standards of new generating capacity. As a consequence, electric utilities have the incentive to keep older coal plants running far beyond what would otherwise be their economic lifespan (e.g., Ackerman et al., 1999; Maloney and Brady, 1988; Nelson et al., 1993). As a result, while 90% of new electric generating capacity is produced with gas turbines, coal still produces 55% of the electricity in the United States (Moomaw, 2002). And it was estimated that, in 1998, coal-fired power plants were responsible for nearly 90% NOx emissions from electric power generation (USEPA, 2000b, 2006a). About a quarter of the coal-fired electric generating capacity in 1996 was constructed prior to 1965, and almost one-half was constructed prior to 1975 (Ackerman et al., 1999).

Considerable reductions in NOx emissions can be achieved with existing commercial technologies by replacing outdated coal-fired capacity with modern gas-fired combined-cycle power plants (Howarth et al., 2005). Existing coal plants can also be retrofitted with new control technologies, such as low-NOx burners (Ackerman et al., 1999; Bradley and Jones, 2002). Other promising technologies for reduction emissions from coal-fired power plants include fluidized-bed boilers (Cogeneration Technologies, 2006) and gasified coal combined-cycle power plants (U.S. DOE, 2006).

For the most part, NOx emissions in the United States are regulated because of concerns over formation of smog and ozone and seldom because of water quality concerns (Melillo and Cowling, 2002; Moomaw, 2002). Since smog and ozone pollution occur mostly in summer months, regulation of NOx emissions from stationary sources has often focused on summertime only regulation (Howarth et al., 2005). Since the largest cost of controlling NOx from power plants is the capital cost of building scubber systems, the additional cost of requiring year-round NOx control from power plants is small compared to that for summertime only controls. Thus, year-round operation of existing control technologies represents a cost-effective approach for reducing NOx emissions. Some local and state governments, such as New York State, have recently moved toward year-round regulation of NOx because of concern over coastal nitrogen pollution (Ron Entringer, NY State DEC, personnel communication).

4.5.7.2 Residential and Urban Sources

Urban and suburban runoff comes from a variety of sources, including impervious surfaces like roads, rooftops, and parking lots, as well as pervious surfaces like lawns. Urban and suburban runoff can be important sources of pollutants, especially for local water quality effects. For example, the *National Water Quality Inventory: 2000 Report to Congress* concluded that urban runoff is a major source of water quality impairment in surface waters (USEPA, 2002). A variety of actions can be used to control nonpoint urban sources, including both structural and nonstructural practices (e.g., USEPA, 2005).

Although controlling urban nonpoint sources can provide significant benefits from improvements to local water quality, these nonpoint sources are not significant determinants of hypoxia in the Gulf of Mexico, both because concentrations tend to be lower than those from agricultural sources and because the urban land comprises less than 1% of the Mississippi River basin (e.g., Mitsch et al., 1999). Thus, although actions to reduce urban nonpoint sources may be justified, these control actions will not likely contribute significantly to reductions in the size of the Gulf of Mexico hypoxic zone. Since control of urban nonpoint sources will not have an important role in reducing hypoxia, we do not focus on actions to reduce urban nonpoint sources of nutrients in this book.

> **Key Findings and Recommendations**
>
> Atmospheric deposition is a small but significant (8% in Upper Mississippi and 16% in Ohio River subbasins) contribution to N inputs in the Mississippi River basin. Opportunities exist to lower NO_x emissions in a number of ways, but it is not likely that hypoxia will drive most of these regulatory decisions. Rather, hypoxia reduction and other water quality benefits should be incorporated in a number of regulatory decisions regarding air pollution. Based on these findings, the Study Group offers the following recommendations.
>
> - Water quality benefits and effects on hypoxia should be incorporated into decisions involving retirement or retrofitting of old coal-fired power plants; NO_x controls, such as the extension of the current summertime NO_x standards to a year-round requirement; and emissions standards; and mileage requirements for sport utility vehicles, heavy trucks, and buses.

4.5.8 Most Effective Actions for Industrial and Municipal Sources

Sewage treatment plants and industrial dischargers represent a more significant source of N and P in the MARB than was originally identified in the *Integrated Assessment*. Although most point sources in the MARB do not have permits that require removal of N or P from discharged effluent, as local water quality standards for these nutrients have not yet been developed, states are charged with developing water quality criteria for achieving and maintaining designated beneficial uses of surface waters, including those waters that receive sewage treatment plant effluent. However, the process by which these criteria are translated into quantitative and enforceable nutrient limits from regulated point sources remains unclear.

Based on data from the recent MART (2006b) report, the Study Group has estimated that permitted point source discharges represented approximately 22 and 34% of the average annual total N and total P flux to the Gulf, respectively, for the 2001–2005 water years (for a detailed discussion see Appendix D). These point sources represent a significant opportunity to reduce N and P loadings that should be fully evaluated in the context of other potential management changes in the MARB.

Encouraging behavioral changes of nondomestic sewer users as well as increasing capital investments in sewage treatment and industrial treatment plant upgrades have proven to be effective approaches to managing nutrient discharges in other areas of the United States (USEPA, 2004b, 2003a; Chesapeake Bay Commission, 2004). The use of Biological Nutrient Removal and Enhanced Nutrient Removal technologies for N and P removal is being implemented to reduce N and P concentrations in sewage treatment plant effluent discharge by 50–80% (Maryland Department of Environment, 2005; USEPA, 2004b). Sewage treatment plant upgrades designed to remove phosphorus typically include enhanced chemical precipitation applied alone or in combination with biological phosphorus treatment and membrane filtration. These types of sewage treatment plant unit operations, which can achieve effluent discharge phosphorus concentrations as low as 0.1 mg/L total phosphorus or less, now constitute the BMP for phosphorus removal at sewage treatment plants. Removing P to a 0.1 mg P/L limit is most commonly implemented where there is a market for water recycling, such as in communities located in the desert Southwest, and the increased cost can be justified. In locations where there is no market for recycled water, higher limits for P (for example, 0.3 or < 1.0 mg P/L) will be more cost-effective.

The Study Group presents an example calculation to demonstrate the magnitude of reduction possible in riverine total N and P fluxes to the NGOM if technology for N and P removal from sewage effluent were implemented for large sewage treatment plants (0.5 million gallons per day and above) across the MARB. Based on the Study Group's adjustment to the MART report's estimates of N and P effluent from sewage treatment plants (MART, 2006b), the Study Group has calculated that upgrades for large sewage treatment plants in the MARB to achieve total N concentration limits of 3 mg/L could create reductions in N flux from sewage treatment plants from 192,000 metric tonne N/year (212,000 ton N/year) to 70,000 metric tonne N/year (77,000 ton N/year), about a 64% reduction in annual N flux from sewage treatment plants. This translates into a reduction of total annual N flux to the Gulf by about 10% and the total spring N flux by about 6%. Upgrading to achieve P concentrations of 0.3 mg/L would create reductions in P fluxes from sewage treatment plants from 41,000 metric tonne P/year (45,000 ton P/year) to 10,500 metric tonne P/year (11,600 ton P/year) or about a 75% reduction in annual flux from sewage treatment plants to the MARB. These reductions, in turn, would translate into a decrease in the total annual P flux to the Gulf by about 20% and the total spring P flux by about 15%. It is important to recognize that these estimates assume that the changes in biosolids quality and production rates resulting from the capital improvements to the sewage treatment plant do not adversely impact nutrient management procedures implemented at biosolids land application sites.

In the Chesapeake Bay watershed, nutrient reductions from sewage treatment plant upgrades were determined to be as cost-effective as, and more predictable than, the estimated reductions achieved through implementation of agricultural nonpoint source BMPs. The Chesapeake Bay Commission (2004) found average point source costs to remove N and P to be within the range of most widely implemented agricultural BMPs (USEPA, 2003b). The Commission stated that "this technology-based approach provides the highest degree of confidence for consistent, long-term reductions. Furthermore, the cost of this technology has continued to decline in recent years."

However, there are many differences in point source distribution, population, and income in various subbasins of the MARB compared to other areas of the country where point sources have had total N and P reductions (such as the Chesapeake Bay or Long Island Sound). Therefore, a cost-effectiveness analysis of point source controls of N and P in the MARB is needed to fully evaluate this particular method of reducing nutrient inputs to rivers in the context of nonpoint source control costs. A part of that analysis should consider the cost of N and P removal that could be optimized by establishing loading caps for individual treatment plants and/or groups of plants within river basins and by allowing nutrient credit trades between the plants. This "point-to-point" trading allows those plants that can most efficiently achieve reductions to sell nutrient reduction credits to plants that would incur much higher costs to achieve their loading cap. This approach is being used in Long Island Sound and in the Chesapeake Bay watershed within Virginia. These point-to-point trading programs are consistent with an overall cap and trade program as discussed in Section 4.4.3.

Another potential approach for reducing the nutrient discharge from sewage treatment plants, which could be applied alone or in combination with plant upgrades, is to encourage local sewer districts to establish more stringent nutrient pretreatment standards for private industries and other nondomestic sewer users. Meat packing, chemical manufacturing, and food processing are examples of the types of industries that generate wastewater containing large amounts of N and P. Through the regulatory authority granted to them under the National Pollutant Discharge Elimination System (NPDES) program, sewer districts can encourage industries to reduce their nutrient discharge to sewage treatment plants through the establishment of local sewer discharge nutrient limits as well as by the judicious development of technology-based wastewater surcharge rates.

The overall decrease in the mass of nutrients discharged into the local sewer system due to pretreatment will improve the quality of both the sewage treatment plant effluent and biosolids and will result in a net reduction of nutrients entering the MARB. A feasibility study is needed to evaluate the regulatory and economic options that could be applied to provide incentives for major industries to identify and implement pollution prevention measures to reduce and/or recycle nutrients that would otherwise be discharged into the local sewer system.

In addition, industrial treatment plant upgrades designed to remove nutrients can also reduce nutrients that are directly discharged to the MARB and the Gulf. Industrial discharges account for about 28% of the point source N flux and 23% of the point source P fluxes, or 75,000 metric tonne N/year (83,000 ton N/year) and 17,000 metric tonne P/year (18,700 ton P/year). Experience in other regions has shown that industrial sources could be targeted on a permit-by-permit basis since frequently a limited number of permitted facilities are responsible for a large part of the load. This approach could be recommended for the MARB. It would be useful to design initial efforts to focus on discharge categories likely to have high nutrient discharges. Examination of discharge information (Table 3.2, MART, 2006b) reveals that two categories (industrial organic chemicals and plastic materials/synthetic resins) account for about half of industrial N discharges, about 45,000 metric tonne N/year (50,000 ton N/year). For P, four categories (crude petroleum and natural gas, electrical services, refuse systems, and wet corn milling) account for about 40% of the industrial load or about 5,500 metric tonne P/year (6,000 ton P/year). Industries in these categories should be evaluated for opportunities to reduce N and P discharges through pollution prevention, process modification, or treatment.

While P removal is technologically feasible and widely implemented elsewhere, advanced treatment increases the amount of biosolids generated and, therefore, the land area needed to manage a given amount of biosolids based on P and N needs of the crop, rather than just the N requirements. This will create additional costs for biosolids-management programs in the MARB and needs to be considered when evaluating the total cost of implementing P removal at sewage and industrial treatment plants in the basin.

Unlike nitrogen, which can be biochemically transformed and removed from the sewage treatment plant as a volatile gas (N_2 and/or N_2O) through the nitrification/denitrification process, phosphorus is simply moved from the liquid to solid phases and accumulates in the biosolids. Physical upgrades in sewage treatment plants specifically aimed at reducing the phosphorus concentration in the effluent discharge typically include substantial additions of precipitating chemicals (e.g., alum) alone, or in combination with, higher efficiency membrane filtration. The net effect of these capital improvements is a significant increase in the mass of biosolids requiring handling and management. Most biosolids are beneficially used in crop production on land located as near to the treatment facility as feasible to minimize transportation costs. Transportation distances range from essentially zero to several 100 km depending on plant location, size, and the amount of biosolids or biosolid nutrient content. Phosphorus removal will increase both the mass of biosolids and the P content of the biosolids.

Biosolids application to agricultural land is regulated through the NPDES permit of the treatment facility. In many places in the MARB, land application of biosolids is based on the N needs of the crop. As with animal manures, biosolids application to meet crop N needs results in overapplication of P and buildup of bioavailable P in the soil surface. Research during the past two decades has indicated that soil P levels substantially in excess of crop needs can cause elevated P concentrations in runoff, particularly from critical source areas within fields. As a result, recommendations for application of organic nutrient sources, such as manure or biosolids, suggest that applications be limited based on P where the risk of loss is moderate to high. This will minimize the opportunity for P removed from discharged effluent to be lost in runoff when biosolids are land applied. All states now have a tool to estimate the potential for P loss from application of manure or biosolids. Nearly all states use a locally adapted version of the Phosphorus Site Index (PSI) to estimate P loss risk. Since biosolids currently contain more P relative to N than crops require, land application of biosolids should routinely involve an evaluation of the risk of P loss using the PSI or another risk assessment tool.

> **Key Findings and Recommendations**
>
> Sewage treatment plants and industrial dischargers represent a more significant source of N and P in the MARB than was originally identified in the *Integrated Assessment*. Tightening effluent limits on large sewage treatment plants together with establishing more stringent pretreatment nutrient standards on nondomestic sewer users may offer some of the most certain short-term and cost-effective opportunities for substantial nutrient reductions, particularly for P, but a full analysis of costs needs to be conducted in the context of nonpoint source reduction costs. Based on these findings, the Study Group offers the following recommendations.
>
> - Tighter limits on N and P effluent discharge concentrations for major sewage treatment plants, together with concomitant reductions in nutrient discharges from nondomestic sewer users, should be considered, following an analysis of the cost and technical feasibility for a particular basin.
> - A review of discharge data, including N and P loads, for industrial dischargers could identify possible industrial facilities to target for cost-effective reductions.
> - Regulatory authorities should encourage or require sewage treatment plants to utilize phosphorus-based biosolids land application rates rather than the nitrogen-based rates in beneficial-use programs.

4.5.9 Ethanol and Water Quality in the MARB

The production of renewable fuels has been of interest since the 1973 oil price shocks, and technologies for the conversion of crops into ethanol and bio-diesel have existed since the 1940s. Currently about 99% of renewable transportation fuel produced domestically is ethanol from grains and oil crops, primarily corn (Institute for Agricultural and Trade Policy [IATP] 2006). This section focuses on the potential water quality implications of both ethanol production from corn and its potential production from lignocellulosic feedstocks.

The rapid growth in corn prices is primarily a result of increased energy prices (Kline et al., 2009). Increased ethanol production is only a minor contributor even though it is projected to rise from less to 2 billion gallons in 2001 to more than 19 billions gallons in 2009, a 950% increase (IATP, 2006). Current estimates are that about 75% of that production will be in the nine Upper Mississippi River Corn Belt states (IATP, 2006). The Food and Agricultural Policy Institute (FAPRI) projects that ethanol production from corn will increase from about 6.8 billion gallons in 2007 to over 14 billion gallons by 2012. Associated with this increase in ethanol production, FAPRI projects an increase in corn acreage from about 80 million acres to about 94 million acres in the same time period (www.fapri.missouri.edu). This growth of grain-based ethanol production may have major water quality implications for the MARB and the country.

Cellulosic ethanol is an alternative fuel made from a variety of nonfood feedstocks (such as agricultural residuals like corn stover and cereal straws, industrial plant byproducts like saw dust and paper pulp, and crops grown specifically for fuel production like switchgrass, *Panicum virgatum*). By using a variety of regional feedstocks for refining cellulosic ethanol, the fuel can be produced in nearly every region of the country. Though it requires a more complex refining process, cellulosic ethanol produces less impacts on water quality, contains more net energy, and results in lower greenhouse emissions than traditional corn-based ethanol (Mclaughlin and Walsh, 1998). One of the challenges for wider use of cellulosic ethanol is that the cost of production is higher than current prices for corn ethanol and gasoline. Another challenge is that technology has not yet developed the fermentation efficiency for conversion of cellulosic feedstocks to the level at which it is commercially viable. Contributing to the high cost is the need to consolidate enough feedstock close to the plant to produce an adequate supply as well as the cost of transporting the heavy and bulky feedstock (Perlack and Turhollow, 2003).

Many hope that the heightened interest in biofuels will lead to a more sustainable mode of energy production by reducing impacts on water quality, recycling biomass residuals and emitting little, if any, greenhouse gases. The vision is that future biorefineries will use tailored perennial plants in increasing amounts (Perlack et al., 2005). Integration of agroenergy plant resources and biorefinery technologies can lead to a new manufacturing paradigm (Ragauskas et al., 2006). While these possibilities exist, much is unknown concerning how this future might develop and whether it is economically and technically viable.

4.5.9.1 Water Quality Implications of Projected Grain-Based Ethanol Production Levels

The Study Group could find no published estimates of the likely impact of the consequences of expanded corn-based ethanol production on nutrient flows from the MARB. To characterize the short-term potential impact, a set of simple calculations is reported in Table 4.8 that combine acreage projections from the FAPRI baseline for CRP and three major field crops in the Unites States with estimates of the per acre nutrient losses from these crops (CEAP, 2007). The second and third columns in the table report the projected nationwide acreage for the years 2007 and 2013 for corn, soybeans, wheat, and CRP and the fourth column reports the projected change in acreage for each. As can be seen, the FAPRI baseline projects a sizable increase in corn acreage, with that increase coming largely from soybeans and the CRP (totals do not add up since other cropland is omitted).

The fifth column estimates per acre N loss for corn, soybeans, and winter wheat based on the sum of waterborne losses reported in the CEAP assessment (http://www.nrcs.usda.gov/technical/nri/ceap/croplandreport/table 36, page 117) for the Upper Midwest region. The CEAP report did not estimate N loss from CRP, but for the current analysis, losses from CRP are assumed to be 10% of the average loss

Table 4.8 Estimated changes in N losses from cropping changes predicted by FAPRI from 2007 to 2013

	2007 FAPRI baseline (million acres)	2013 Acreage projections, FAPRI[a] (million acres)	Projected change in acreage[b] (million acres)	N Loss estimate per acre[c] (lbs./acre)	Difference in total N losses – million lbs[d]
Corn	78.3	93.7	15.4	28.1	431.6
Soybeans	75.5	67.9	−7.6	17.7	−134.2
Wheat	57.3	58.3	0.9	12.9	11.7
CRP	36.0	30.0	−6.0	2	−12
Total	247.2	249.9			297

[a]These projections are from the August, 2007 baseline http://www.fapri.missouri.edu/outreach/publications/2007/FAPRI_MU_Report_28_07.pdf
[b]This column is the difference between columns 1 and 2.
[c]Per acre estimates of N loss for corn, soybeans, and winter wheat are the sum of waterborne losses reported in the CEAP assessment (http://www.nrcs.usda.gov/technical/nri/ceap/croplandreport/ table 36, page 117) for the Upper Midwest region. The CEAP report did not estimate N loss from CRP, but for the current analysis, losses from CRP are assumed to be 10% of the average loss from cropland. The CEAP N loss rates are based on simulations using the Erosion Productivity Impact Calculator (EPIC) model. The CEAP estimates tend to overestimate surface losses and underestimate subsurface losses because EPIC does not estimate tile drainage losses that increase the dissolved subsurface loss of nitrate.
[d]The difference in total N losses is computed by multiplying the projected changes in acreage (column 3) by the N loss estimate per acre (column 4).

from cropland. The sixth column reports the estimated change in total N losses due to the change in acreage of CRP and each respective crop, with the sum in the bottom row representing the total projected increase in N loss. By this calculation, N losses nationwide could increase by 297 million pounds N/year between 2007 and 2013. Implications for nutrient loads to the Gulf of course depend on how much of the predicted acreage change will occur in the MARB. Assuming the MARB accounts for 80% of the change in cropping systems, additional losses of 238 million pounds N/year could be expected for the MARB.

While these estimates are rough and omit numerous factors that could affect the nutrient loss from these lands (policy changes like higher mandates for the ethanol content of gasoline, farming practices, energy prices, and climate change), they provide an idea of the magnitude of the possible short-term nutrient consequences from increased corn-based ethanol production.

4.5.9.2 Impacts on Nutrient Application to Corn

In the simple calculations made in Table 4.8, it was implicitly assumed that N application rates will remain unchanged. However, reductions in N application rates have been identified as one tool to reduce N loss from corn (CERN, 2000). The level of nitrogen application that maximizes farm profits for a given soil and climate is a function of price and input costs. Corn price has increased, but fertilizer N costs have also skyrocketed in recent years so it is not possible, without further analysis, to determine the net effects of these two price trajectories on fertilizer application rates. Further, as Laboski et al., (2008) point out, simply applying N at economically optimal rates will not resolve the issue of nitrate movement from fields in subsurface drainage, for nitrate losses occur in corn production systems even when no N is applied.

High corn prices associated with market impacts of increased ethanol production will make it less profitable for farmers to manage N conservatively. Higher corn prices are likely to reinforce the perception that assurance of adequate N is worth the cost, since farmers are more likely to be adverse to risks of yield loss when corn prices are high. Based on economic optimum yield and historic response to high corn prices by farmers, $4/bushel corn may tend to increase N application rates to levels where N use efficiency is lower. High corn prices also provide a disincentive for cropland retirement or conversion to perennials.

Finally, it is worth noting that a large literature exists on the likely magnitude of yield drag associated with continuous corn and other crop rotations. These effects may also mean higher fertilization over the levels assumed in the CEAP study used in Table 4.8. See Katsvairo and Cox (2000a, 2000b) and Pikul et al. (2005).

4.5.9.3 Grain Versus Cellulosic Ethanol and Water Quality

Cellulosic ethanol produced from perennial grasses, fast-growing woody species, manures, and other biomass residuals such as corn stover could allow the United States to meet renewable transportation fuel goals while improving water quality (Mann and Tolbert, 2000; Perlack et al., 2005). Yet the rapid expansion of grain-based ethanol products may be a disincentive to development of perennial crops or crop residual-based ethanol. The technology to produce ethanol from cellulosic materials is rapidly improving but is not yet operational. The production, storage, and handling infrastructure are in place for grain but not for perennial crops or residuals. Cellulosic material is harder to handle, and only biomass sources such as forestry residuals and corn stover are in sufficient abundance to provide reliable supplies.

Grain-based ethanol producers are interested in the development of technology using corn stover and other crop residue as feedstock. Crop residues represent the largest potential source of feedstock, projected to be 354 million metric tonne/year (390 million ton/year). Graham et al. (2007) estimated about 58 million dry metric tonne/year (64 million dry ton/year) could be removed with soil loss at "tolerable levels" (T) levels, but at $\frac{1}{2}$ T soil loss removals could only be about 18 million metric tonne/year (19.8 million ton/year) (at 1995–2000 corn production levels). However, soil losses could increase 2–20 fold and still be below T. Therefore harvesting corn stover to keep soil losses just below T would result in substantial increases in erosion and associated N and P losses compared to current conservation or no-till production.

English et al. (2006) proposed that corn stover may be the largest potential source of cellulosic materials for ethanol production once cellulosic technologies are cost competitive. However, the contribution of returning stover to soil quality and quantity has long been recognized. Wilhelm et al. (2004) conclude that corn stover can be harvested for ethanol production, but recommendations for removal vary depending on regional yield, climatic conditions, and cultural practices.

Perennial grasses, including switchgrass and high biomass-producing trees, are currently considered the most promising energy crops (Kurt et al., 1998; McLaughlin and Kszos, 2005; Tolbert, 1998). Miscanthus and sweet sorghum have also been suggested as possible perennial feedstocks. This discussion focuses on switchgrass, which is a warm-season perennial native prairie grass that produces high biomass in its above ground growth and in deep roots. Switchgrass requires some N and P for optimal production, but less than corn. Switchgrass normally requires two growing seasons to become fully productive, but then it can grow for 20 years or more without replanting. Thus, either expected profitability from switchgrass production must be large enough to overcome early lower yields, or an incentive program will be needed to compensate the farmer during the 2-year transition. As mentioned previously, the transport and storage infrastructure needed to handle the large quantities of materials for an ethanol facility will need to be developed.

The evidence thus far suggests that switchgrass is a more favorable energy crop for reducing impacts on the land and climate; however, the technology for converting switchgrass to ethanol is not yet commercially viable. The fermentation co-product is a lignocellulosic material that can be dried and burned to provide part of the energy for the facility with net positive energy returns (Farrell et al., 2006). Switchgrass requires few nutrient additions, is not suited as a feed amendment, and can enhance threat to water quality. If it is grown instead of corn on productive soils, N and P losses are expected to be reduced by over 50% (Chesapeake Bay Program, 2003). Switchgrass will also sequester carbon, increase soil organic matter, and improve soil quality through its extensive, deep root system. These positive environmental attributes have substantial potential to provide multiple revenue streams. Lower production cost, greater net energy production, multiple revenue streams, and environmental benefits of switchgrass all favor its long-term use as a dedicated energy crop. However, the lag in development of fermentation technology and the lack of existing infrastructure prevent it from replacing corn as the major ethanol feedstock for the near future.

Increasing grain prices have increased the relative economic advantage that row crops, particularly corn, have over switchgrass. Substantial incentives will be needed before farmers would convert row crop land to switchgrass or other perennials at current market conditions. Babcock et al. (2007) estimated that the magnitude of subsidies would be significant and that conversion of all cropland to switchgrass in a watershed in northeastern Iowa would result in an 84, 83, 44, and 53% reduction, respectively, in sediment, total phosphorus (TP), nitrate (NO3), and total nitrogen (TN) at the watershed outlet compared to existing conditions. Model results also indicated that conversion of all cropland in the watershed to continuous corn would increase sediment, TP, NO3, and TN from current levels by 23, 128, 147, and 150%, respectively. They also evaluated the impact of growing switchgrass on all Highly Erodible Land (HEL) and continuous corn on other cropland. Careful placement of the switchgrass on other sensitive landscapes and as a buffer on non-HEL land could provide additional water quality benefits.

> **Key Findings and Recommendations**
>
> Expansion and intensification of corn production to support grain-based ethanol production and impacts of ethanol co-products from the animal production sector are likely to cause major increases in N and P losses in the MARB. The opportunity still exists to make choices that result in a renewable energy strategy that achieves energy goals with a reduced impact on the environment. Grain-based ethanol production is rapidly expanding, and the Study Group's preliminary calculations demonstrate a significant short-run increase in N and P losses to water resulting from current market incentives favoring corn.
>
> Cellulosic ethanol production can be less environmentally detrimental, but current technology and infrastructures do not make it competitive with grain-based ethanol. Harvesting corn stover as a feedstock for cellulosic ethanol has water and soil quality implications. Switchgrass or other perennial grasses or woody biomass provide greater net energy and lower production costs and potentially higher total revenue with substantial environmental benefits when compared to corn and could become the dominant feedstock if investment, policy, and market conditions do not keep renewable energy policy focused on grain feedstocks. Based on these findings, the Study Group offers the following recommendations regarding biofuel production.
>
> - Life-cycle analysis, examining all impacts to air, water, and climate, is needed to compare the various feedstocks for ethanol production.
> - Research and development should focus on biofuel production systems that are both economically viable and ecologically desirable.
> - If research continues to support the potential of cellulosic materials to meet energy and environmental goals, incentives (or the removal of disincentives) should be provided to promote ethanol production with more environmentally benign feedstocks.

4.5.10 Integrating Conservation Options

The previous sections have described land-management and conservation practices that can enhance nutrient loss reduction and water quality locally and in the

Gulf. As discussed, these practices vary, sometimes substantially, in their effectiveness among watersheds and subbasins in the MARB. Furthermore, there can be synergistic effects on nutrient loss reductions, where combinations of these practices can produce more (or less) than the sum of their individual reductions. In evaluating suites of management options, it is crucial to determine whether the nutrients that are not released to waters are being lost instead to other systems so that reactive N and P are not actually removed from the environment but just redistributed. These facts are an important part of the basis for our recommendation that watershed-based modeling approaches continue to be developed and that they be explicitly used to design optimal land-management systems within an adaptive management context. As noted in Sections 2.1.9 and 3.4, watershed-based models can be a key source of information for considering alternative sets of conservation practices and implementation approaches. Ideally, integrated modeling systems would be used to evaluate whether it is more cost-effective to reduce nutrient loadings with targeted nutrient management practices on the farm, to subsidize edge-of-field buffers in targeted watersheds, to change cropping patterns or to focus financing on well-placed off-site freshwater wetlands, or to implement some carefully chosen combination of these practices. However, while such models exist and are continuously being further improved, there remain limitations of these models in their current state (see Sections 2.1.9 and 3.4).

In Tables 4.9, 4.10, and 4.11, we provide a summary of the potential total nitrogen (TN) and phosphorus (TP) reduction efficiencies (percent, %) in surface runoff, subsurface flow, and tile drainage that can be realized where the various conservation practices could be implemented within the MARB. The cost-effectiveness of these measures will vary from site to site and with current and future land- and water-use designations. To a large extent, these estimates are based on relevant sections of this book and on reports by Devlin et al. (2003), Dinnes (2004), and Gitau et al. (2005). Where numeric values for reduction efficiency were not included in these reports, relative effects of practices were estimated based on expert opinion as negative (−, indicating increased export expected), positive (+, indicating reduced export expected), or neutral (±, indicating no significant effect expected).

Values for percent nutrient loss reductions are basin-scale averages, derived from edge-of-field and small watershed studies and not from widespread implementation. It must be emphasized that there is a great deal of site-specificity (spatial and temporal), which results in a wide range in observed conservation practice efficiency. While some of the conservation practices detailed have large, local, water quality benefits, they may not have a major impact on nutrient loss to the Gulf. To help facilitate implementation of practices that reduce nutrient loads to the Gulf, local water quality benefits are essential to MARB-wide adoption of these strategies. Estimates of N and P reductions are only appropriate to areas where a specific conservation practice can be implemented. For instance, it would not be effective to implement surface runoff control practices, such as sedimentation basins on flat lands with no concentrated surface flow of water. To a certain extent, N and P risk assessment tools that identify and quantify site vulnerability to N or P loss should be used at a local or field level to effectively target practices and to maximize reduction.

4.5 Options for Managing Nutrients, Co-benefits, and Consequences

Table 4.9 Potential total nitrogen (TN) and phosphorus (TP) efficiencies (percent change) produced by *nutrient-use* conservation practices on surface runoff, subsurface flow, and tile drainage. Estimates are average values for a multiple-year basis, and some of the numbers in this table are based on a very small amount of field information. Shading highlights the methods producing the greatest reduction efficiencies within the three types of N and P loss (surface runoff, subsurface, and tile drainage)

Conservation practice	Surface runoff		Subsurface flow		Tile drainage	
	TN	TP	TN	TP	TN	TP
Nitrification and urease inhibitors	+[a]	±	+	±	1–21[b]	±
Nitrogen: spring versus fall application	+	±[d]	0–25[c]	±	10–30 %[b]	±
Nitrogen: Recommended rate versus above-recommended rate	28–44[b]	+[d]	+	±	27–50[b]	±
Nitrogen: Subsurface versus surface broadcast	50[e]	20[f]	–	±	16[b]	±
Phosphorus: Avoid runoff producing rainfall	±	28–57[b]	±	+	±	+
Phosphorus: Rate balanced to crop use versus above-recommended rate	0–25[e]	15–47[b]	±	36[b]	±	25[b]
Phosphorus: Subsurface versus surface broadcast	±	8–92[a,b]	±	–	±	–
Manure: bioenergy, treatment, alternative use, transport to nutrient-deficit areas	+[f]	+[f]	+[f]	+[f]	+[f]	+[f]
Adoption of comprehensive farm nutrient management plan	0–65[e,g]	0–45[e,g]	+[f]	+[f]	+[f]	+[f]

NOTE: For references, see Table 4.11.

Implementation of any one of the tabulated conservation practices can positively or negatively influence the effectiveness of another.

Awareness of the weather forecast in planning any nutrient application or tillage operation is important to avoiding rainfall-induced runoff of applied nutrients and erosion. The conversion of cropped acres to perennial crops is distinguished from conversion to CRP lands, in that perennial crops will include grasses harvested for cellulosic biofuel production, which may receive maintenance or low fertilizer N and P inputs. The conversion of lands to CRP and from annual cropping to perennials is expected to decrease N and P loss because of reduced fertilizer and manure nutrient inputs and reduced erosion afforded by increased vegetative cover. Improved N use efficiency via appropriate timing, rate, and method of application is expected to benefit P loss reductions by increasing crop P uptake and removal (if harvested).

Table 4.10 Potential total nitrogen (TN) and phosphorus (TP) efficiencies (percent change) produced by *in-field* conservation practices on surface runoff, subsurface flow, and tile drainage. Estimates are average values for a multiple-year basis, and some of the numbers in this table are based on a very small amount of field information. Shading highlights the methods producing the greatest reduction efficiencies within the three types of N and P loss (surface runoff, subsurface, and tile drainage)

Conservation practice	Surface runoff		Subsurface flow		Tile drainage	
	TN	TP	TN	TP	TN	TP
No-till versus conventional tillage	0–25[b,e]	35–70[b,e]	–	±	–	±
Cover crops	50[b]	7–63[b]	+	48[b]	13–50[b]	+
Diverse cropping systems and rotations within row cropping [g,h]	25–70[b,e,g]	25–88[b]	±	±	52–93[b]	±
Contour plowing and terracing	20–55[c,f]	30–75[e,g]	–	±	±	±
Standard tile drainage versus undrained	25[g]	70[g]	+	±	–	–
Water table management versus uncontrolled drainage	–	–	+	+	25–54[b]	+
Shallow and/or wide versus standard tile placement	–	–	+	+	39[b]	25–42[b]
Conversion to CRP	40[b]	+	40[b]	+	40–97[b]	+
Conversion to perennials crops	+60–90[h]	+75–95[h]	+90[i]	+	+	+
Livestock exclusion from streams versus constant intensive grazing	10–80[b,g]	32–76[g,i,k]	+	75[b]	±	±
Managed grazing versus constant intensive grazing	–100–80[b,g]	0–78[b,g]	+	+	+	±
In-field vegetative buffers	12–51[b,e,g]	4–67[b,e,g]	±	±	–	–

NOTE: For references, see Table 4.11.

The estimated reduction efficiencies in Tables 4.9, 4.10, and 4.11 are based on edge-of-field losses for studies conducted within the MARB and do not represent expected whole-basin reductions. These values represent potential reductions only for those areas where the particular practices could be implemented and do not address how broadly a practice could be applied. The shaded areas indicate those practices expected to have the greatest impact on reducing nutrient export from the MARB as a whole: red shading indicates conservation practices that translate into N loss reduction in tile drainage, green shading is for surface runoff of N and P, and blue shading for nutrient loss in subsurface flow. It is clear that where edge-of-field loss estimates are available, there is a large variability in reduction efficiencies, which is both temporally and spatially dependent. This inherent variability must be recognized when developing conservation or remedial strategies for the MARB,

4.5 Options for Managing Nutrients, Co-benefits, and Consequences

Table 4.11 Potential total nitrogen (TN) and phosphorus (TP) efficiencies (percent change) produced by *off-site* conservation practices on surface runoff, subsurface flow, and tile drainage. Estimates are average values for a multiple-year basis, and some of the numbers in this table are based on a very small amount of field information. Shading highlights the methods producing the greatest reduction efficiencies within the three types of N and P loss (surface runoff, subsurface, and tile drainage)

Conservation practice	Surface runoff		Subsurface flow		Tile drainage	
	TN	TP	TN	TP	TN	TP
Sedimentation basins	55[g]	65[g]	±	±	±	±
Riparian buffers: total n total P	50 82[g,i]	40–93[g,i,k]	+	+	±	±
Riparian buffers: nitrate-N and dissolved P	41–92[i,k]	28–85[i,k]				
Wetlands: total P	61–92[b,g]	0–79[g,i,k]	9–74[b]	+	20–90[i,k]	+
Wetlands: dissolved P		22–86[g,i,k]				

[a] Relative effects of practices estimated based on expert opinion as negative (−, indicating increased export expected), positive (+, indicating reduced export expected), or neutral (±, indicating no significant effect expected).
[b] From Dinnes (2004) report or from Study Group report. Values from IA, IL, MO, MN, NE, OH, and OK are included.
[c] From Randall and Sawyer (2005), nitrogen application timing, forms, and methods. pp. 73–84. Session 6, UMRSHNC (2006) report.
[d] Increased crop yields afforded with N fertilizer is likely to increase P uptake by crop and P removal if harvested.
[e] From Devlin et al. (2003).
[f] Improved manure management leads to lower land application and thereby less potential for loss in any pathway.
[g] Values based on data included in Gitau et al. (2005).
[h] Studies with only corn–soybean systems are not included, although they were included in Dinnes (2004).
[i] Values from Smith et al. (1992).
[j] Values from Randall et al. (1997).
[k] Values are modifications of values in Dinnes (2004) based on values in Study Group report.

in the context of probability of expected outcomes. It is also a key component of the conservation premise that there is no "one size fits all" rationale for adaptive management.

As a complement to the information summarized in Tables 4.9, 4.10, and 4.11 a second summary of the likely environmental benefits is provided in association with the conservation and land management. In Tables 4.12 and 4.13, the focus is on the broader contribution these practices can have with respect to a wide variety of environmental services including local water quality, carbon sequestration in agricultural soils, wildlife habitat, biodiversity, general recreational activities, and air pollution. These effects are based on the scientific literature and professional judgment, and potential repercussions are indicated only as being positive (+) or negative (−) or having no effect (0).

Table 4.12 Anticipated benefits associated with different agricultural management options

Agricultural management option	Reduce N load to Gulf	Reduce P load to Gulf	Local surface water quality - N	Local surface water quality - P and sediments	Ground-water quality	Carbon sequestration	Local wildlife habitat[a]	Biodiversity[a]	Recreational activities	Air pollution reduction	Soil quality
Decrease drainage intensity	+	−	+	−	0	0	+	+	+	0	0
Increase freshwater wetlands	+	+/?	+	+/?	0	+	+	+	+	−	0
Forested riparian buffers	+	+	+	+	+	+	+	+	+	+	+
Herbaceous riparian buffers	+	+	+	+	+	+	+	+	+	+	+
Improve manure mgmt.	+	+	+	+	+	0	0	0	0	+	+
Increase acreage of perennials	+	+	+	+	+	+	+	+	+	+	+
Increase acres of farmland retired	+	+	+	+	+	+	+	+	+	+	+
Reduce fertilizer N and/or P application	+	+	+	+	+	0	+	+	+	+	0
Spring fertilizer N and/or P application	+	0	+	0	+	0	0	0	0	0	0
Expand corn-based ethanol production	−	−	−	−	−	−	−	−	−	−	−
Expand cellulosic ethanol production	+	+	+	+	+	+	+	+	+	+	+

+ = will lead to improvements in conditions; − = likely to be further degraded; 0 = will have little effect; ? = effect unknown.

4.5 Options for Managing Nutrients, Co-benefits, and Consequences

Table 4.13 Anticipated benefits associated with other management options

Management option	Reduce N load to Gulf	Reduce P load to Gulf	Local surface water quality		Groundwater quality	Carbon sequestration	Local wildlife habitat	Biodiversity	Recreational activities	Air pollution reduction	Soil quality
			N	P and Sediments							
Decrease NO$_x$ emissions	+	0	+	0	0	0	0	0	0	+	0
Reduce point source loads	+	+	+	+	0	0	+	+	+	0	0
Reduce urban nonpoint source loads	+	+	+	+	+	0	+	+	+	0	+
Enhance floodplain connectivity	+	+	+	+	0	+	+	+	+	0	0
Atchafalaya diversion	?	?	?	?	0	0	0	0	?	0	0
Increase coastal wetlands	?	+	?	?	0	+	+	+	+	0	+

+ = will lead to improvements in conditions; − = likely to be further degraded; 0 = will have little effect; ? = effect unknown.

In each of these tables, the effects predicted assume that conservation practices are implemented and managed (maintained) as designed to maximize effectiveness and life expectancies. Inadequate implementation and maintenance can lead to poor performance of such systems. Further, these strategies need to be carefully targeted at an appropriate level of intensity and over sufficient time in order to effectively reduce nutrient export.

Finally, when considering these tables, it is important to note there are synergistic effects of combinations of conservation practices that result in greater nutrient loss reductions than do individual practices (Table 4.10). For example, N application management that minimizes the potential for excess N available to be leached (nutrient management, Table 4.9) should be combined with efforts to reduce the potential off-site movement of water (infield management, Table 4.9). Conversely, there are potential trade-offs. For example, reduced-till, no-till, and tile drainage can decrease runoff, erosion, and P loss but can enhance NO_3 nitrate leaching potential. As another example, while N-based manure application can be a cost-effective N source to meet crop N needs, P may be overapplied, increasing the potential for increased runoff and loss of P.

Key Findings and Recommendations

A number of conclusions concerning the appropriate use of conservation practices can be drawn from these tables. First, there is no "one size fits all" land-use or conservation practice strategy that will be cost-effective in all locations. Rather, site-specific and regional optimization of conservation practices and appropriate targeting of conservation practices and measures will be needed and will include a broad range of alternative practices and land uses, such as crop, animal, fertilizer, and drainage management measures targeted to appropriate areas. The reduction efficiencies of these practices are spatially and temporally variable, making it impossible to assign a specific reduction efficiency for any given conservation practice. As information from ongoing monitoring of nutrient loss reduction efficiencies becomes available, we will be better be able to determine what major factors influence reduction efficiencies. This learning and integration of new knowledge is important and will enhance the process of adaptive management.

Second, practices that are likely to address NGOM hypoxia effectively in tile-drained landscapes can differ markedly from those appropriate in nontiled lands. Further, while there are no-one-size-fits-all strategies, there are some approaches that appear particularly promising. For example, inter-seeding of leguminous cover or relay crops within corn and other grain rotations can decrease fertilizer N requirements, reduce soil profile N at critical loss times of the year, and mine excess soil P. Reconnecting the floodplain with managed agricultural lands, by managing hydrology to increase the amount of time water is retained on the land (wetland) prior to entering the major fluvial systems, should be considered an important part of an adaptive management plan to reduce NGOM hypoxia.

Third, practices that are likely to be cost-effective in addressing NGOM hypoxia may not be the same that yield the highest benefits in other environmental dimensions. This has important planning and implementation implications, for it suggests that, when considering implementation strategies, the optimal set of conservation practices and sinks needs to be considered with respect both to NGOM hypoxia and to the suite of other environmental concerns that are likely to vary regionally.

Finally, in considering information from the tables and "optimal" sets of practices, the principles of adaptive management imply that approaches need to be changed and updated with time to maximize overall efficiency. In the process, more information can and will be learned about the effectiveness of these practices. This information can be used both to improve the performance of water quality models to aid in better implementation strategies and directly to improve targeting of conservation practices and actions. Based on these findings, the Study Group offers the following recommendations.

- There is great temporal and spatial variability in nutrient loss reduction efficiencies of the various conservation practices available. Thus, continued, new, and enhanced small watershed-based studies of suites of conservation practices as applied in the real-world are necessary and should be set in a context of research, monitoring, and demonstration to stakeholders so that progress (or lack thereof) in response to management change can be assessed. A variety of response measures relevant to different watershed scales and environmental concerns should be monitored. These measures should include both performance measures (e.g., nutrient loading at subwatershed levels, estimates of carbon sequestered on the landscape), and practice-based measures (e.g., number of acres of wetlands installed, miles of conservation buffers installed).
- To reduce spring nitrate loss from tile-drained regions, alternative and more complex cropping systems (including perennials) are thought to be the most effective method of reducing losses. However, given current constraints in cropping systems, the Study Group recommends reducing or discontinuing fall N application for corn, improved N fertilizer management techniques, use of cover crops, wetland establishment, and drainage management where appropriate.
- For P loss reduction, the Study Group again finds that alternative and complex cropping systems are most effective. For current cropping systems, the Study Group recommends that riparian buffer strips; improved P fertilizer and manure management; and, where appropriate, cover crops be implemented.

- Where appreciable drainage occurs in the fall and winter, controlled drainage could significantly reduce nitrate losses but can be expected to increase surface runoff and losses of particulate contaminants.
- If precision agriculture and controlled-release fertilizer technologies are proven to provide reductions in losses of N and P to water resources, then incentives should be considered to stimulate their adoption.
- Incentives for conversion to perennials, which have potential future use as cellulosic biofuels production, should be established to promote the co-benefit of greatly reduced nitrate and P loss from agricultural systems.
- There should be a focus on conservation practices and implementation strategies that appropriately match the nutrient reduction strategies with the goals of reducing NGOM hypoxia as well as local/regional environmental goals (carbon sequestration, wildlife, air quality, local water quality, etc.). Given the breadth and magnitude of these additional environmental goals, these "co-benefits" should be incorporated in the planning process.
- Information on effectiveness and geographic appropriateness of various conservation practices and nutrient reduction strategies should be used in conjunction with formal models to plan implementation strategies for conservation measures that effect a reduction in nutrient loading to the NGOM.

Chapter 5
Summary of Findings and Recommendations

This book responds to questions in three general areas: characterization of hypoxia; characterization of nutrient fate, transport and sources; and the scientific basis for goals and management options. In the sections below, these questions (shown in italics bellow) are addressed very briefly with references to those sections of this book where more detailed science on that particular question may be found.

5.1 Characterization of Hypoxia

I. Characterization of Hypoxia: *The development, persistence, and areal extent of hypoxia is thought to result from interactions in physical, chemical, and biological oceanographic processes along the northern Gulf continental shelf; and changes in the Mississippi River basin that affect nutrient loads and freshwater flow.*

   A. *Address the state-of-the-science and the importance of various processes in the formation of hypoxia in the Gulf of Mexico. These issues include*

   i. *Increased volume and/or funneling of freshwater discharges from the Mississippi River*
   ii. *Changes in hydrologic or geomorphic processes in the Gulf of Mexico and the Mississippi River basin*
   As discussed in Section 2.1, the hydrologic regime of the Mississippi River and spatial distribution and timing of freshwater inputs to the Gulf of Mexico relative to the occurrence of energetic currents and waves are critical to vertical mixing intensity, stratification, and hypoxia in the Gulf. Alteration of the hydrologic regime of the Mississippi and Atchafalaya Rivers from the 1920s to 1960s has likely increased the residence time of freshwater on the Louisiana–Texas shelf as well as the area of the NGOM shelf that is conducive to hypoxia.
   iii. *Increased nutrient loads due to coastal wetlands losses, upwelling, or increased loadings from the Mississippi River basin*

As discussed in Section 2.1, increased nutrient loadings from the Mississippi River basin have triggered hypoxia by stimulating in situ phytoplankton production of labile organic matter in shallow near-shore receiving waters of the Gulf. Nutrients also enter this region of the Gulf by advective transport from deeper offshore sources and from atmospheric deposition. However, advective imports and atmospheric deposition are relatively minor sources of nutrients in comparison with those from the Mississippi River basin. The extent to which coastal wetland losses have changed nutrient processing and loading to the Gulf of Mexico is a subject of continued study but is largely believed to be of secondary importance.

iv. *Increased stratification and seasonal changes in magnitude and spatial distribution of stratification and nutrient concentrations in the Gulf*

As discussed in Section 2.1, increased phytoplankton production, coupled with stratification and suppressed vertical mixing associated with fresh water discharge, has caused hypoxia in bottom waters of the northern Gulf of Mexico. However, historic analyses indicate a great deal of variability in seasonal, interannual, and decadal-scale patterns of primary productivity, phytoplankton biomass, and the amounts of freshwater and nutrients discharged to the Gulf. Therefore, trends for nutrient-driven eutrophication and hypoxia on these timescales have been difficult to interpret.

v. *Temporal and spatial changes in nutrient limitation or co-limitation, for nitrogen or phosphorus, as significant factors in the development of the hypoxic zone*

As discussed in Section 2.1.3, studies of waters overlying the hypoxic region of the northern Gulf of Mexico indicate that N limitation characterizes offshore waters, but inshore productivity appears to be P limited and P and N co-limited. This is particularly true from February to May when peak phytoplankton productivity and biomass formation coincide with peak freshwater discharge and nutrient loading. Inshore primary productivity shifts to an N-limited mode during the drier (lower freshwater discharge) summer and fall seasons, and there are likely to be periods when both N and P are supplied at low levels and co-limit phytoplankton production during the spring to summer transition.

vi. *The implications of reduction of phosphorus or nitrogen without concomitant reduction of the other*

As discussed in Section 2.1, the Study Group finds ample evidence to conclude that N loading from the Mississippi–Atchafalaya River basin is the significant factor driving the timing and extent of hypoxia in the northern Gulf of Mexico. However, P supplies also play a significant role in controlling primary production. Therefore, as discussed in Section 2.1.8, reducing the size of the hypoxic zone requires both N and P discharge reductions.

B. Comment on the state-of-the-science for characterizing the onset, volume, extent, and duration of the hypoxic zone

Section 2.1.9 describes modeling approaches that have been used to characterize the onset, volume, extend, and duration of the hypoxic zone. Simple linear and multiple regression models that use nutrient loadings to predict hypoxic zone area have been constructed. Other models have included some consideration of processes and mechanisms.

5.2 Nutrient Fate, Transport, and Sources

II. Characterization of Nutrient Fate, Transport, and Sources: Nutrient loads, concentrations, speciation, seasonality, and biogeochemical recycling processes have been suggested as important causal factors in the development and persistence of hypoxia in the Gulf. The Integrated Assessment (CENR, 2000) presented information on the geographic locations of nutrient loads to the Gulf and the human and natural activities that contribute nutrient loadings.

A. Given the available literature and information (especially since 2000), data and models on the loads, fate and transport, and effects of nutrients, evaluate the importance of various processes in nutrient delivery and effects. These may include the following:

    i. The pertinent temporal (annual and seasonal) characteristics of nutrient loads/fluxes throughout the Mississippi River basin and, ultimately, to the Gulf of Mexico

    Total annual N flux discharged to the Gulf of Mexico, primarily nitrate-N and particulate/organic N, has decreased during the past 25 years, as has the spring (April–June) flux. Neither total P nor SRP fluxes show major annual or seasonal trends during the same period.

    As discussed in Section 3.1, the upper Mississippi and Ohio–Tennessee River subbasins contribute about 82% of the annual nitrate-N flux, 69% of the TKN flux, and 58% of the total P flux to the Gulf of Mexico while representing only 31% of the drainage area of the MARB. When the upper Mississippi River basin is further divided, the subbasin contributing to the upper Mississippi River between Clinton, IA, and Grafton, IL (only 7% of the drainage area) contributes about 29% of the total annual nitrate-N flux to the Gulf. Perhaps more importantly, the upper Mississippi and Ohio–Tennessee River subbasins currently contribute nearly all the spring N flux to the Gulf. These subbasins represent the tile-drained, corn–soybean landscape of Iowa, Illinois, Indiana, and Ohio and illustrate that corn–soybean agriculture with tile drainage leaks considerable N under the current management system. The source of riverine P is more diffuse, although these subbasins are also the largest sources of P.

ii. *The ability to determine an accurate mass balance of the nutrient loads throughout the basin*

Estimates of mass balances for nutrient inputs during the period since the *Integrated Assessment* have been recalculated and are discussed in Section 3.2, but the research needs described in the *Integrated Assessment* remain unresolved. Therefore, the Study Group's ability to determine an accurate mass balance of nutrient inputs to the MARB is limited by the available information and understanding. For example, some components of the N mass balance (e.g., denitrification, N_2 fixation, manure N, soil N pool processes such as mineralization and immobilization) are not measured each year. N_2 fixation and manure N are the only two of these components that can be estimated. There are too few data available for the remaining processes to allow calculations. There also is still a disconnect between estimates of inputs to the land (i.e., fertilizer and manure use) and estimates of the proportion of N and P from those inputs that reach the riverine system and contribute to the nutrient flux. Point sources discharge N and P directly to rivers and are estimated by this Study Group to contribute about 22 and 34% of the annual riverine N and P flux, respectively, yet their contributions continue to be estimated from permit limits and are not actually measured. Better point source data are needed to improve mass balance estimates of nutrient loads.

iii. *Nutrient transport processes (fate/transport, sources/sinks, transformations, etc.) through the basin, the deltaic zone, and into the Gulf*

As discussed in Section 3.3, the percentage of annual N and P inputs removed by in-stream processes varies by MARB subbasin and ranges from 20 to 55% for N and 20–75% for P based on model estimates. Denitrification can be a significant pathway for N removal in small streams during low flow, warm periods, thereby enhancing local water quality. However, most nitrate-N is exported to the Gulf during high flows in the period from January to June, when denitrification is not an effective removal process. Although current estimates of denitrification rates in coastal wetlands are higher than the estimates used in the *Integrated Assessment*, current studies still conclude that river diversions to coastal wetlands would remove only small amounts of nutrients relative to the total fluxes. However, better estimates of nutrient and organic matter loss rates (denitrification; long-term burial of C, N, and P; and plant uptake) are needed to better understand observed differences between wetland inputs and outputs in coastal areas.

B. Given the available literature and information (especially since 2000) on nutrient sources and delivery within and from the basin, evaluate capabilities to

i. *Predict nutrient delivery to the Gulf, using currently available scientific tools and models and*

ii. *Route nutrients from their various sources and account for the transport processes throughout the basin and deltaic zone, using currently available scientific tools and models*

In Section 3.4, the Study Group singled out three models for discussion: SPARROW, SWAT, and IBIS/THMB. Each is capable of N and P load estimation on the scale of the MARB, yet each has strengths and weaknesses requiring further development. The uncertainty of results from each model reflects the uncertainty of the model structure and algorithms, as well as that propagated by the input data, user parameterization, the calibration process, other user-defined conditions, and the skill of the model user. Even though the capability to predict and route nutrients throughout the MARB has improved since the *Integrated Assessment*, future adaptive management will require a smooth interface between watershed, economic, and Gulf of Mexico hypoxia models that will allow resource managers the capability to assess the effects of policy decisions and management practices on the sources, fate, and transport of nutrients from the MARB to the Gulf of Mexico.

5.3 Goals and Management Options

III. Scientific Basis for Goals and Management Options: The Task Force has stated goals of reducing the 5-year running average areal extent of the Gulf of Mexico hypoxic zone to less than 5,000 km^2 by the year 2015, improving water quality within the basin and protecting the communities and economic conditions within the basin. Additionally, nutrient loads from various sources in the Mississippi River basin have been suggested as the major driver for the formation, extent, and duration of the Gulf hypoxic zone.

A. *Are these goals supported by present scientific knowledge and understanding of the hypoxic zone, nutrient loads, fate and transport, sources, and control options?*

The Study Group affirms the major findings of the *Integrated Assessment*. Although the 5,000 km^2 target remains a reasonable end point for continued use in an adaptive management context, it may no longer be possible to achieve this goal by 2015. Accordingly, it is even more important to proceed in a directionally correct fashion to manage factors affecting hypoxia than to wait for greater precision in setting the goal for the size of the zone.

i. *Based on the current state-of-the-science, should the reduction goal for the size of the hypoxia zone be revised?*

No. As discussed in the Executive Summary, it is more important to begin to move in a directionally correct fashion than to refine the goal for the exact size of the hypoxic zone.

ii. *Based on the current state-of-the-science, can the areal extent of Gulf hypoxia be reduced while also protecting water quality and social welfare in the basin?*

Social welfare can be protected by choosing policies that incorporate targeting, provide economic incentives and maximize co-benefits. As discussed in Section 4.3, improvements in large-scale integrated economic and biophysical models are needed to better capture system-wide response and effects.

B. *Based on the current state-of-the-science, what level of reduction in causal agents (nutrients/discharge) will be needed to achieve the current reduction goal for the size of the hypoxic zone?*

As discussed in Section 4.2, to reduce the size of the hypoxic zone, the Study Group recommends an adaptive management approach targeting at least a 45% reduction in discharges of total N and total P from the 1980 to 1996 fluxes.

C. *Given the available literature and information (especially since 2000) on technologies and practices to reduce nutrient loss from agricultural, runoff, from other nonpoint sources, and from point source discharges, discuss options (and combinations of options) for reducing nutrient flux in terms of cost, feasibility, and any other social welfare considerations.*

In general, the social costs of reducing nutrients will vary widely with the policy chosen, hence overall cost-effectiveness is largely a function of policy. Policies that target and provide economic incentives are essential to minimize costs. A wide range of policy options are discussed in Section 4.4, while management options are covered extensively in Section 4.5.

These options may include the following:

i. *The most effective agricultural practices, considering maintenance of soil sustainability and avoiding unintended negative environmental consequences*

The cost and reduction efficiency rankings of agricultural management practices will vary by site and region, historic land use and management, crops grown, local soil conditions, distance to waterway, field slopes and configuration, presence of buffers, drainage structures, and so forth. Table 4.8 in Section 4.5.10 provides the Study Group's summary of the evidence comparing the relative effectiveness of nutrient (N and P) reduction options in agriculture. Section 4.5.6 discusses management options for in-field nutrients. A targeted and adaptive management framework will maximize local and regional water quality benefits in the MARB and Gulf.

ii. *The most effective actions for other nonpoint sources*

As discussed in Section 4.5.7, there are significant policy opportunities to reduce atmospheric deposition of N; however, a detailed examination of air pollution control policy options was beyond the

Study Group's scope. Nonetheless, the Group strenuously recommends incorporating water quality benefits and effects on hypoxia in air pollution control decisions.

iii. *The most effective technologies for industrial and municipal point sources*

As discussed in Section 4.5.8, a targeted permit-by-permit approach to industrial point source discharges could yield significant opportunities for nutrient (N and P) reduction since frequently a limited number of permitted facilities are responsible for a large part of the N and P loads. Municipal point sources are also discussed in Section 4.5.8, where the Study Group recommends an analysis to assess the cost and feasibility of tightening limits on N and P concentrations in discharges for large sewage treatment plants.

In all three areas, please address research and information gaps (expanded monitoring, documentation of sources and management practices, effects of practices, further model development and validation, etc.) that should be addressed prior to the next 5-year review.

Recommendations for monitoring and research are found in nearly every section of the book and are included below in the summary of the Study Group's recommendations.

5.4 Conclusion

This book constitutes the Study Group's response to questions posed by the USEPA Office of Water. This Study Group reaffirms the major findings of the *Integrated Assessment*, while pointing out the need for economic incentives to encourage conservation in the Mississippi–Atchafalaya River basin. Although the science has grown, actions to control hypoxia have lagged. The Study Group urges the USEPA and other agencies to act on the recommendations of this Study Group and move ahead with implementing programs, strategies, and policies to reduce the size of the hypoxic zone and improve water quality in the Mississippi–Atchafalaya River basin.

Most of the research and monitoring needs identified in the *Integrated Assessment* have not been met, and fewer rivers and streams are monitored today than in 2000. The majority of monitoring recommendations in the *Integrated Assessment* remain relevant and should be heeded, specifically the CENR's call to improve and expand monitoring of the temporal and spatial extent of hypoxia and the processes controlling its formation; the flux of nutrients, carbon, and other constituents from nonpoint sources throughout the MARB and to the NGOM; and measured (rather than estimated) nitrogen and phosphorus fluxes from municipal and industrial point sources. Echoing the CENR, the Study Group affirms the need for research on the ecological effects of hypoxia; watershed nutrient dynamics; effects of different agricultural practices on nutrient losses from land, particularly at the small watershed scale; nutrient cycling and carbon dynamics; long-term changes in hydrology and climate; and economic and social impacts of hypoxia. A suite of models is needed to simulate the processes and linkages that regulate the onset, duration, and extent of hypoxia. Emerging coastal ocean observation and prediction systems should be encouraged to monitor dissolved oxygen and other physical and biogeochemical parameters needed to continue improving hypoxia models.

Although there are over 90 recommendations in this book, the following major recommendations reflect the Study Group's consideration of the *new* science that has emerged since the *Integrated Assessment*.

To advance the science characterizing hypoxia and its causes, the Study Group finds that research is needed to

- collect and analyze additional sediment core data needed to develop a better understanding of spatial and temporal trends in hypoxia;
- investigate freshwater plume dispersal, vertical mixing processes, and stratification over the Louisiana–Texas continental shelf and Mississippi Sound, and use three-dimensional hydrodynamic models to study the consequences of past and future flow diversions to NGOM distributaries;
- advance the understanding of biogeochemical and transport processes affecting the load of biologically available nutrients and organic matter to the Gulf of Mexico and develop a suite of models that integrate physics and biogeochemistry;
- elucidate the role of P relative to N in regulating phytoplankton production in various zones and seasons and investigate the linkages between inshore primary production, offshore production, and the fate of carbon produced in each zone;
- improve models that characterize the onset, volume, extent, and duration of the hypoxic zone and develop modeling capability to capture the importance of P, N, and P–N interactions in hypoxia formation.

With respect to advancing the science on sources, fate, and transport of nutrients, the Study Group finds that research is needed to

- develop models to simulate fluvial processes and estimate N and P transfer to stream channels under different management scenarios;
- improve the understanding of temporal and seasonal nutrient fluxes and develop nutrient, sediment, and organic matter budgets within the MARB;

To enhance the scientific basis for implementation of management options, the Study Group finds that research is needed to

- examine the efficacy of dual nutrient control practices;
- determine the extent, pattern, and intensity of agricultural drainage as well as opportunities to reduce nutrient discharge by improving drainage management;
- integrate monitoring, modeling, experimental results, and ongoing management into an improved conceptual understanding of how the forces at key management scales influence the formation of the hypoxia zone; and

5.4 Conclusion

- develop integrated economic and watershed models to support adaptive management at multiple scales.

To reduce the size of the hypoxic zone, the Study Group recommends at least a 45% reduction in N accompanied by a comparable reduction in P. Five areas offer the most significant opportunities for N and P reductions:

- promotion of environmentally sustainable approaches to biofuel production and associated cropping systems (e.g., perennials);
- improved management of nutrients by emphasizing infield nutrient management efficiency and effectiveness to reduce losses;
- construction and restoration of wetlands, as well as criteria for targeting those wetlands that may have a higher priority for reducing nutrient losses;
- introduction of tighter N and P limits on municipal point sources; and
- improved targeting of conservation buffers, including riparian buffers, filter strips, and grassed waterways to control surface-borne nutrients.

Appendices

Appendix A: Studies on the Effects of Hypoxia on Living Resources

The abstracts in this appendix all came from a workshop sponsored by the NOAA Center for Sponsored Coastal Ocean Research held at Tulane University, New Orleans, LA held September 25–26, 2006.

Brouwer, Marius, 2006. "Changes in Gene and Protein Expression and Reproduction in Grass Shrimp, Palaemonetes pugio, Exposed to Chronic Hypoxia" Presentation at "Hypoxia Effects on Living Resource in the Gulf of Mexico" NOAA Center for Sponsored Coastal Ocean Research, Tulane University, New Orleans, LA. September 25–26, 2006.

Abstract: Hypoxic conditions in estuaries are one of the major factors responsible for declines in habitat quality. Previous studies examining the effects of hypoxia on crustacea have focused on individual/population-level, physiological or molecular responses but have not considered more than one type of response in the same study. The objective of this study was to integrate disciplines by examining the responses of grass shrimp to chronic hypoxia both at the molecular and whole animal level. Hypoxia-induced alterations in gene expression were screened using custom cDNA macroarrays containing 78 clones from a hypoxia-responsive suppression subtractive hybridization (SSH) cDNA library. Grass shrimp respond differently to moderate (2.5 ppm DO) versus severe (1.5 ppm DO) chronic hypoxia. The initial response to moderate hypoxia was down-regulation of genes coding for ribosomal proteins, HSP 70 and MnSOD. The initial response after short-term (3 d) exposure to severe hypoxia was upregulation of genes involved in oxygen uptake/transport and energy production, such as hemocyanin and ATP synthases. The major response by day 7 was an increase of transcription of genes present in the mitochondrial genome, together with upregulation of a putative heme binding protein and the iron storage protein, ferritin. By day 14 a dramatic reversal was seen, with a significant downregulation of transcription of genes in the mitochondrial genome. Both ferritin and the heme binding protein were downregulated as well. Levels of Hypoxia Inducible Factor (HIF1-alpha) remained unchanged. The macroarray data were validated using real-time qPCR. Changes in mitochondrial proteins were examined by separating proteins in 2 dimensions (IEF and reverse phase) followed by MS. At the organismal level, hypoxia exposure resulted in marked effects on shrimp egg production and larval survival, suggesting population-level implications of long-term hypoxia.

Baltz, Donald M., Hiram W. Li, Philippe A. Rossignol, Edward J. Chesney and Theodore S. Switzer, 2006. "A Qualitative Assessment of the Relative Effects of Bycatch Reduction, Fisheries and Hypoxia on Coastal Nekton Communities in the Gulf of Mexico", Presentation at "Hypoxia Effects on Living Resource in the Gulf of Mexico" NOAA Center for Sponsored Coastal Ocean Research, Tulane University, New Orleans, LA. September 25–26, 2006.

> Abstract: We applied qualitative mathematical models to develop an understanding of linkages that influence shrimp, fishes, and fisheries in coastal Louisiana where biotic communities face many natural and anthropogenic stressors, one of which is fishing activities related to the harvest of shrimp. Shrimp trawling ranks high in terms of impact on nekton and their habitats, and like most fishing gears catches nontarget species or sizes that are not marketed. These individuals, termed 'bycatch', are often returned to the water in dead or dying condition. Numerous other individuals are not 'caught' per se but also suffer the 'effects of fishing', that can degrade habitats or cause injuries leading to mortality. Modeling was used to examine the effects of fishing and bycatch mortality on community structure in the 'Fertile Fisheries Crescent' and how major stressors interact with hypoxia to influence fisheries. We explored direct and indirect interactions between shrimp, their predators, bycatch species, and shrimp landings. A major finding was that bycatch reduction efforts may feedback on fisheries and shrimp populations in an unexpectedly negative manner. Another was that changes in community structure that might be attributed to hypoxia are also possible from fishing alone. To corroborate our models, we analyzed 15 years of quantitative data on National Marine Fisheries Service shrimp landings, Louisiana Department of Wildlife and Fisheries (LDWF) gillnet surveys, and LDWF shrimp trawl surveys from central Louisiana. Abundant bycatch and other species were summarized into several functional groups including small and large shrimp predators, nonshrimp predators, major bycatch consumers, minor bycatch consumers, and non-bycatch consumers. Factor and correlation analyses of quantitative data for functional groups on a bimonthly basis corroborated results from the qualitative models, and combined indicated that shrimp abundance and shrimp landings would likely suffer from increased natural mortality if the shrimp-fishery bycatch was substantially reduced.

Craig, J. Kevin and Larry B. Crowder, 2005. "Hypoxia-induced habitat shifts and energetic consequences in Atlantic croaker and brown shrimp on the Gulf of Mexico shelf" Marine Ecology Progress Series, Vol. 294, pp 79–94.

> Abstract: This paper evaluates the effects of hypoxia-induced habitat loss on Atlantic croaker and brown shrimp. The compare spatial distributions and the relationship to abiotic factors, including temperature, dissolved oxygen and salinity across years with differing levels of hypoxia using 14 years of fishery-independent trawl data. They find that hypoxia results in considerable shifts in temperature and oxygen conditions that croaker and brown shrimp experience. Croaker typically occupy relative warm, inshore waters. During periods of hypoxia, croaker remain in the warmest inshore waters, but are also displaced to cooler offshore waters. Brown shrimp typically are distributed more broadly and further offshore. During periods of hypoxia, brown shrimp shift to warm inshore waters and cooler waters near the offshore edge of the hypoxic zone. The shifts in spatial distribution are reflected in decreases in water temperature for croaker that are displaced offshore the hypoxic region, and increases in water temperature for brown shrimp that are displace inshore of the hypoxic zone. Both species also face increased variance in water temperatures due to hypoxia-induced habitat displacement. Despite avoidance of the lowest oxygen waters, high densities of croaker and brown shrimp occur in areas of 1.6–3.7 mg/l near the offshore hypoxic edge. Shifts in spatial distribution during severe hypoxia may impact organism energy budgets. For example, laboratory studies indicate low oxygen impacts individual movement, growth,

and mortality (Taylor and Miller, 2001; Wannamaker and Rice, 2000; Wu, 2002). High croaker and shrimp densities near the hypoxic edge likely have implications for trophic interactions as well as the harvest of both target (brown shrimp) and nontarget (croaker) species by the commercial shrimp fishery. Croaker may benefit from high concentrations of brown shrimp at the edge of the hypoxic zone, while brown shrimp may become more susceptible to predation by croaker.

Craig, J. Kevin, Larry B. Crowder, and Tyrrell A. Henwood, 2005. "Spatial distribution of brown shrimp (Farfantepenaeus aztecus) on the northwestern Gulf of Mexico shelf: effects of abundance and hypoxia" Canadian Journal of Fisheries and Aquatic Science. Vol. 62, pp 1295–1308.

> Abstract: This paper uses fishery-independent hydrographic and bottom trawl surveys from 1983–2000 used to test for density dependence and effects of hypoxia on spatial distribution of brown shrimp. The spatial distribution of shrimp was found to be positively related to abundance on the Texas shelf, but negatively related to abundance on the Louisiana shelf. Density dependence was weak, and may have been due to factors other than habitat selection. Large-scale hypoxia (up to ~20 000 km2) on the Louisiana shelf occurs in regions of typically high shrimp density, resulting in loss of up to 25% of shrimp habitat on the Louisiana shelf. They also find shifts in distribution and densities both inshore and offshore of the hypoxic region. Results placed in terms of the generality of density-dependent spatial distributions in marine populations. Potential consequences of habitat loss and associated shifts in distribution due to low dissolved oxygen. They note that shifts in spatial distribution may precede major stock declines, and thus could potentially serve as an early warning sign of future declines in abundance (Overholtz, 2002; Rose et al., 2000).

Diaz, Robert, 2001. "Overview of Hypoxia around the World" Journal of Environmental Quality. Vol. 30, No. 2, (March–April) pp 275–281.

> Abstract: This paper summarizes effects of hypoxia in various locations around the world, which provides lessons for potential consequences of hypoxia in the Gulf of Mexico. They note that hypoxia was probably not a prominent feature of the shallow continental shelf in the Northern Gulf of Mexico prior to the 1920's through 1950's based on geo-chronology of sediment cores. A longer, 2000-year chronology in the Chesapeake indicates that early European settlement of the watershed was a key feature that set the stage for current oxygen problems. Improved water quality in Lake Erie is the best example in the US that large ecosystems can respond positively to nutrient regulation, but the time interval for recovery can be long. In Lake Erie, the extent of hypoxia was similar between 1970 and 1990 despite reduced nutrient loads. Delayed improvements in oxygen levels are argued to be consistent with mechanisms and processes that contribute to ecosystem's resilience (Charlton et al., 1993), and as a consequence improvements in oxygen may not be noticed for decades following implementation of management actions.

Hendon, Laura A. Erik A. Carlson, Steve Manning, and Marius Brouwer, 2006 "Cross-talk between Pyrene and Hypoxia Signaling Pathways in Embryonic Cyprinodon variegates" Presentation at "Hypoxia Effects on Living Resource in the Gulf of Mexico" NOAA Center for Sponsored Coastal Ocean Research, Tulane University, New Orleans, LA. September 25–26, 2006.

> Abstract: The aryl hydrocarbon nuclear translocator (ARNT) is a general dimeric partner for the aryl hydrocarbon receptor (AhR) and hypoxia-inducible factor one alpha (HIF1-α). The AhR/ARNT complex binds to promoters in target genes, such as CYP1A1, resulting in alterations in gene expression, while the HIF1-α/ARNT heterodimer binds to hypoxia response elements in target genes, such as VEGF. While AhR is activated by PAHs, such as

pyrene, HIF1-α is activated by hypoxia. Since ARNT is a general dimeric partner for both AhR and HIF1-α, possible cross-talk may exist between the two pathways in which the activation of one results in inhibition of the other. The objective of this study was to determine if pyrene-activation of AhR2, or hypoxia-activation of HIF1-α could sequester the ARNT protein away from HIF1-α and AhR2, respectively, resulting in reduced developmental toxicity associated with hypoxia or pyrene alone in embryonic *Cyprinodon variegatus*. As a first step to examine this hypothesis, we cloned AhR2, CYP1A1 (PAH-activated gene) and VEGF (HIF-activated gene). Next, pyrene (20, 60, and 150 ppb) and hypoxia's (1–2 ppm) individual developmental toxicity endpoints were determined, together with CYP1A1 and VEGF expression levels using real-time quantitative RT-PCR. Combined treatments of pyrene and hypoxia were examined in order to determine sequestration of the ARNT protein and developmental toxicity endpoints. Results demonstrate that pyrene-treated embryos alone develop toxicity endpoints such as pericardial edema and dorsal body curvature. Hypoxia-treated embryos alone display delayed hatching and less-developed characteristics in comparison to normoxic treatments. Under hypoxic conditions alone, real-time quantitative RT-PCR determined that VEGF was down-regulated significantly at 24 hpf, while at 14 dph, the HIF-activated gene was significantly up-regulated. Pyrene-treated embryos showed a dose-dependent and time-dependent response in CYP1A1 regulation with increasing expression over time of exposure. The combined effects of pyrene and hypoxia appeared to alter VEGF expression, while CYP1A1 remained unaffected in *C. variegatus*.

Montagna, Paul, Ben Hodges, David Maidment and Barbara Minsker, 2006. "Long-Term Studies of Hypoxia in Corpus Christi Bay: The Cybercollaboratory Testbed" Presentation at "Hypoxia Effects on Living Resource in the Gulf of Mexico" NOAA Center for Sponsored Coastal Ocean Research, Tulane University, New Orleans, LA. September 25–26, 2006.

Abstract: Corpus Christi Bay is a shallow (~3.2 m) enclosed bay with a level bottom. It experiences high wind speeds, temperatures, and receives a low amount of fresh water inflow. Hypoxia has been documented in the southeastern region of Corpus Christi Bay every summer since 1988. Hypoxia found in bottom waters, usually within 1 m from bottom, when the bay is stratified. Over the last 20 years, there has been increased surface water temperatures, but no change in nutrient concentrations, which are low. Ecosystem processes during salinity stratification likely drive the hypoxia, because respiration is stimulated and the surface and bottom water masses are not mixing. Hypoxia causes reduced benthos abundance, biomass, and diversity. The reduction is due to loss of deeper-dwelling organisms, and is likely a direct effect (stress or death), and not an indirect effect (increased predation by exposure to the surface). There is increased interest in developing real-time environmental forecasting and management to better monitor and understand large-scale, event-based environmental phenomena, e.g., hypoxia and flooding. A new project focuses on creating a new Corpus Christi Bay Observatory Testbed Project to demonstrate how cyberinfrastructure can enable real-time forecasting from a hydrographic information system. Although only a few months old, the testbed project has already created a few simple models and visualization tools that improved sampling designs to better identify hypoxic events, extent, and intensity.

O'Connor, Thomas and David Whitall, 2007. "Linking Hypoxia to Shrimp Catch in the Northern Gulf of Mexico", Marine Pollution Bulletin Vol. 54, No. 4 (April), pp 460–463.

Abstract: This study updates the statistical analysis of Zimmerman and Nance (2001) of the effect of hypoxia on commercial shrimp landings data for 1985 through 2004. This study uses commercial landings data, not the interview data, and therefore does not use spatial data on the location of catch. The paper confirms the results of Zimmerman and Nance that there is no correlation of hypoxic area with landings of white shrimp or with landings of brown shrimp in Louisiana, but there is a significant correlation with the total combined landings in Texas and Louisiana. Unlike Zimmermann and Nance, they find a significant relationship between the hypoxic area and brown shrimp landings in Texas alone. Hypoxia explains about 32% of the variance in catch using data for catch in July and August, and about 27% of the variance in catch using annual data.

Perez, Amy N., Leon Oehlers and Ronald B. Walter, 2006. "Detection of Hypoxia-related Proteins in Medaka (*Oryzias latipes*) by Difference Gel Electrophoresis and Identification by Sequencing of Peptides using MALDI-TOF Mass Spectrometry" Presentation at "Hypoxia Effects on Living Resource in the Gulf of Mexico" NOAA Center for Sponsored Coastal Ocean Research, Tulane University, New Orleans, LA. September 25–26, 2006.

Abstract: Multidimensional separation techniques combined with matrix-assisted laser desorption/ionization tandem time-of-flight mass spectrometry (MALDI-TOF/TOF-MS) were used to identify hypoxia-related biomarker proteins in tissues of medaka fish (*Oryzias latipes*) and medaka cultured cells. The multidimensional protein/peptide separation methods used included two-dimensional difference gel electrophoresis (2D-DIGE) using fluorescent cyanine dyes, and gel electrophoresis combined with reversed phase liquid chromatrography of tryptic peptides isotopically labeled with ^{16}O or ^{18}O (geLC-MS). In both methods, control and hypoxia-treated tissue or cell protein extracts were differentially labeled, combined in 1:1 mass ratios, and subjected to separation and MALDI-TOF/TOF-MS analysis of tryptic peptides derived from proteins exhibiting significant changes in expression upon hypoxia exposure. Prior to MALDI-TOF/TOF-MS analysis, the peptides were N-terminally sulfonated using the derivatizing reagent 4-sulfophenyl isothiocyanate (SPITC) to enhance the post-source decay (PSD) fragmentation spectra of the peptides in MALDI-TOF/TOF-MS, which was shown to dramatically improve de novo sequencing of labeled peptides. The methods described here were used to monitor and analyze the changes in protein resulting from exposures of both cultured medaka cells and medaka fish to hypoxic conditions (0.8–1.0 mg/L dissolved oxygen) for periods up to 120 h. We have identified a number of potential candidate biomarker proteins differentially-regulated upon exposure to hypoxia, including carbonic anhydrase, hemoglobin, calbindin, aldolase, glutathione-S-transferase, succinate dehydrogenase, and lactate dehydrogenase.

Rabalais, Nancy N, 2006. "Benthic Communities and the Effects of Hypoxia in Louisiana Coastal Waters" Presentation at "Hypoxia Effects on Living Resource in the Gulf of Mexico" NOAA Center for Sponsored Coastal Ocean Research, Tulane University, New Orleans, LA. September 25–26, 2006.

Abstract: The responses of the benthic fauna to decreasing concentration of dissolved oxygen follow a fairly consistent pattern of progressive stress and mortality as the oxygen concentration decreases from 2 mg L^{-1} to anoxia (0 mg L^{-1}). Motile organisms (fish, portunid crabs, stomatopods, penaeid shrimp and squid) are seldom found in bottom waters with oxygen concentrations less than 2 mg L^{-1}. Below 1.5–1 mg L^{-1} oxygen concentration, less motile and burrowing invertebrates exhibit stress behavior, such as emergence from the sediments, and eventually die if the oxygen remains low for an extended period. At minimal concentrations just above anoxia, sulfur-oxidizing bacteria form white mats on the sediment surface, and at 0 mg L^{-1}, there is no sign of aerobic life, just black anoxic sediments. The composition of the benthic communities reflects differences in sedimentary regime, seasonal input of organic material and seasonally severe hypoxia/anoxia. Decreases in species richness, abundance and biomass of organisms are dramatic when bottom-waters are affected by severe hypoxia/anoxia. Some macroinfauna, the polychaetes Ampharete and Magelona and a sipuculan Aspidosiphon, are capable of surviving extremely low dissolved oxygen concentrations and/or high hydrogen sulfide concentrations. Macroinfauna, primarily opportunistic polychaetes, increase in the spring following flux of primary produced carbon, and increase to a lesser extent in the fall following the dissipation of hypoxia. Fewer taxonomic groups characterize the severely affected benthos, and long-lived, higher biomass and direct-developing species are mostly excluded. Suitable feeding habitats (in terms of severely reduced populations of macroinfauna that may characterize substantial areas of the seabed) are frequently removed from the foraging base of demersal organisms, including the commercially important penaeid shrimps.

Switzer, Theodore S., Edward J. Chesney, and Donald M. Baltz, 2006. "Habitat Selection by Flatfishes along Gradients of Environmental Variability: Implications for Susceptibility to Hypoxia in the Northern Gulf of Mexico" Presentation at "Hypoxia Effects on Living Resource in the Gulf of Mexico" NOAA Center for Sponsored Coastal Ocean Research, Tulane University, New Orleans, LA. September 25–26, 2006.

Abstract: Although eutrophication in the northern Gulf of Mexico contributes to the high fisheries productivity characteristic of the region, nutrient over-enrichment leads to the seasonal formation of hypoxic (< 2 mg L^{-1} O_2) bottom water along the Louisiana-Texas continental shelf. Despite an increase in the magnitude and duration of hypoxic episodes in recent decades, fisheries landings have remained high; nevertheless, hypoxia remains a persistent threat to the long-term sustainability of regional fisheries production. The greatest threat to mobile nekton is likely the influence of reduced dissolved oxygen concentrations on habitat quality, potentially forcing the movement of individuals and/or prey from generally favorable habitats. At the population level, these movements may result in altered spatial distributions that reflect selection of resources along gradients of environmental variability. To unravel the potential influence of hypoxia on the distribution of nekton, we examined patterns of habitat use by several abundant flatfishes based on data collected during summer SEAMAP groundfish surveys from 1987 to 2000. Results from habitat suitability analyses indicated that most flatfishes selected a restricted range of suitable depths, temperatures, and salinities. Although most flatfishes were tolerant of moderately-low dissolved oxygen concentrations, hypoxic environments were generally avoided, indicating that hypoxia likely renders large areas of the Gulf of Mexico unsuitable. In comparisons of spatial habitat suitabilities between years of moderate (< 15,000 km^2) and severe hypoxia (> 15,000 km^2), all flatfishes exhibited a reduction in the suitability of areas immediately west of the Mississippi River and a concomitant increase in suitability within adjacent areas. Altered spatial distributions corresponded to species-specific suitabilities along depth, temperature, and salinity gradients, indicating that habitat suitability analyses may be effective in predicting population-level responses to hypoxic episodes.

Wells, Melissa C., Zhenlin Ju, Sheila J. Heater and Ronald B. Walter, 2006. "Microarray Gene Expression Analyses in Medaka (Oryzias latipes) Exposed to Hypoxia" Presentation at "Hypoxia Effects on Living Resource in the Gulf of Mexico" NOAA Center for Sponsored Coastal Ocean Research, Tulane University, New Orleans, LA. September 25–26, 2006.

Abstract: We are investigating the genomic and proteomic effects of hypoxia exposure using the Japanese medaka (Oryzias latipes) aquaria fish model as a tool for biomarker discovery. We have developed a hypoxia exposure system allowing programmable exposure scenarios and have initiated experimental assessment of changes in gene expression and protein abundance using microarray and 2D-DIGE gel analyses of hypoxia exposed fish. We present the design, construction, validation, and subsequent use of a medaka 8,046 (8 K) unigene oligonucleotide microarray to begin the study of hypoxia exposure. Array performance was validated via self-self hybridization. Optimization of sample size needed for robust array data, based upon the number features detected and the signal intensity, suggest 2 μg total RNA as a starting template for amplification is sufficient. For treatment, adult medaka are exposed to a hypoxic environment of 4% dissolved oxygen (DO) for 2 days and then the DO lowered to 2% for an additional 5 days. Upon sacrifice, changes in gene expression in brain, liver, skin, and gill tissues of these fish were assessed in conjunction with matched control fish exposed similarly to 18% DO. Analyses of array results identified 501 features from brain, 442 from gill, and 715 features from liver that exhibit statistically significant changes in transcript abundance upon hypoxia exposure. Nine features were found to exhibit common expression patterns between all three tissues. Data mining of the array results suggest hypoxic exposure results in a general slowdown of metabolic function. Real-time PCR was then employed to support the microarray results and this independent validation agreed well with the microarray findings. Overall these results indicate the medaka microarray will be a sound diagnostic tool for changes in gene expression due to hypoxia exposure.

Zimmerman, Roger J. and James M. Nance, 2001. "Effects of Hypoxia on the Shrimp Industry of Louisiana and Texas" Chapter 15 in Rabalais, N.N. and R.E. Turner, Coastal Hypoxia: Consequences for Living Resources Coastal and Estuarine Studies, Vol. 58, pp 293–310.

Abstract: This study carries out a statistical test for effects of hypoxia on commercial catch of shrimp in the Gulf of Mexico for 1985–1997. The analysis combines landings data and interview data on fishing effort, catch and location of each trip. The analysis is spatially explicit, based on catch in 9 statistical subareas in Louisiana and Texas, with each subarea divided into 10 depth zones. Zimmerman and Nance found no correlation of hypoxic area with landings of white shrimp or with landings of brown shrimp in Louisiana, but they found a statistically significant relationship between hypoxia and combined landings in Texas and Louisiana. The finding of no relationship for white shrimp is consistent with prior expectations, because white shrimp are less sensitive to hypoxia (Renaud, 1986), and because white shrimp habitat is mostly in-shore the hypoxic region. In comparison, brown shrimp travel from inshore areas to offshore in order to spawn. Since brown shrimp migrate through the hypoxic region, they are more likely to be effected by hypoxia. The absence of a significant relationship between the size of the hypoxic region and catch of brown shrimp in Louisiana may be explained by the fact that much of the catch in Louisiana occurs in-shore of the hypoxic region, while catch in Texas occurs offshore.

Zou, Enmin, 2006. "Impacts of Hypoxia on Physiology and Toxicology of the Brown Shrimp Penaeus aztecus" Presentation at "Hypoxia Effects on Living Resource in the Gulf of Mexico" NOAA Center for Sponsored Coastal Ocean Research, Tulane University, New Orleans, LA. September 25–26, 2006.

Abstract: The brown shrimp, Penaeus aztecus, in the northern Gulf of Mexico is faced with dual stresses of environmental hypoxia, which occurs as a result of oxygen depletion from microbial decomposition of organic materials from algal blooms, and pollution from polycyclic aromatic hydrocarbons (PAHs) from petroleum and gas production on the continental shelf of the northern Gulf of Mexico. This study aimed to address the questions of (1) whether the presence of PAH contamination makes penaeid shrimps more susceptible to hypoxia and (2) whether hypoxia can promote PAH bioaccumulation in penaeid shrimps. The susceptibility of shrimps to hypoxia was represented by the oxyregulating capacity, a physiological parameter that describes how well an animal regulates its oxygen consumption when subjected to hypoxia. It was found that acute exposure to naphthalene significantly reduced the oxyregulating capacity of Penaeus aztecus. An ensuing consequence of a decrease in oxyregulating ability is that the stress from the lack of oxygen would set in sooner in the presence of PAH contamination than when shrimps are in the clean environment. Hypoxia was found to have no significant effect on naphthalene bioaccumulation in Penaeus aztecus. The absence of a significant effect was attributed to increased naphthalene metabolism in the brown shrimp subjected to hypoxia.

Appendix B: Flow Diagrams and Mass Balance of Nutrients

Global Material Cycles

For the reader's information, the following flow diagrams of the global nitrogen, phosphorus, and silicon cycles (Figs. B-1, B-2, and B-3) are taken from the *Encyclopedia of Earth*, a new electronic reference about the Earth, its natural environments, and their interaction with society. The Encyclopedia is a free, fully searchable collection of articles written by scholars, professionals, educators, and experts (http://www.eoearth.org/eoe/about). Its contents may be freely copied and distributed with proper attribution. These diagrams are not the deliberative products of the Study Group but are provided to illustrate important processes discussed in this book including: fertilization, nitrogen fixation, nitrification, denitrification, ammonification, nutrient assimilation, sedimentation, recycling from sediment, and weathering of rocks. Chemical equations representing processes depicted in the flow diagrams are available from many sources in the published literature, including standard textbooks on biogeochemistry, limnology, and oceanography.

Atmospheric Deposition

The *Integrated Assessment* concluded that atmospheric deposition as a new nitrogen input to the Mississippi River basin was not as important as agricultural sources but that deposition nonetheless was a significant source (Goolsby et al., 1999). Atmospheric deposition of nitrogen generally shows a trend of increasing from west to east in the Mississippi basin, and deposition was a particularly important source of nitrogen in the Ohio River basin (Goolsby et al., 1999). The *Integrated Assessment* followed the net anthropogenic nitrogen input (NANI) budgeting approach established by the International SCOPE Nitrogen Project in assuming that deposition of oxidized nitrogen (NO_y) is a new input of nitrogen while the deposition of ammonium is not but rather is a recycling of nitrogen emitted to the atmosphere from agricultural sources within the basin (Howarth et al., 1996). The oxidized nitrogen is presumed to come largely from fossil–fuel combustion and, thus, is not accounted for in any other input to the budget (Goolsby et al., 1999; Howarth et al., 1996). The

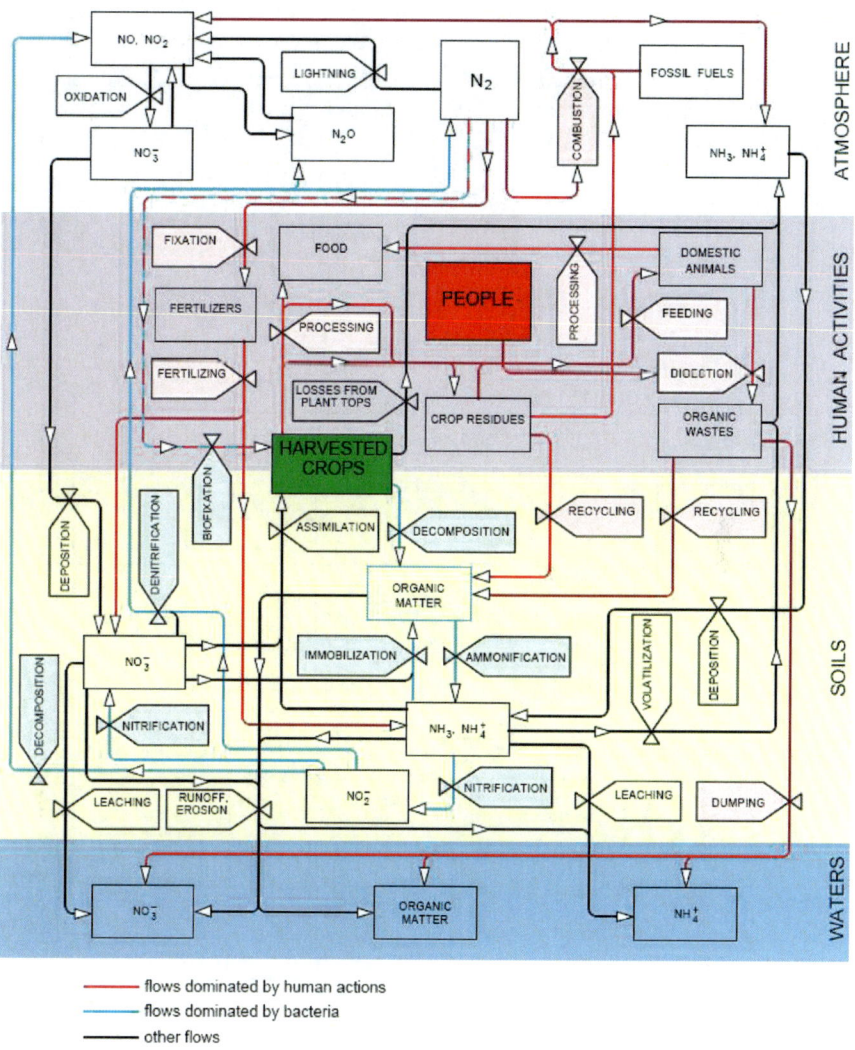

Fig. B.1 Nitrogen cycle flow diagram. Taken from Encyclopedia of Earth (2007) at http://www.eoearth.org/global_material_cycles

Integrated Assessment further considered that the deposition of organic nitrogen was a new input of nitrogen (Goolsby et al., 1999).

The *Integrated Assessment* used monitoring data to estimate NOy deposition and made a very rough guestimate for the magnitude of deposition of organic nitrogen. They used data from the NADP for wet deposition and from CASTnet for dry deposition. This yielded an average estimate of NOy deposition for the Mississippi

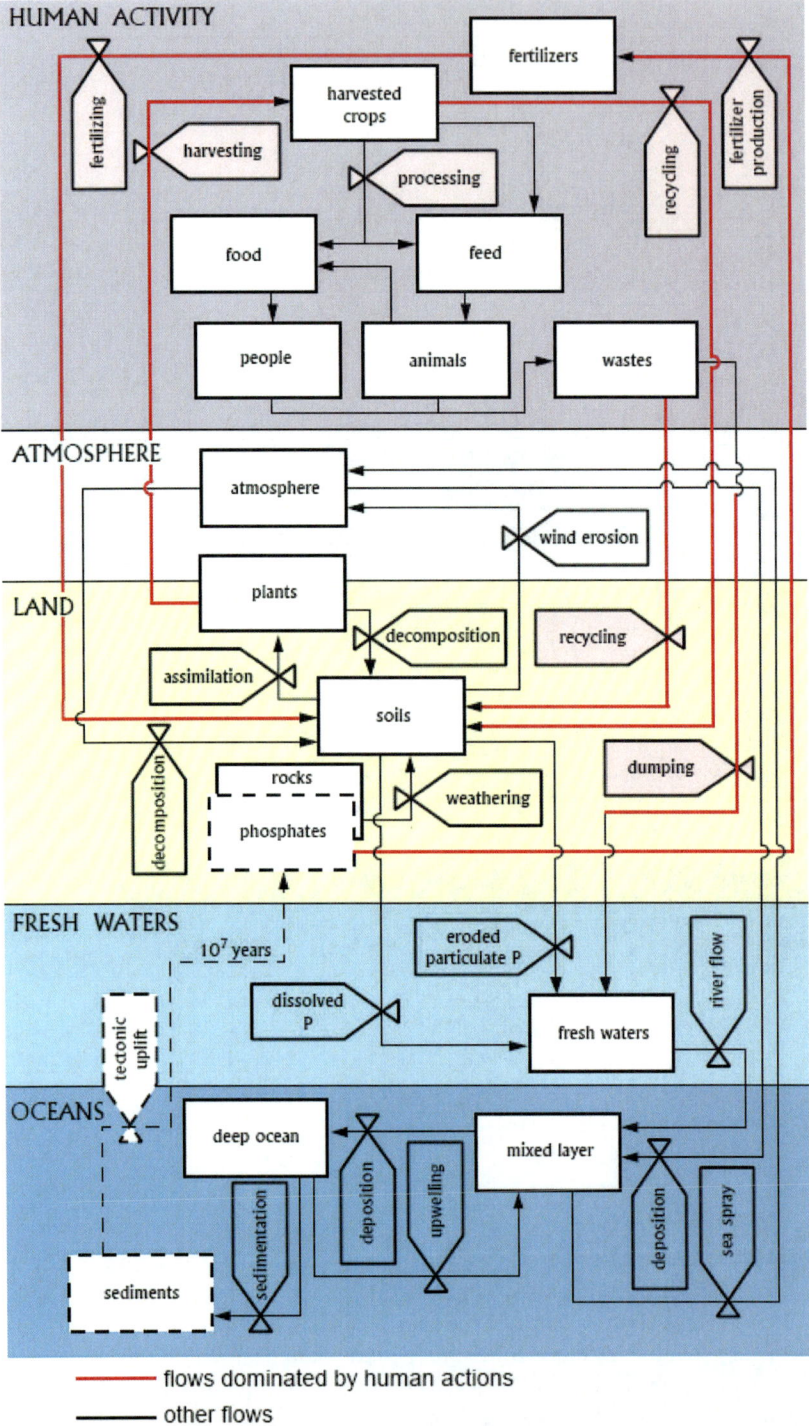

Fig. B.2 Phosphorus cycle flow diagram. Taken from Encyclopedia of Earth (2007) at http://www.eoearth.org/global_material_cycles

Appendices

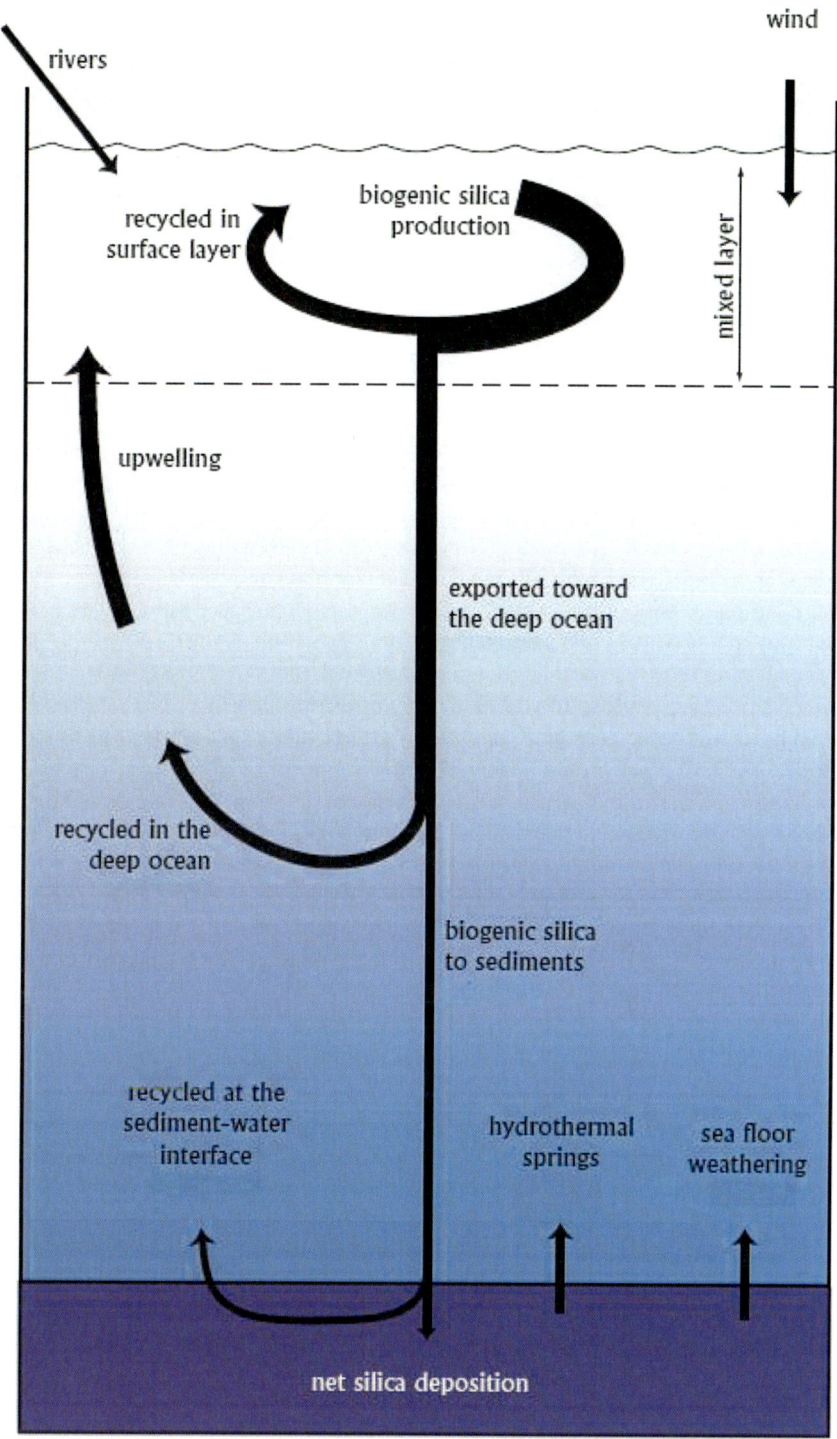

Fig. B.3 Silicon cycle flow diagram. Taken from Encyclopedia of Earth (2007) at http://www.eoearth.org/global_material_cycles

River basin for the time period 1988–1994 of 3.4 kg N/ha-year (3 lb N/ac-year), of which 2 kg N/ha-year (1.8 lb N/ac-year) was nitrate in wet deposition and 1.4 kg N/ha-year (1.25 lb N/ac-year) was NOy dry deposition (Goolsby et al., 1999). The assessment estimated the deposition of organic nitrogen as 1 kg N/ha-year (0.89 lb N/ac-year), yielding a total estimate for new nitrogen deposition of 4.4 kg N/ha-year (3.9 lb N/ac-year) (Goolsby et al., 1999). This can be compared with an estimate for NOy deposition derived from the GCTM model, which estimates deposition rates from data on emissions to the atmosphere and on rates of reaction and advection within the atmosphere (Prospero et al., 1996). For the Mississippi River basin for essentially the same time period used in the *Integrated Assessment*, the GCTM model suggested a total NOy deposition of 6.6 kg N/ha-year (5.9 lb N/ha-year), with 6.2 kg N/ha-year (5.5 lb N/ac-year) of this input being attributable to new inputs from fossil–fuel burning and 0.4 kg N/ha-year (0.36 lb N/ac-year) originating from natural sources (Howarth et al., 1996).

Holland et al. (1999, 2005) noted that deposition estimates based on monitoring data are typically lower than those from emission-based models across most of the United States. For the northeastern United States from Maine through Virginia, the estimates from the GCTM model (Howarth et al., 1996) are again almost twice as high as are estimates from NADP and CASTnet monitoring data (Boyer et al., 2002). There are many possible reasons for this discrepancy, but probably at least part of the problem lies with an underestimation of dry deposition by the CASTnet program (Holland et al., 1999; Howarth, 2006; Howarth et al., 2006b). Most CASTnet monitoring stations are purposefully located away from emission sources, and deposition is likely to be higher near these emission sources, creating a bias in the network. Further, the CASTnet program only estimates deposition of nitrogen in particles and deposition of nitric acid vapor. The deposition of several other gases (including NO, NO$_2$, and nitrous acid vapor) is not measured. Deposition of these gases, which would be included in the estimates from the emission-based models, is likely to be particularly high near emission sources (Howarth, 2006). Both the GCTM and the TM3 models only estimate deposition at coarse spatial scales, but a new emission-based model (CMAQ) shows promise for estimation at relatively fine spatial scales (Robin Dennis, NOAA, personal communication). This model suggests very high NOy deposition rates near urban centers in the eastern United States and associated with power plant emissions in the Ohio River basin.

In the mass balance presented in Sect. 3.2, deposition was estimated as in Goolsby et al. (1999). ***Organic N was not included, however, as it was unclear what the importance of this form of N was, or what an appropriate estimate would be (Keene et al., 2002). A comparison by region for 2001 was made of deposition inputs from the NOy estimate used in the mass balance and from the CMAQ model. For the upper Mississippi basin, NOy deposition was 4.2 kg N/ha-year (3.8 lb N/ac-year), the same as the CMAQ model[1]. For the Missouri basin, both methods

---

[1] CMAQ model unpublished results courtesy of Robin Dennis, NOAA, with analysis by states provided by Dennis Swaney, Cornell University; unpublished.

again gave similar estimates, with NOy deposition of 2.2 N/ha/year (2 lb N/ac/year) and CMAQ modeled deposition of 2.1 kg N/ha/year (1.9 lb N/ac/year). For regions with more fuel combustion, the pattern was different, with an Ohio basin NOy estimate of 5.0 N/ha/year (4.5 lb N/ac/year) and the CMAQ model estimate of 8.8 kg N/ha/year (7.8 lb N/ac/year). For the lower Mississippi River basin, NOy was 3.7 kg N/ha/year (3.3 lb N/ac/year) and the CMAQ estimated was 5.1 kg N/ha/year (4.6 lb N/ac/year). Overall, this supports the assertion of the mass-balance analysis that, for the upper Mississippi basin, atmospheric deposition is a small component about 8% of N inputs and is more important in the Ohio region (about 16% of N inputs according to the CMAQ model for 2001).

Appendix C: Animal Production Systems

Intensification of Animal Feeding Operations

Current census information shows that there has been an 18% increase in the number of pigs in the United States during the past 10 years along with a 72% decrease in the number of farms. Over the same 10 years, the number of dairies has decreased by 40%, but herd size has increased by 50%. A similar trend in the poultry and beef industries has also occurred, with 97% of poultry production in the United States coming from operations with more than 100,000 birds and over a third of beef production from <2% of the feedlots (Gardner, 1998). Fattened cattle numbers remained fairly constant from 1982 to 1997, but the number of fattening operations decreased more than 50% (Kellogg et al., 2000). Overall, cattle, pig, and poultry numbers have increased 10–30%, while the number of farms on which they were reared has decreased 40–70% during the past 10 years (Gardner, 1998).

Nutrient Budgets

The large-scale consolidation has created much larger animal feeding operations, which makes economical utilization and re-distribution of manure to croplands difficult and has profound consequences for farm and regional nutrient transfer and management within the MARB. For example, the accumulation of nutrients is first evident at the farm scale, where N and P management is affected by daily operation decisions and the long-term goals of each farmer. For example, the potential for P and N surplus on farms with AFOs can be much greater than in cropping systems where nutrient inputs become dominated by feed rather than fertilizer (Table C.1). With a greater reliance on imported feeds, only 30% of N and 29% of P in purchased feed for a 1280-hog operation on a 30-hectare farm could be accounted for in farm outputs. These nutrient budgets clearly show that animal feed is the largest input of nutrients to farms with AFOs and thus is the primary source of on-farm nutrient excess, for which a resolution will require innovative management. Current animal

Table C.1 Farming system and nutrient budget; amounts given in kg ha^{-1} year^{-1}

Farming system	Nutrient input in		Output in produce	Surplus	Nutrient utilization (%)
	Feed	Fertilizer			
Nitrogen budget					
Cash crop[a]	–	95	92	3	97
Dairy[b]	155	40	75	120	38
Hog[c]	390	10	120	280	30
Poultry[d]	5800	–	1990	3810	34
Phosphorus budget					
Cash crop[a]	–	22	20	2	91
Dairy[b]	30	11	15	26	37
Hog[c]	105	–	30	75	29
Poultry[d]	1560	–	440	1120	28

[a] 30-hectare cash-crop farm growing corn and alfalfa.
[b] 40-hectare farm with 65 dairy Holsteins averaging 6600 kg milk cow^{-1} year^{-1}, 5 dry cows and 35 heifers. Crops were corn for silage and grain and alfalfa and rye for forage.
[c] 30-hectare farm with 1280 hogs; surplus includes 36 kg P and 140 kg N ha^{-1} year^{-1} manure exported from the farm.
[d] 12-hectare farm with 74,000 poultry layers; surplus includes 180 kg P and 720 kg N ha^{-1} year^{-1} manure exported from the farm.

Table C.2 Number of animals and amount of manure produced and N and P excreted within the MARB states based on information from the 1997 US Census of Agriculture (data obtained from USDA-ERS, http://ers.usda.gov/data/MANURE/)

Animal type	Number of farms	Animal units	Manure excreted[a] (Mg)	Manure N excreted (million kg)	Manure P excreted (million kg)
Beef	837,972	52,627,536	71,354,009	2,712	864
Dairy	77,363	5,944,742	13,287,687	439	79
Poultry	45,870	3,044,000	8,743,736	433	149
Swine	84,717	6,591,998	7,310,054	419	124

[a] Manure in dry state, as excreted adjusted for water content.

number and estimated manure N and P production within the MARB is given in Table C.2.

Nutrient Surpluses

USDA (2003) estimated the amount of manure produced from animal distribution numbers from the 1997 US Census of Agriculture, using standard values of manure production and nutrient concentration for each animal type. Estimates of excess N and P were calculated based on crop N and P removal and the assumption that all suitable crop and pasture land was available for manure application. Most areas with CAFOs have some excess N and P. These distributions demonstrate that within the MARB, regional excesses were similar for N and P.

Targeting Remedial Strategies Within the MARB

The importance of targeting nutrient management within a watershed is shown by several MARB studies. In the early 1980s, conservation practices were installed on about 50% of the Little Washita River watershed (54,000 hectares or 133,000 acres) in central Oklahoma. Practices included construction of flood control impoundments, eroding gully treatment, and conservation tillage (Sharpley et al., 1996, Sharpley and Smith, 1994). Although conservation measures decreased N and P export about 80 to 92%, there was no effect on P concentration in flow at the outlet of the main Little Washita River watershed. Thus, a lack of effective targeting of nutrient management and control of major sources of nutrient export contributed to field or subwatershed scale responses not being translated to reductions in nutrient export from the main Little Washita River watershed.

Managing Manures

Manure application timing and method relative to rainfall influences the concentration of N and P in runoff (Dampney et al., 2000; Sims and Kleinman, 2005). For example, several studies have shown a decrease in N and P loss with an increase in the length of time between manure application and surface runoff (Djodjic et al., 2000; Edwards and Daniel, 1993a; Sharpley, 1997; Westerman et al., 1983). This decrease can be attributed to the reaction of added P with soil and dilution of applied P by infiltrating water from rainfall that did not cause surface runoff.

The incorporation of manure into the soil profile either by tillage or by subsurface placement decreases the potential for P loss in surface runoff. Rapid incorporation of manure also reduces NH_3 volatilization and potential loss in runoff as well as improving the N:P ratio for crop growth. Mueller et al. (1984) showed that incorporation of dairy manure by chisel plowing reduced total P loss in runoff from corn by about 95%, compared to no-till areas receiving surface applications. In fact, P loss in runoff was decreased by a lower concentration of P at the soil surface and a reduction in runoff with incorporation of manure (Mueller et al., 1984; Pote et al., 1996). As with fertilizer application methods, other factors are important in selecting or recommending the most appropriate application method. Equipment availability, whether the soil is sufficiently free of rocks to allow subsurface application, labor requirements, product availability, and availability of operating capital all affect the application method decision.

Crop Selected to Receive Manure Application

Manure has traditionally been applied for corn or other grass production. However, corn acreage to which manure is applied has not expanded proportionally to animal operation expansions; thus the risks increase for applying manure in excess of the amount necessary to meet crop nutrient requirements (Dou et al., 1998; Schmitt et al., 1996). One solution to minimize these risks, and the subsequent potential risk of NO_3 leaching to ground water, is to select alternative crops to receive manure applications. Although legumes are not usually considered for manure application, soybean can annually remove as much as 385 kg N/ha (344 lb N/ac) (Shibles, 1998) and alfalfa as much as 500 kg N/ha (446 lb N/ac) (Russelle et al., 2001), compared to less than 200 kg N/ha (179 lb N/ac) for corn. Schmidt et al. (2000) demonstrated that nodulation in soybean effectively compensated with additional N when manure N was insufficient to meet crop demands; so if necessary, manure could be applied conservatively without risk of applying too little to meet crop needs.

Rate and Frequency of Application

As might be expected, N and P loss in runoff increases with greater frequency and rates of applied manure (Edwards and Daniel, 1993b; McDowell and McGregor, 1984). Although rainfall intensity and duration, as well as when rainfall occurs relative to applied manure, influence the concentration and overall loss of manure N and P in runoff, the relationship between potential loss and application rate is critical to establishing environmentally sound nutrient management guidelines. Also evident is that the effect of applied manure on increasing the concentration of P in surface runoff can be long lasting. For instance, Pierson et al. (2001) found that a poultry litter application tailored to meet pasture N demands elevated surface runoff P for up to 19 months after application. Although few studies have evaluated the loss of P in surface runoff as a function of application frequency, more frequent manure applications can be expected to rapidly increases soil P (Haygarth et al., 1998; Sharpley et al., 1993; 2005; Sims et al., 1998), with a concomitant increase in runoff P loss.

Intensity and Duration of Grazing

As beef grazing of pastures is an important component of animal production in many regions of the MARB, careful management of grazing is needed to minimize P loss and water quality impacts. The localized accumulations of P where manure is deposited can saturate the P sorption capacity of a soil, increasing the potential for P loss from grazed pastures in runoff or drainage waters. However, at a field and watershed scale, it is likely that critical stocking factors, such as density and duration, will influence both hydrologic and chemical factors controlling P transport. For example, Owens et al. (1997) found that decreasing grazing density and duration dramatically reduced runoff and erosion from a pastured watershed in Ohio. Clearly, increased runoff and erosion with grazing will enhance the potential for P loss. In Oklahoma, Olness et al. (1975) found that P losses were greater from continuously (4.6 kg P/ha-year or 4.1 lb P/ha-year) than rotationally grazed pastures (1.3 kg P/ha-year or 1.2 lb P/ha-year). In fact, P losses with continuous grazing were greater than from alfalfa or wheat [2.7 kg P/ha-year (2.4 lb P/ha-year); Olness et al., 1975]. However, the work of Owens et al. (1997) does show that, when management is changed, the impacts of the previous grazing impacts were not long lasting, changing within a year. Even so, there is a need to determine critical stocking densities and durations as a function of grazing management.

Stream-Bank Fencing

By observing four pastured dairy herds with stream access over four intervals during the spring and summer of 2003 in the Cannonsville Watershed south central, New York, James et al. (2007) were able to estimate fecal P contributions to streams. In the herds observed, on a per cow basis, cattle were especially likely to defecate in the stream, although they spent a small proportion of their time there. On average, approximately 30% of all fecal deposits expected from a herd were observed to fall on land within 40 m of a stream, and 7% fell directly into streams. Although amenities in pasture (such as water troughs, feeders, salt, and shade located away from the stream) did affect where cattle congregated, the stream demonstrated a consistent draw.

Using spatial databases of streams, pasture boundaries, and animal characteristics (i.e., number of cattle, time in pasture, and type of cattle [heifers versus milk cows]) for 90% of the dairy farms in the Cannonsville watershed, approximately 3,600 kg (7,940 lb) of manure P are estimated as deposited directly into streams with 7,650 kg (16,900 lb) deposited in pasture near streams (<10 m) from the 11,000 dairy cattle in the watershed. At this magnitude, P loadings represent a significant environmental concern, with in-stream deposits equivalent to approximately 12% of watershed-level P loadings attributed to agriculture (Scott et al., 1998). Riparian shade can also attract grazing cattle and influence P loss in stream flow.

Appendix D: Calculation of Point Source Inputs of N and P

As discussed in Sect. 3.2 and 4.5.8, estimates of N and P fluxes from sewage treatment plants from the MART (2006b) report were much lower for both total N and total P than in the *Integrated Assessment*. As pointed out in the MART report, much of this decline is thought to be due to the values assigned for total N and P concentrations in sewage treatment plants effluent. Few measured data were used, but rather estimated values were applied to most. Most estimates were made using a "typical pollutant concentration" (TPC) for N or P based on the level of treatment. These TPC's were from an update of values compiled in a report by Tetra Tech (1998). The 2006 MART report assumed that sewage treatment plants with advanced wastewater treatment had TPC values of 5.6 and 0.82 mg/L for total N and total P, respectively. The MART report then applied these assumed values to estimated daily discharges to calculate an estimated daily flux. The MART report further assumed that plants that had less than advanced wastewater treatment had TPC values of 11.2 and 2.02 mg/L for total N and total P, respectively, applied to estimated daily discharges to calculate an estimated daily flux. The Study Group is not comfortable with these assumptions and instead believes that most wastewater treatment plants in the MARB had TPCs applied that were too low. The Study Group, therefore, adjusted the database, by using TPCs of 11.2 and 2.02 mg/L for total N and total P, respectively, for plants with advanced wastewater treatment, and TPCs of 15 and 4 mg/L for total N and total P, respectively, for plants with less than advanced treatment.

As an example of how these adjustments changed estimates, the Study Group examined seven Chicago plants (Stickney is the largest sewage treatment plant in the basin) and one in Champaign-Urbana, IL, where measured flux data were available (daily to weekly measurements of total N and total P and flow were made at each plant). From this analysis, it is clear that the TPCs used in the MART report were not appropriate and gave substantially lower flux estimates (Table D.1) than the actual measured values. The MART report indicated that each of these plants had advanced treatment, and therefore applied their estimated TPCs of 5.6 mg/L total N and 0.82 mg/L total P, respectively. Most plants in the MARB do not have treatment processes (either biological or chemical) to remove P, and much of the advanced treatment is to nitrify ammonium to nitrate, because most are permitted for ammonia in effluent.

Table D.1 Comparison of MART estimated sewage treatment plant annual effluent loads of total N and P and values from measurements at each plant for 2004

Plant	MART (tons N/year)	Measured (tons N/year)	Difference (%)	MART (tons P/year)	Measured (tons P/year)	Difference (%)
Stickney	6,282	9,850	64	921	1,105	83
Calumet	1,799 (3,599)	3,243	55	264 (650)	1,065	25
Lemont	13 (26)	51	25	2 (5)	8	23
Northside	2,259 (4,518)	3,161	71	331 (207)	441	75
Egan	209 (418)	386	54	31 (75)	99	31
Hanover Park	70 (139)	144	48	10 (73)	33	31
Kirie	201 (402)	331	61	29 (73)	44	67
Champaign-Urbana	77 (155)	310	25	11 (28)	58	20

*All plants are in Illinois. Also shown in parentheses is the recalculated MART value as described below, except for Stickney, where actual values were used because of plant size and concentration considerations.

This analysis supports the Study Group's use of increased TPCs when estimating point source loads. Therefore, all plants that were labeled as advanced treatment and used the Clean Water Needs Survey data for load estimates were recalculated using total N and P concentrations of 11.2 and 2.02 mg/L, respectively (this included most plants, and some were estimated using the permit compliance system data and were not recalculated). These concentrations were much closer to the values reported by the plants in Table D.1 (parenthetical values in table), although there still was considerable variability and included 2,080 point sources (the total database has 33,302 point sources of all types). For plants identified as receiving secondary treatment, total N and P concentrations of 15 and 4 mg/L, respectively, were applied (there were 4,480 plants of this type). For the seven plants in Table D.1 recalculated this way, total N and P fluxes were 113 and 91% of measured values, respectively, much closer to the measured values than the original MART values. The Study Group's discussion of point sources in the MARB utilizes these adjustments to the MART values. Finally, the Study Group again emphasizes that measured data are generally not available in these large databases, so that many assumptions need to be made.

Appendix E: USUSEPA's Guidance on Nutrient Criteria

In 2000, USEPA recommended criteria to states and tribes for use in establishing their water quality standards consistent with Section 303(c) of the Clean Water Act (CWA). Under Section 303(c) of the CWA, States and authorized tribes have the primary responsibility for adopting water quality standards as state or tribal law or regulation. The standards must contain scientifically defensible water quality criteria that are protective of designated uses. On its web site at http://www.epa.gov/waterscience/criteria/nutrient/ecoregions/, USEPA provides recommended criteria for nutrients in four major types of waterbodies – lakes and reservoirs, rivers and streams (USEPA, 2006b), estuarine and coastal areas, and wetlands across 14 major ecoregions of the United States. The Study Group asked USEPA for a comparison of the Study Group's proposed 45% reductions for TN and TP flux to the nutrient levels that would correspond to USEPA's recommended ecoregional criteria.

Before presenting that preliminary analysis, the following caveats are stressed:

- USEPA's recommended ecoregional nutrient criteria are not laws or regulations; they are guidance that states and tribes may use as a starting point for developing criteria for their water quality standards. Final criteria developed by states and tribes may have concentrations higher or lower than USEPA ecoregional recommendations, or, if scientifically defensible, not include a nutrient if an impact on "designated use" was not found.
- USEPA's recommended ecoregional nutrient criteria do not take into account local site-specific conditions and "designated uses" for particular water bodies (e.g., recreation, water supply, aquatic life, agriculture).

- USEPA's guidance for ecoregional nutrient criteria is based on ambient concentrations of nutrients (expressed in mg/L or µg/L) in various ecoregions. By contrast, the Study Group's recommended reductions of TN and TP are based on flux (expressed in million metric tons of TN and TP discharged at the mouth of the Mississippi River). A direct comparison of concentrations to flux necessitates the simplifying assumption that percentage reductions in concentrations have a one-to-one correspondence with percentage reductions in flux.
- USEPA's guidance for ecoregional criteria is based on estimated "reference conditions," i.e., reference sites chosen to represent the least culturally impacted waters of the class existing at the present time. The estimated reference conditions are based on the 25th percentile of the frequency distribution of nutrient concentration data available for each ecoregion. This assumption lends uncertainty to USEPA's guidance for ecoregional nutrient criteria.

Given these caveats, the following analysis by USEPA's Office of Water's Office of Science and Technology and USEPA's Office of Research and Development allows some comparison between USEPA's guidance for ecoregional nutrient criteria and the Study Group's proposed 45% nutrient reductions.

Comparison of SAB Nitrogen and Phosphorus Recommendations with USEPA Nitrogen and Phosphorus Criteria Recommended Reference Conditions – Submitted by USEPA's Office of Water, 8-24-07.

Question: How do the 45% recommended reductions in nitrogen (N) and phosphorus (P) at the mouths of the Mississippi and Atchafalaya Rivers compare with the 25th percentile of TN and TP concentration data from ecoregions draining the Mississippi-Atchafalaya River Basin (MARB)?

Answer: This question is addressed with a preliminary approach. A more thorough approach is needed, but this would require a longer period of time.

The preliminary approach was developed by staff from the USEPA Office of Research and Development's Gulf Breeze Lab and the USEPA Office of Water's Office of Science and Technology using USGS loading estimates from the lower Mississippi River at St. Francisville, LA, and the Atchafalaya River at Melville, LA, over the past 20 years. This approach compares the 45% reduction in nitrogen and phosphorus recommended by the SAB to the 25th percentiles of the distribution of data in USEPA's National Nutrient Database for total nitrogen (TN) and total phosphorus (TP) in each aggregate nutrient ecoregion of the MARB. These 25th percentiles represent USEPA's approximated reference conditions for those ecoregions.

It is important to note that these 25th percentile values are not intended to be implemented or promulgated directly as criteria. Rather, USEPA developed the nutrient criteria recommendations with the intent that they serve as a starting point for states and tribes to develop more refined criteria, as appropriate, to reflect local conditions. States and tribes may adopt criteria that are higher or lower than these 25th percentiles. Text in two USEPA documents helps to clarify the use of the ecoregional reference condition values. See introductions to the ecoregional criteria documents at http://www.epa.gov/waterscience/criteria/nutrient/ecoregions/rivers/index.html and USEPA's Nutrient Criteria Technical Guidance Manual for Rivers and Streams (http://www.epa.gov/waterscience/criteria/nutrient/guidance/rivers/chapter_1.pdf).

Given this description, one can compare a 45% reduction in N and P measured in two locations to the estimated reference conditions in each of the MARB ecoregions to obtain a rough estimate of whether a 45% reduction could be more or less stringent than what could result if USEPA's recommended reference conditions were adopted without further modification, as state water quality standards.

Data Sources: River flow and nutrient flux are monitored and computed by the US Geological Survey's (USGS) National Stream Quality Accounting Network (NASQAN) program at numerous river gauge stations in the Mississippi River Basin. A description of the USGS NASQAN program, the flux estimation methodology, and the downloadable data records are available at http://toxics.usgs.gov/hypoxia/. Monthly average nutrient concentrations were calculated from the USGS data as

The monthly average nutrient concentration = USGS monthly load/monthly average discharge rate, where monthly average discharge rates for the mainstem Mississippi River were calculated from daily discharge rates obtained at the Tarbert Landing, MS, gauge (ID = 01100).

Nitrogen. The median monthly nitrate concentration for the combined Mississippi River at St. Francisville and Atchafalaya River at Melville over the period 1979–2007 is 1.24 mg/L. In comparison, historical data from the Mississippi River at St. Francisville indicate that the median nitrate concentrations during the period 1955–1970 was 0.6 mg/L.

Nitrate, as a component of TN is about 60% on an average (based on USGS nutrient load data); thus 1.24 mg/L nitrate would extrapolate to 2.07 mg/L TN.

A proposed 45% reduction of 2.07 mg/L TN would yield a concentration of 1.14 mg/L TN.

The relevant USEPA-recommended ecoregional reference conditions for TN are

Ecoregion IV – 0.56 mg/L
Ecoregion V – 0.88 mg/L

Ecoregion VI – 2.18 mg/L
Ecoregion VII – 0.54 mg/L
Ecoregion IX – 0.69 mg/L
Ecoregion X – 0.76 mg/L
Ecoregion XI – 0.31 mg/L

These values range from 27 to 191% of the estimated 1.14 mg/L TN that would result from a 45% reduction, with all but one value below 100% (the Corn Belt and northern Great Plains ecoregion VI). This suggests that a 45% reduction of estimated median monthly TN concentrations to 1.14 mg/L would likely be less stringent than could be obtained if states adopted USEPA's recommended reference condition values into state water quality standards for TN.

Phosphorus. Using the same data (Mississippi River at St. Francisville and Atchafalaya River at Melville, 1979–2007, monthly means), the median monthly concentration of TP is 202 µg/L. Thus a 45% reduction of 202 µg/L TP would yield a concentration of 111 µg/L.

The relevant USEPA recommended ecoregional reference conditions for TP are:

Ecoregion IV – 23.00 µg/L
Ecoregion V – 67.00 µg/L
Ecoregion VI – 76.00 µg/L
Ecoregion VII – 33.00 µg/L
Ecoregion IX – 36.00 µg/L
Ecoregion X – 128.00 µg/L
Ecoregion XI – 10.00 µg/L

These values range from 9 to 115% of the estimated 111 µg/L TP that would result from a 45% reduction, with all but one value below 100% (the Texas–Louisiana Coastal and Mississippi Alluvial Plains ecoregion X). This also suggests that a 45% reduction of estimated median monthly TP concentrations to 111 µg/L would likely be less stringent than could be obtained if states adopted USEPA's recommended reference condition values into state water quality standards for TP.

A More Comprehensive Approach

A thorough comparison of the distribution approach to reference condition estimation and the 45% reduction in TN and TP could be made by calculating the nutrient concentrations from the USGS loading estimates at river gauge stations at each of the nine subbasins. The USGS provides monthly or annual nutrient flux estimates and river flow data from which nutrient concentration data can be derived (http://toxics.usgs.gov/hypoxia/). These data provide values over many years for nine subbasins located within the MARB. The data could be used in the following steps to compare the two sets of values:

- use the USGS nutrient loading data to compile a TN and TP concentration dataset for each subbasin;
- calculate the median TN and TP concentrations at each of the nine subbasin river gauge stations;
- overlay nutrient ecoregions on subbasins and extract nutrient ecoregional data from subbasins. From this refined data set, calculate the median value of the seasonal 25th percentiles of TN and TP for the ecoregion–subbasin.

These data can be used for the following comparisons:

(1) Calculate the concentrations resulting from a 45% reduction in the median concentration for each subbasin.
(2) Compare these to the USEPA 25th percentiles (ecoregional reference conditions) in each subbasin, or specific subbasins of interest.

References

Ackerman, F., Biewald, B., White, D., Woolf, T., and Moomaw, W., 1999, Grand-fathering and coal plant emissions—The cost of cleaning up the Clean Air Act: Energy Policy, v. 27, p. 929–940.

Adviento-Borbe, M.A.A., Haddix, M.L., Binder, D.L., Walters, D.T., and Dobermann, A., September 2007, Soil greenhouse gas fluxes and global warming potential in four high-yielding maize systems: Global Change Biology, v. 13, no. 9, p. 1972–1988, listed online as Accepted Articles at: http://www.blackwell-synergy.com/doi/abs/10.1111/j.1365-2486.2007.01421.x, last accessed August 15, 2007.

Aillery, M., Gollehon, N., Johansson, R., Kaplan, J., Key, N., and Ribaudo, M., 2005, Managing manure to improve air and water quality: Washington, D.C., U.S. Government Printing Office, U.S. Department of Agriculture, Economic Research Service, Economic Research Report No. 9, 65 p., available online at: http://www.ers.usda.gov/publications/err9/err9.pdf.

Alexander, R.B., Boyer, E.W., Smith, R.A., Schwarz, G.E., and Moore, R.B., 2007, The role of headwater streams in downstream water quality: Journal of the American Water Resources Association, v. 43, no. 1, p. 41–59.

Alexander, R.B., Elliott, A.H., Shankar, U., and McBride, G.B., 2002a, Estimating the sources and transport of nutrients in the Waikato River basin, New Zealand: Water Resources Research, v. 38, p. 1268–1290.

Alexander, R.B., Johnes, P.J., Boyer, E.W., and Smith, R.A., 2002b, A comparison of models for estimating the riverine export of nitrogen from large watersheds: Biogeochemistry, v. 57/58, p. 295–339.

Alexander, R.B., Slack, J.R., Ludtke, A.S., Fitzgerald, K.K., and Schertz, T.L., 1998, Data from selected U.S. Geological Survey national stream water-quality monitoring networks: Water Resources Research, v. 34, no. 9, p. 2401–2405.

Alexander, R.B., and Smith, R.A., 2006, Trends in the nutrient enrichment of U.S. rivers during the late 20th century and their relation to changes in probable stream trophic conditions: Limnology and Oceanography, v. 51, p. 639–654.

Alexander, R.B., Smith, R.A., and Schwarz, G.E., 2000, Effect of stream channel size on the delivery of nitrogen to the Gulf of Mexico: Nature, v. 403, p. 758–761.

Alexander, R.B., Smith, R.A., and Schwarz, G.E., 2004, Estimates of diffuse phosphorus sources in surface waters of the United States using a spatially referenced watershed model: Water Science and Technology, v. 49, no. 3, p. 1–10.

Alexander, R.B., Smith, R.A., Schwarz, G.E., Boyer, E.W., Nolan, J.V., and Brakebill, J.W., 2008, Differences in phosphorus and nitrogen delivery to the Gulf of Mexico from the Mississippi River basin: Environmental Science and Technology, v. 42, no. 3, p. 822–830.

Alexander, R.B., Smith, R.A. Schwarz, G.E., Preston, S.D., Brakebill, J.W., Srinivasan, R., and Pacheco, P.A., 2001, Atmospheric nitrogen flux from the watersheds of major estuaries of the United States: An application of the SPARROW watershed model, *in* Valigura, R., Alexander, R., Castro, M., Meyers, T., Paerl, H., Stacey, P., and Turner, R.E., eds., Nitrogen loading in coastal water bodies—An atmospheric perspective: American Geophysical Union Monograph: Madison, WI, American Society of Agronomy, Volume 57, 119–170 p.

Aller, R.C., 1998, Mobile deltaic and continental shelf muds as sub-oxic, fluidized bed reactors: Marine Chemistry, v. 61, p. 143–155.

Aller, R.C., Heilbrun, C., Panzeca, C., Zhu, Z.-B., and Baltzer, F., 2004. Coupling between sedimentary dynamics, early diagenetic processes, and biogeochemical cycling in the Amazon-Guianas mobile mud belt: Coastal French Guiana: Marine Geology, v. 208, p. 331–360.

Altieri, A.H., 2008, Dead zones enhance key fisheries species by providing predation refuge: Ecology, v. 89, no. 10, p. 2808–2818.

Ammerman, J.W., and Sylvan, J.B., 2004, Phosphorus limitation of phytoplankton growth in the Mississippi River plume—A case for dual nutrient control?: EOS Transactions AGU, v. 85, no. 47, Fall Meeting Supplement, Abstract OS11B-07.

Anand, S., Mankin, K.R., McVay, K.A., Janssen, K.A., Barnes, P.L., and Pierzynski, G.M., 2007, Calibration and validation of ADAPT and SWAT for field-scale runoff prediction: Journal of the American Water Resources Association, v. 43, no. 4, p. 899–910.

Anderson, C.J., Nairn, R.W., and Mitsch, W.J., 2005, Temporal and spatial development of surface soil conditions at two created riverine marshes: Journal of Environmental Quality, v. 34, p. 2072–2081.

Anderson, D.M., and Garrison, D.J., eds., 1997, The ecology and oceanography of harmful algal blooms: American Society of Limnology and Oceanography Special Issue, v. 42, no. 5, p. 1009–1305.

Andraski, T.W., and Bundy, L.G., 2003, Relationships between phosphorus levels in soil and in runoff from corn production systems: Journal of Environmental Quality, v. 32, p. 310–316.

Arnold, J.G., Srinivasan, R., Muttiah, R.S., and Allen, P.M., 1999, Continental scale simulation of the hydrologic balance: Journal of the American Water Resources Association, v. 35, no. 5, p. 1037–1051.

Arnold, J.G., Srinivasan, R., Muttiah, R.S., and Williams, J.R., 1998, Large area hydrologic modeling and assessment—Part I, Model development: Journal of the American Water Resources Association, v. 34, no. 1, p. 73–89.

Atwood, J.D., Benson, V.W., Srinivasan, R., Walker, C., and Schmid, E., 2001, Simulated nitrogen loading from corn, sorghum, and soybean production in the Upper Mississippi Valley, in Stott, D.E., Mohtar, R.H., and Steinhardt, G.C., eds., Sustaining the Global Farm, 10th International Soil Conservation Organization Meeting, Purdue University, IN, May 24–29, 1999, p. 344–348.

Aulenbach, B.T., and Hooper, R.P., 2006, The composite method—An improved method for stream-water solute load estimation: Hydrological Processes, v. 20, p. 3029–3047.

Aulenbach, B.T., Buxton, H.T., Battaglin, W.A., and Coupe, R.H., 2007, Streamflow and nutrient fluxes of the Mississippi-Atchafalaya River basin and subbasins for the period of record through 2005: U.S. Geological Survey Open-File Report No. 2007-1080, available online at: http://toxics.usgs.gov/pubs/of-2007-1080/index.html.

Babcock, B.A., Gassman, P.W., Jha, M., and Kling, C.L., 2007, Adoption subsidies and environmental impacts of alternative energy crops: Iowa State University, Center for Agricultural and Rural Development (CARD) Briefing Paper 07-BP 50, 15 p., available online at: http://www.card.iastate.edu/publications/DBS/PDFFiles/07bp50.pdf.

Baker, J.L., David, M.B., and Lemke, D.W., 2008, Understanding nutrient fate and transport, including the importance of hydrology in determining losses, and potential implications on management systems to reduce those losses, in Final Report: Gulf Hypoxia and Local Water Quality Concerns Workshop. Saint Joseph, MI: American Society of Agricultural and Biological Engineers.

Baker, J.L., Melvin, S.W., Lemke, D.W., Lawlor, P.A., Crumpton, W.G., and Helmers, M.J., 2004, Subsurface drainage in Iowa and the water quality benefits and problem, in Cooke, R., ed., Proceedings of the 8th International Drainage Symposium, Sacramento, CA, March 21, 2004, ASAE Pub #701P0304, p. 39–50.

Baker, J.L., Mickelson, S.K., and Crumpton, W.G., 1997, Integrated crop management and off-site movement of nutrients and pesticides, in Hatfield, J.C., Buhler, D.B., and Stewart, B.A., eds., Weed biology, soil management, and weed management—Advances in soil science: Boca Raton, CA, CRC Press, 135–160 p.

References

Baker, D.B., and Richards, P.R., 2002, Phosphorus budgets and riverine phosphorus export in northwestern Ohio watersheds: Journal of Environmental Quality, v. 31, p. 96–108.

Baltz, D.M., Hiram, W.L., Rossignol, P.A., Chesney, E.J., and Switzer, T.S., 2006, A qualitative assessment of the relative effects of bycatch reduction of fisheries and hypoxia on coastal nekton communities in the Gulf of Mexico: Paper presented at Hypoxia Effects on Living Resources in the Gulf of Mexico, September 25–26, 2006: New Orleans, Louisiana, Tulane University, sponsored by National Oceanic and Atmospheric Administration Center for Sponsored Coastal Ocean Research.

Barker, D.W., Sawyer, J.E., and Al-Kaisi, M.M., 2006a, Assessment of the amino sugar-nitrogen test on Iowa soils—I. Evaluation of soil sampling and corn management practices: Agronomy Journal, v. 98, p. 1345–1351.

Barker, D.W., Sawyer, J.E., and Al-Kaisi, M.M., 2006b, Assessment of the amino sugar-nitrogen test on Iowa soils—II. Field correlation and calibration: Agronomy Journal, v. 98, p. 1352–1358.

Batie, S.S., Gilliam, J.W., Groffman, P.M., Hallberg, G.R., Hamilton, N.D., Larson, W.E., Lee, L.K., Nowak, P.J., Renard, K.G., Rominger, R.E., Stewart, B.A., Tanji, K.K., Van Schilfgaarde, J., Wagenet, R.J., and Young, D.L., 1993, Soil and water quality: An agenda for agriculture: National Academy of Sciences, Board of Agriculture, p. 278.

Battaglin, W., 2006, Streamflow and nitrogen, phosphorus, and silica flux at selected sites in the Mississippi River basin, 1980–2005: Presented at Science Symposium: Sources, Transport and Fate of Nutrients in the Mississippi and Atchafalaya River basins, November 7–9, 2006, Minneapolis, MN.

Baumol, W., and Oates, W., 1988, The theory of environmental policy, 2nd edition: Cambridge, UK, Cambridge University Press, 299 p.

Baustian, M.M., and Rabalais, N.N., 2009, Seasonal composition of benthic macroinfauna exposed to hypoxia in the Northern Gulf of Mexico: Estuaries and Coasts, v. 32, no. 5, p. 975–983.

Beegle, D.B., 2005, Assessing soil phosphorus for crop production by soil testing, *in* Sims, J.T., and Sharpley, A.N., eds., Phosphorus: Agriculture and the environment: Madison, WI, American Society of Agronomy Monograph Series No. 46, 123–144 p.

Belabbasi, L., 2006, Examination of the relationship of river water to occurrences of bottom water with reduced oxygen concentrations in the northern Gulf of Mexico: College Station, Texas, Texas A&M University, Ph.D. thesis, xii + 119 p.

Benner, R., and Opsahl, S., 2001, Molecular indicators of the sources and transformations of dissolved organic matter in the Mississippi River plume: Organic Geochemistry, v. 32, p. 597–611.

Bermudez, M., and Mallarino, A.P., 2007, Impacts of variable-rate phosphorus fertilization based on dense grid soil sampling on soil-test phosphorus and grain yield of corn and soybean: Agronomy Journal, v. 99, p. 822–832.

Bernot, M.J., and Dodds, W.K., 2005, Nitrogen retention, removal, and saturation in lotic ecosystems: Ecosystems, v. 8, p. 442–453.

Besiktepe, S.T., Lermusiaux, P.F.J., and Robinson, A.R., 2003, Coupled physical and biogeochemical data-driven simulations of Massachusetts Bay in late summer: Real-time and postcruise data assimilation: Journal of Marine Systems, v. 40–41, p. 171–212.

Beven, K.J., 2001, Rainfall-runoff modeling: The primer: Chichester, UK, Wiley, 360 p.

Bharati, L., Lee, K.H., Isenhart, T.M., and Schultz, R.C., 2002, Soil-water infiltration under crops, pasture and established riparian buffer in Midwestern USA: Agroforestry Systems, v. 56, p. 249–257.

Bianchi, T.S., Allison, M.A., Canuel, E.A., Corbett, D.R., McKee, B.A., Sampere, T.P., Wakeham, S.G., Waterson, E., 2006, Rapid export of organic matter to the Mississippi canyon, Mississippi: EOS, Transactions of the American Geophysical Union, v. 87, no. 50, p. 572–574.

Bianchi, T.S., Filley, T., Dria, K., and Hatcher, P.G., 2004, Temporal variability in sources of dissolved organic carbon in the lower Mississippi River: Geochimica et Cosmochimica Acta, v. 68, no. 5, p. 959–967.

Bianchi, T.S., Galler, J.J., and Allison, M.A., 2007, Hydrodynamic sorting and transport of terrestrially derived organic carbon in sediments of the Mississippi and Atchafalaya Rivers: Estuarine, Coastal and Shelf Science, v. 73, nos. 1–2, p. 211–222.

Bianchi, T.S., Mitra, S., and McKee, B., 2002, Sources of terrestrially-derived carbon in the lower Mississippi River and Louisiana shelf—Implications for differential sedimentation and transport at the coastal margin: Marine Chemistry, v. 77, p. 211–223.

Bierman, V.J., Jr., Hinz, S.C., Zhu, D.-W., Wiseman, W.J., Jr., Rabalais, N.N., and Turner, R.E., 1994, A preliminary mass balance model of primary productivity and dissolved oxygen in the Mississippi River Plume/Inner Gulf Shelf region: Estuaries, v. 17, no. 4, p. 886–899.

Blaylock, A.D., 2006, Review of enhanced-efficiency nitrogen fertilizers, in Proceedings of Southern Plant Nutrient Management Conference, Olive Branch, MS, October 3–4, 2006, p. 4–10.

Blomqvist, S., Gunnars, A., and Elmgren, R., 2004, Why the limiting nutrient differs between temperate coastal seas and freshwater lakes—A matter of salt: Limnology & Oceanography, v. 49, p. 2236–2241.

Bode, A., and Dortch, Q., 1996, Uptake and regeneration of inorganic nitrogen in coastal waters influenced by the Mississippi River—Spatial and seasonal variations: Journal of Plankton Research, v. 18, no. 12, p. 2251–2268.

Boesch, D.F., 2002, Challenges and opportunities for science in reducing nutrient over-enrichment of coastal ecosystems: Estuaries, v. 25, p. 744–758.

Boesch, D.F., 2003, Continental shelf hypoxia: Some compelling answers: Comments on "Continental shelf hypoxia: Some nagging questions": Gulf of Mexico Science, v. 21, no. 2, p. 202–205, available online at: http://goms.disl.org/tocpages/december2003vol21no2.htm.

Boicourt, W.C., 1992, Influences of circulation processes on dissolved oxygen in the Chesapeake Bay, in Smith, D.E., Leffler, M., and Mackiernan, G., eds., Oxygen dynamics in the Chesapeake Bay—A synthesis of recent research: College Park, MD, Maryland Sea Grant College, 7–79 p.

Booth, M.S., and Campbell, C., 2007, Spring nitrate flux in the Mississippi River basin: A landscape model with conservation applications: Environmental Science and Technology, v. 41, no. 15, p. 5410–5418.

Boyer, E.W., Goodale, C.L., Jaworski, N.A., and Howarth, R.W., 2002, Anthropogenic nitrogen sources and relationships to riverine nitrogen export in the northeastern U.S.A.: Biogeochemistry, v. 57/58, p. 137–169.

Boyer, E.W., and Howarth, R.W., 2008, Nitrogen fluxes from rivers to the coastal oceans, in Capone, D., Mulholland, M., and Carpenter, E., eds., Nitrogen in the marine environment, 2nd edition: New York, NY, Academic Press, p. 1565–1587.

Boynton, W.R., and Kemp, W.M., 2000, Influence of river flow and nutrient loads on selected ecosystem processes: A synthesis of Chesapeake Bay data, in Hobbie, J.E., ed., Estuarine science—A synthetic approach to research and practice: Washington, D.C., Island Press, 269–298 p.

Bradley, M.J., and Jones, B.M., 2002, Emissions—Developing advanced energy and transportation technologies: AMBIO, A Journal of the Human Environment, v. 31, no. 2, p. 141–149.

Bratkovich, A., Dinnel, S.P., and Goolsby, D.A., 1994, Variability and prediction of freshwater and nitrate fluxes for the Louisiana-Texas shelf—Mississippi and Atchafalaya river source functions: Estuaries, v. 17, p. 766–778.

Breetz, H., Fisher-Vander, K., Garzon, L., Jacobs, H., Droetz, K., and Terry, R., 2004, Water quality trading and offset initiatives in the U.S.—A comprehensive survey: Dartmouth College, Hanover, NH, 337 p., available online at: http://www.dartmouth.edu/~kfv/waterqualitytradingdatabase.pdf.

Breitburg, D.L., Hondorp, D.W., Davias, L.A., and Diaz, R.J., 2009a, Hypoxia, nitrogen, and fisheries: Integrating effects across local and global landscapes: Annual Review of Marine Science, v. 1, p. 329–349.

Breitburg, D.L., Craig, J.K., Fulford, R.S., Rose, K.A., Boynton, W.R., Brady, D.C., Ciotti, B.J., Diaz, R.J., Friedland, K.D., Hagy, J.D., Hart, D.R., Hines, A.H., Houde, E.D., Kolesar, S.E.,

Nixon, S.W., Rice, J.A., Secor, D.H., and Targett, T.E., 2009b, Nutrient enrichment and fisheries exploitation: Interactive effects on estuarine living resources and their management: Hydrobiologia, v. 629, no. 1, p. 31–47.

Brezonik, P.L., Bierman, V.J., Jr., Alexander, R., Anderson, J., Barko, J., Dortch, M., Hatch, L., Hitchcock, G.L., Keeney, D., Mulla, D., Smith, B., Walker, C., Whitledge, T., and Wiseman, W.J., Jr., 1999, Effects of reducing nutrient loads to surface waters within the Mississippi River basin and the Gulf of Mexico: Topic 4 report for the integrated assessment of hypoxia in the Gulf of Mexico: Silver Spring, MD, National Oceanic and Atmospheric Administration Coastal Ocean Program Decision Analysis Series No. 18, 158 p., available online at: http://oceanservice.noaa.gov/products/hypox_t4final.pdf.

Bridgham, S.D., Johnston, C.A., Schubauer-Berigan, J.P., and Wesihampel, P., 2001, Phosphorus sorption dynamics in soils and coupling with surface and pore water in riverine wetlands: Soil Sciences Society of America Journal, v. 65, p. 577–588.

Broshears, R.E., Clark, G.M., and Jobson, H., 2001, Simulation of stream discharge and transport of nitrate and selected herbicides in the Mississippi River Basin: Hydrological Processes, v. 15, p. 1157–1167.

Brouwer, M., 2006, Changes in gene and protein expression and reproduction in grass shrimp, *Palaemonetes pugio*, exposed to chronic hypoxia: Paper presented at Hypoxia Effects on Living Resources in the Gulf of Mexico, September 25–26, 2006, Tulane University, New Orleans, Louisiana, sponsored by National Oceanic and Atmospheric Administration Center for Sponsored Coastal Ocean Research.

Bruland, G.L., and Richardson, C.J., 2006, An assessment of the phosphorus retention capacity of wetlands in the Painter Creek Watershed, Minnesota, USA: Water, Air, and Soil Pollution, v. 171, p. 169–184.

Bundy, L.G., 1998, Corn fertilization: University of Wisconsin Cooperative Extension Service Publication AA3340.

Cai, W.J., and Lohrenz, S.E., 2005, Carbon, nitrogen, and phosphorus fluxes from the Mississippi River and the transformation and fate of biological elements in the river plume and the adjacent margin, *in* Liu, K.K., Atkinson, L., Quinones, R., Talaue-McManus, L., eds., Carbon and nutrient fluxes in continental margins—A global synthesis: New York, NY, Springer-Verlag.

Cai, W.J., 2003, Riverine inorganic carbon flux and rate of biological uptake in the Mississippi River plume: Geophysical Research Letters, v. 30, no. 2, doi:10.1029/2002GL016312, p. 1032.

Caraco, N., Cole, J.J., and Likens, G.E., 1990, A comparison of phosphorus immobilization in sediments of freshwater and coastal marine systems: Biogeochemistry, v. 9, p. 277–290.

Caraco, N.F., Cole, J.J., and Likens, G.E., 1989, Evidence for sulfate controlled phosphorus release from sediments of aquatic systems: Nature, v. 341, p. 316–318.

Carter, G.S., Gregg, M.C., and Lien, R.C., 2005, Internal waves, solitary like waves, and mixing on the Monterey Bay shelf: Continental Shelf Research, v. 25, p. 1499–1520.

CENR, 2000, Integrated assessment of hypoxia in the northern Gulf of Mexico: National Science and Technology Council, Committee on Environmental and Natural Resources, May 2000, 66 p., available online at: http://oceanservice.noaa.gov/products/hypox_final.pdf.

Cerco, C.F., and Cole, T., 1993, Three-dimensional eutrophication model of Chesapeake Bay: Journal of Environmental Engineering, v. 119, no. 6, p. 1006–1025.

Charlton, M.N., Milne, J.E., Booth, W.G., and Chiocchio, F., 1993, Lake Erie offshore in 1990: Restoration and resilience in the central basin: Journal of Great Lakes Research, v. 19, p. 291–309.

Chen, N., Bianchi, T.S., and McKee, B.A., 2005, Early diagnosis of chloropigment biomarkers in the lower Mississippi River and Louisiana shelf—Implications for carbon cycling in a river-dominated margin: Marine Chemistry, v. 93, p. 159–177.

Chen, N., Bianchi, T.S., McKee, B.A., and Bland, J.M., 2001, Historical trends of hypoxia on the Louisiana shelf—Applications of pigments as biomarkers: Organic Geochemistry, v. 32, p. 543–561.

Chen, R.F., and Gardner, G.B., 2004, High resolution measurements of chromophoric dissolved organic matter in the Mississippi and Atchafalaya river plume regions: Marine Chemistry, v. 89, p. 103–125.

Chen, X., Lohrenz, S.E., and Wiesenburg, D.A., 2000, Distribution and controlling mechanisms of primary production over the Louisiana-Texas continental shelf: Journal of Marine Systems, v. 25, p. 179–207.

Chesapeake Bay Commission, 2004, Cost effective strategies for the Bay: Annapolis, MD, 14 p.

Chesapeake Bay Program, 2003, Recommendations for the 2003 Directive on Expanded Riparian Forest Buffer Goals in the Chesapeake Watershed, Annapolis, MD: Chesapeake Bay Program.

Chesapeake Bay Program, 2006, Watershed model progress scenario results for 2005: Available online at: http://www.chesapeakebay.net/tribtools.htm.

Chesney, E.J., and Baltz, D.M., 2001, The effects of hypoxia on the northern Gulf of Mexico coastal ecosystem—A fisheries perspective, in Rabalais, N.N., and Turner, R.E., eds., Coastal hypoxia—Consequences for living resources and ecosystems: Washington, D.C., American Geophysical Union, Coastal and Estuarine Studies, Volume 58, 321–354 p.

Childs, C.R., Rabalais, N.N., Turner, R.E., and Proctor, L.M., 2002, Sediment denitrification in the Gulf of Mexico zone of hypoxia: Marine Ecology Progress Series, v. 240, p. 285–290.

Childs, C.R., Rabalais, N.N., Turner, R.E., and Proctor, L.M., 2003, Erratum—Sediment denitrification in the Gulf of Mexico zone of hypoxia: Marine Ecology Progress Series, v. 247, p. 310.

Chow, T.L., Rees, H.W., and Daigle, J.L., 1999, Effectiveness of terraces/grassed waterway systems for soil and water conservation—A field evaluation: Journal of Soil and Water conservation, v. 54, no. 3, p. 577–583.

Chua, T.T., Bronson, K.F., Booker, J.D., Keeling, J.W., Mosier, A.R., Bordovsky, J.P., Lascano, R.J., Green, C.J., and Segarra, E., 2003, In-season nitrogen status sensing in irrigated cotton— I. Yields and nitrogen-15 recovery: Soil Science Society of America Journal, v. 67, p. 1428–1438.

Chung, S.W., Gassman, P.W., Gu, R., and Kanwar, R.S., 2002, Evaluation of EPIC for assessing tile flow and nitrogen losses for alternative agricultural management systems: Transactions of the American Society of Agricultural Engineers, v. 45, no. 4, p. 1135–1146.

Claassen, R., 2000, Agricultural resources and environmental indicators—Compliance provisions for soil and wetland conservation: U.S. Department of Agriculture, Economic Research Service, Agricultural Resources and Environmental Indicators, chap. 6.3, 20 p.

Claassen, R., Breneman, V., Bucholtz, S., Cattaneo, A., Johansson, R., and Morehart, M., 2004, Environmental compliance in U.S. agricultural policy—Past performance and future potential: U.S. Department of Agriculture, Economic Research Service, Agricultural Economic Report No. 832, p. 52.

Clapp, C.E., Allmaras, R.R., Layese, M.F., Linden, D.R., and Dowdy, R.H., 2000, Soil organic carbon and ^{13}C abundance as related to tillage, crop residue, and nitrogen fertilization under continuous corn management in Minnesota: Soil & Tillage Research, v. 55, no. 3, p. 127–142.

Clark, C., and Russell, C., 2005, Public information provision as a tool of environmental policy?, in Krarup, S., and Russell, C., eds., Environment, information, and consumer behaviour: Cheltenham, UK, Edward Elgar Publishing, chap. 6.

Cogeneration Technologies, 2006, Clean coal technology & the President's clean coal power initiative: Available online at: http://www.cogeneration.net/IntegratedGasificationCombinedCycle.htm.

Cohn, T.A., Caulder, D.L., Gilroy, E.J., Zynjuk, L.D., and Sommers, R.M., 1992, The validity of a simple statistical model for estimating fluvial constituent loads: An empirical study involving nutrient loads entering Chesapeake Bay: Water Resources Research, v. 28, p. 2352–2363.

Collie, J.S., Richardson, K., and Steele, J.H., 2004, Regime shifts—Can ecological theory illuminate the mechanisms?: Progress in Oceanography, v. 60, p. 281–302.

Conley, D.J., 2002, Terrestrial ecosystems and the global biogeochemical silica cycle: Global Biogeochemical Cycles, v. 16, p. 1121, doi:10.1029/2002GV001894.

Conley, D.J., Carstensen, J., Ærtebjerg, G., Christensen, P.B., Dalsgaard, T., Hansen, J.L.S., and Josefson, A.B., 2007, Long-term changes and impacts of hypoxia in Danish coastal waters: Ecological Applications, Supplement, v. 17, no. 5, p. S165–S184.

Conley, D.J., Humborg, C., Rahm, L., Savchuk, O.P., and Wulff, F., 2002a, Hypoxia in the Baltic Sea and basin-scale changes in phosphorus biogeochemistry: Environmental Science and Technology, v. 36, p. 5315–5320.

Conley, D.J., Markager, S., Andersen, J., Ellermann, T., and Svendsen, L.M., 2002b, Coastal eutrophication and the Danish National Aquatic Monitoring and Assessment Program: Estuaries, v. 25, p. 706–719.

Cooke, R.A., Sands, G.R., and Brown, L.C., 2008, Drainage water management: a practice for reducing nitrate loads from subsurface drainage systems, *in* Final Report: Gulf Hypoxia and Local Water Quality Concerns Workshop. Saint Joseph, MI: American Society of Agricultural and Biological Engineers.

Corbett, D.R., McKee, B.A., and Allison, M.A., 2006, Nature of decadal-scale sediment accumulation in the Mississippi river deltaic region: Continental Shelf Research, v. 26, p. 2125–2140.

Corbett, D.R., McKee, B.A., and Duncan, D., 2004, An evaluation of mobile mud dynamics in the Mississippi River deltaic region: Marine Geology, v. 209, p. 91–112.

Costello, C., Griffin, W.M., Landis, A.E., and Matthews, H.S., 2009, Impact of biofuel crop production on the formation of hypoxia in the Gulf of Mexico: Environmental Science & Technology, v. 43, no. 20, p. 7985–7991.

CPRA, 2007, Integrated ecosystem restoration and hurricane protection: Louisiana's comprehensive master plan for a sustainable coast: Coastal Protection and Restoration Authority (CPRA) of Louisiana, Office of the Governor (Louisiana), 140 p., available online at: http://www.lacpra.org/assets/docs/cprafinalreport5-2-07.pdf.

Craig, J.K., and Crowder, L.B., 2005, Hypoxia-induced habitat shifts and energetic consequences in Atlantic croaker and brown shrimp on the Gulf of Mexico shelf: Marine Ecology Progressive Series, v. 294, p. 79–94.

Craig, J.K., Crowder, L.B., and Henwood, T.L., 2005, Spatial distribution of brown shrimp (*Farfantepenaeus aztecus*) on the northwestern Gulf of Mexico shelf—Effects of abundance and hypoxia: Canadian Journal of Fisheries and Aquatic Sciences, v. 62, p. 1295–1308.

Craig, J.K., Gray, C.D., McDaniel, C.M., Henwood, T.L., and Hanifen, J.G., 2001, Ecological effects of hypoxia on fish, sea turtles, and marine mammals in the northwestern Gulf of Mexico, *in* Rabalais, N.N., and Turner, R.E., eds., Coastal hypoxia—Consequences for living resources and ecosystems: Washington, D.C., American Geophysical Union, Coastal and Estuarine Studies, Volume 58, 269–291 p.

Crespi, J., and Marette, S., 2005, Eco-labeling economics—Is public involvement necessary?, *in* Krarup, S., and Russell, C.S., eds., Environment, information, and consumer behaviour: North Hampton, NH, Edward Elgar Publishing, chap. 5.

Crumpton, W.G., Kovacic, D., Hey, D., and Kostel, J., 2008, Potential of wetlands to reduce agricultural nutrient export to water resources in the corn belt, *in* Final Report: Gulf Hypoxia and Local Water Quality Concerns Workshop. Saint Joseph, MI: American Society of Agricultural and Biological Engineers.

Crumpton, W.G., Stenback, G.A., Miller, B.A., and Helmers, M.J., 2006, Potential benefits of wetland filters for tile drainage systems—Impact on nitrate loads to Mississippi river subbasins: U.S. Department of Agriculture, CSREES Project Completion Report, (release pending).

Czapar, G.G., Payne, J., and Tate, J., 2007, An educational program on the proper timing of fall-applied nitrogen fertilizer: On-line Crop Management, doi:10.194/CM-2007-1510-01-RS.

D'Sa, E.J., and Miller, R.L., 2003, Bio-optical properties in waters influenced by the Mississippi River during low flow conditions: Remote Sensing of Environment, v. 84, p. 538–549.

Dagg, M., Benner, R., Lohrenz, S., and Lawrence, D., 2004, Transformation of dissolved and particulate materials on continental shelves influenced by large rivers—Plume processes: Continental Shelf Research, v. 24, p. 833–858.

Dagg, M.J., 1995, Copepod grazing and the fate of phytoplankton in the northern Gulf of Mexico: Continental Shelf Research, v. 15, nos. 11–12, p. 1303–1317.

Dagg, M.J., Ammerman, J.V., Amon, R., Gardner, W., Green, R., and Lohrenz, S., 2007, A review of water-column processes influencing hypoxia in the Northern Gulf of Mexico, Estuaries and Coasts, v. 30, p. 735–752.

Dagg, M.J., and Breed, G.A., 2003, Biological effects of Mississippi River nitrogen on the northern Gulf of Mexico—A review and synthesis: Journal of Marine Systems, v. 43, p. 133–152.

Dagg, M.J., and Brown, S.L., 2005, The potential contribution of fecal pellets from the larvacean *Oikopleura dioica* to vertical flux of carbon in a river dominated coastal margin, *in* Gorsky, G., Youngbluth, M.J., and Deibel, D., eds., Response of marine ecosystems to global change: Ecological impact of appendicularians, New York, NY, Gordon and Breach, 293–307 p.

Dahl, T.E., 1990, Wetlands losses in the United States—1780s to 1980s: Washington, D.C., U.S. Department of Interior, Fish and Wildlife Service, p. 21.

Dale, V.H., Brown, S., Haeuber, R.A., Hobbs, N.T., Huntly, N., Naiman, R.J., Riebsame, W.E., Turner, M.G., and Valone, T.J., 2000, Ecological principles and guidelines for managing the use of land: Ecological Applications, v. 10, p. 639–670.

Dalsgaard, T., Canfield, D.E., Petersen, J., Thamdrup, B., and Acuha-Gonzalez, J., 2003, N_2 projection by the anammox reaction in the anoxic water column of Golfo Dulce, Costa Rica: Nature, v. 422, p. 606–608.

Dalzell, B.J., Gowda, P.H., and Mulla, D.J., 2004, Modeling sediment and phosphorus losses in an agricultural watershed to meet TMDLs: Journal of American Water Resources Association, v. 40, p. 533–543.

Dampney, P.M.R., Lord, E.I., and Chambers, B.J., 2000, Development of advice for farmers and advisors: Soil Use and Management, v. 16, p. 162–166.

Darrow, B.P., Walsh, J.J., Vargo, G.A., Masserini, R.T., Jr., Fanninga, K.A., and Zhang, J.-Z., 2003, A simulation study of the growth of benthic microalgae following the decline of a surface phytoplankton bloom: Continental Shelf Research, v. 23, p. 1265–1283.

David, M.B., and Gentry, L.E., 2000, Anthropogenic inputs of nitrogen and phosphorus and riverine export for Illinois, USA: Journal of Environmental Quality, v. 29, p. 494–508.

David, M.B., Gentry, L.E., Kovacic, D.A., and Smith, K.M., 1997, Nitrogen balance in and export from an agricultural watershed: Journal of Environmental Quality, v. 26, p. 1038–1048.

David, M.B., McIsaac, G.F., Royer, T.V., Darmody, R.G., and Gentry, L.E., 2001, Estimated historical and current nitrogen balances for Illinois: The Scientific World, v. 1, p. 597–604.

David, M.B., Wall, L.G., Royer, T.V., and Tank, J.L., 2006, Denitrification and the nitrogen budget of a reservoir in an agricultural landscape: Ecological Applications, v. 16, p. 2177–2190.

Davis, C.B., Baker, J.L., van der Valk, A.G., and Beer, C.E., 1981, Prairie pothole marshes as traps for nitrogen and phosphorus in agricultural runoff, *in* Richardson, B., ed., Proceedings of the Midwestern Conference on Wetland Values and Management, St. Paul, MN, June 17–19, 1981: Navarre, MN, Fresh Water Society, p. 153–163.

Davis, J.G., Kitchen, N.R., Sudduth, K.A., and Drummond, S.T., 1997, Using electromagnetic induction to characterize soils: Potash & Phosphate Institute, Better Crops, v. 81, no. 4, p. 6–8.

Day, J.W., Jr., Yanez Arancibia, A., Mitsch, W.J., Lara-Dominquez, A.L., Day, J.N., Ko, J.-Y., Lane, R., and Lindsey, J., 2003, Using ecotechnology to address water quality and wetland habitat loss problems in the Mississippi basin—A hierarchical approach: Biotechnology Advances, v. 22, p. 135–159.

Denbaly, M., and Vrooman, H., 1993, Dynamic fertilizer nutrient demands for corn: A co-integrated and error-correcting system: American Journal of Agricultural Economics, v. 75, p. 203–209.

Deutsch, C., Sarmiento, J.L., Sigman, D.M., Gruber, N., and Dunne, J.P., 2007, Spatial coupling of nitrogen inputs and losses in the ocean: Nature, v. 445, p. 163–167, doi:10.1038/nature05392.

Devlin, D., Dhuyvetter, K., McVay, K., Kastens, T., Rice, C., Janssen, K., and Pierznski, G., 2003, Water quality best management practices, effectiveness, and cost for reducing contaminant losses from cropland: Kansas State University Agricultural Experiment Station and Cooperative Extension Service MF-2572, p. 4.

Diaz, R.J., 2001, Overview of hypoxia around the world: Journal of Environmental Quality, v. 30, no. 2, p. 275–281.

Diaz, R.J., Nestlerode, J., and Diaz, M.L., 2003, A global perspective on the effects of eutrophication and hypoxia on aquatic biota, *in* Proceedings of the 7th International Symposium, Fish Physiology, Toxicology, and Water Quality, Tallinn, Estonia, May 12–15, 2003.

Diaz, R.J., and Rosenberg, R., 1995, Marine benthic hypoxia—A review of its ecological effects and the behavioral responses of benthic macrofauna: Oceanography & Marine Biology, An Annual Review, v. 33, p. 245–303.

Diaz, R.J., and Rosenberg, R., 2009, Spreading dead zones and consequences for marine ecosystems: Science, v. 321, no. 5891, p. 926–929.

Diaz, R.J., and Solow, A., 1999, Ecological and economic consequences of hypoxia—Topic 2, Gulf of Mexico hypoxia assessment: Silver Springs, MD, National Ocean and Atmospheric Administration Coastal Ocean Program Decision Analysis Series, 86 p., available online at: http://oceanservice.noaa.gov/products/hypox_t2final.pdf.

Dietz, T., and Stern, P.C., eds., 2002, New tools for environmental protection: Education, information and voluntary measures. Committee on the Human Dimensions of Global Change, Washington, D.C.: National Academy Press.

Dinnel, S.P., and Wiseman, W.J., 1986, Fresh-water on the Louisiana and Texas shelf: Continental Shelf Research, v. 6, p. 765–784.

Dinnes, D.L., 2004, Assessments of practices to reduce nitrogen and phosphorus nonpoint source pollution of Iowa's surface waters: Ames, IA, U.S. Department of Agriculture, Agricultural Research Service, National Soil Tilth Laboratory, 376 p.

Djodjic, F., Ulen, B., and Bergstrom, L., 2000, Temporal and spatial variations of phosphorus losses and drainage in a structured clay soil: Water Research, v. 34, p. 1687–1695.

Dobermann, A., Walters, D.T., and Adviento-Borbe, M.A.A., 2007, Global warming potential of high-yielding continuous corn and corn-soybean systems: Better Crops, v. 91, no. 3, p. 16–19.

Dodds, W.K., Martí, E., Tank, J.L., Pontius, J., Hamilton, S.K., Grimm, N.B., Bowden, W.B., McDowell, W.H., Peterson, B.J., Valett, H.M., Webster, J.R., and Gregory, S., 2004, Carbon and nitrogen stoichiometry and nitrogen cycling rates in streams: Oecologia, v. 140, p. 458–467.

Doering, O., Diaz-Hermelo, F., Howard, C., Heimlich, R., Hitzhusen, F., Kazmierczak, R., Lee, J., Libby, L., Milon, W., Prato, T., and Ribaudo, M., 1999, Evaluation of the economic costs and benefits of methods for reducing nutrient loads to the Gulf of Mexico: Topic 6 report for the integrated assessment of hypoxia in the Gulf of Mexico: Silver Springs, MD, National Oceanic and Atmospheric Administration Coastal Ocean Program, Decision Analysis Series No. 20, 137 p., available online at: http://oceanservice.noaa.gov/products/hypox_t6final.pdf.

Doering, O.C., 2002, Economic linkages driving the potential response to nitrogen over-enrichment: Estuaries, v. 25, no. 4B, p. 809–818.

Donner, S.D., 2006, Surf or turf—A shift from feed to food cultivation could reduce nutrient flux to the Gulf of Mexico: Global Environmental Change, v. 17, no. 1, p. 105–113.

Donner, S.D., Coe, M.T., Lenters, J., Twine, T.E., and Foley, J.A., 2002, Modeling the impact of hydrology on nitrate transport in the Mississippi River System from 1955-1994: Global Biogeochemical Cycles, v. 16(3), 2001GB001396.

Donner, S.D., and Kucharik, C.J., 2003, Evaluating the impacts of land management and climate variability on crop production and nitrogen export across the Upper Mississippi Basin: Global Biogeochemical Cycles, v. 17, doi:10.1028/2001GB1808.

Donner, S.D., and Kucharik, C.J., 2008, Corn-based ethanol production compromises goal of reducing nitrogen export by the Mississippi River, in Proceedings of the National Academy of Sciences of the United States of America, v. 105, no. 11, p. 4513–4518.

Donner, S.D., Kucharik, C.J., and Foley, J.A., 2004, Impact of changing land use practices on nitrate export by the Mississippi River: Global Biogeochemical Cycles, v. 18, no. GB1028, p. 1–21.

Donner, S.D., and Scavia, D., 2007, How climate controls the flux of nitrogen by the Mississippi River and the development of hypoxia in the Gulf of Mexico: Limnology and Oceanography, v. 52, no. 2, p. 856–861.

Dortch, Q., Rabalais, N.N., Turner, R.E., and Qureshi, N.A., 2001, Impacts of changing Si/N ratios and phytoplankton species composition, in Rabalais, N.N., and Turner, R.E., eds., Coastal hypoxia—Consequences for living resources and ecosystems: Washington, D.C., American Geophysical Union, Coastal and Estuarine Studies, Volume 58, 37–48 p.

Dortch, Q., and Whitledge, T.E., 1992, Does nitrogen or silicon limit phytoplankton production in the Mississippi River plume and nearby regions?: Continental Shelf Research, v. 12, p. 1293–1309.

Dosskey, M.G., 2001, Toward quantifying water pollution abatement in response to installing buffers on crop land: Environmental Management, v. 28, no. 5, p. 577–598.

Dou, Z., Lanyon, L.E., Ferguson, J.D., Kohn, R.A., Boston, R.C., and Chalupa, W., 1998, An integrated approach to managing nitrogen on dairy farms—Evaluating farm performance using dairy nitrogen planner: Agronomy Journal, v. 90, p. 573–581.

Du, B., Saleh, A., Jaynes, D.B., and Arnold, J.G., 2006, Evaluation of SWAT in simulating nitrate nitrogen and atrazine fates in a watershed with tiles and potholes: Transactions of the American Society of Agricultural and Biological Engineers, v. 49, no. 4, p. 949–959.

Duan, S., and Bianchi, T.S., 2006, Seasonal changes in the abundance and composition of plant pigments in particulate organic carbon in the lower Mississippi and Pearl Rivers: Estuaries and Coasts, v. 29, no. 3, p. 427–442.

Duan, S., Bianchi, T.S., and Sampere, T.P., 2007, Temporal variability in the composition and abundance of terrestrially-derived dissolved organic matter in the lower Mississippi and Pearl Rivers: Marine Chemistry, v. 103, p. 172–184.

Edwards, D.R., and Daniel, T.C., 1993a, Drying-interval effects on runoff from fescue plots receiving swine manure: Transactions of the American Society of Agricultural Engineers, v. 36, p. 1673–1678.

Edwards, D.R., and Daniel, T.C., 1993b, Runoff quality impacts of swine manure applied to fescue plots: Transactions of the American Society of Agricultural Engineers, v. 36, p. 81–80.

Eldridge, P.M., and Morse, J.W., 2008, Origins and temporal scales of hypoxia on the Louisiana shelf—Importance of benthic and sub-pycnocline water metabolism: Marine Chemistry, v. 108, p. 159–171.

Elobeid, A., Tokgoz, S., Hayes, D.J., Babcock, B.A., and Hart, C.E., 2006, The long-run impact of corn-based ethanol on the grain, oilseed and livestock sectors: A preliminary assessment: Ames, IA, Center of Agriculture and Rural Development Briefing Paper 06-BP 49.

Engelhaupt, E., and Bianchi, T.S., 2001, Sources and composition of high-molecular-weight dissolved organic carbon in a southern Louisiana tidal stream (Bayou Trepagnier): Limnology and Oceanography, v. 46, p. 917–926.

English, B.C., De la Torre Ugarte, D.G., Jensen, K., Hellwinckel, C., Menard, J., Wilson, B., Roberts, R., and Walsh, M., 2006, 25% renewable energy for the Unites States by 2025: Agricultural and economic impacts: Knoxville, TN, University of Tennessee Technical Report.

Etter, P.C., Howard, M.K., and Cochrane, J.D., 2004, Heat and freshwater budgets of the Texas-Louisiana shelf: Journal of Geophysical Research, v. 109, p. C02024, doi:10.1029/2003JC001820.

Farrell, A.E., Plevin, R.J., Turner, B.T., Jones, A.D., O'Hare, M., and Kammen, D.M., 2006, Ethanol can contribute to energy and environmental goals: Science, v. 311, no. 5760, p. 506–508.

Feather, P., Hellerstein, D., and Hansen, L., 1999, Economic valuation of environmental benefits and the targeting of conservation programs—The case of the CRP: U.S. Department of Agriculture, Economic Research Service, Agricultural Economics Report No. 778, 64 p.

Feng, H., Kurkalova, L.A., Kling, C.L., and Gassman, P.W., 2005, Economic and environmental co-benefits of carbon sequestration in agricultural soils—Retiring agricultural land in the upper Mississippi River basin: Iowa State University Center for Agricultural and rural Development (CARD) Publication 05-WP384, 23 p., available online at: http://www.card.iastate.edu/publications/DBS/PDFFiles/05wp384.pdf.

Feng, H., Kurkalova, L.A., Kling, C.L., and Gassman, P.W., 2006, Environmental conservation in agriculture: Land retirement vs. changing practices on working land: Journal of Environmental Economics and Management, v. 52, p. 600–614.

Ferguson, R.B., Lark, R.M., and Slater, G.P., 2003, Approaches to management zone definition for use of nitrification inhibitors: Soil Science Society of America Journal, v. 67, p. 937–947.

Fiener, P., and Auerswald, K., 2003, Effectiveness of grassed waterways in reducing runoff and sediment delivery from agricultural watersheds: Journal of Environmental Quality, v. 32, no. 3, p. 927–936.

Fisher, T.R., and Gustafson, A.B., 2004, Progress Report—Aug. 1990–Dec. 2003, Nutrient-addition bioassays in Chesapeake Bay to assess resources limiting phytoplankton growth: Annapolis, MD, Maryland Department of Natural Resources, 50 p.

Fisher, T.R., Peele, E.R., Ammerman, J.W., and Harding, L.W., 1992, Nutrient limitation of phytoplankton in Chesapeake Bay: Marine Ecology Progress Series, v. 82, p. 51–63.

Forman, R.T.T., 1995, Land mosaics—The ecology of landscapes and regions: Cambridge, UK, Cambridge University Press, 652 p.

Fox, L., Sager, S.L., and Wofsy, S.C., 1985, Factors controlling the concentrations of soluble phosphorus in the Mississippi estuary: Limnology & Oceanography, v. 30, p. 826–832.

Frankenberger, J., Kladivko, E., Sands, G., Jaynes, D.B., Fausey, N.R., Helmers, M., Cooke, R., Strock, J., Nelson, K., and Brown, L., 2006, Drainage water management for the Midwest: Purdue Extension: Knowledge to Go, WQ-44, 8 p., available online at: http://www.ces.purdue.edu/extmedia/WQ/WQ-44.pdf.

Froelich, P.N., 1988, Kinetic control of dissolved phosphate in natural rivers and estuaries—A primer on the phosphate buffer mechanism: Limnology & Oceanography, v. 33, p. 649–668.

Galler, J.J., Bianchi, T.S., Allison, M.A., Campanella, R., and Wysocki, L., 2003, Sources of aged terrestrial organic carbon to the Gulf of Mexico from relict strata in the Mississippi River: EOS, Transactions of the American Geophysical Union, v. 84, p. 469–476.

Gardner, G., 1998, Recycling organic wastes, in Brown, L., Flavin, C., and French, H., eds., State of the world: New York, NY, W.W. Norton, 96–112 p.

Gardner, W.S., Benner, R., Chin-Leo, G., Cotner, J.B., Eadie, B.J., Cavaletto, J.F., Lansing, M.B., 1994, Mineralization of organic material and bacterial dynamics in Mississippi River plume water: Estuaries, v. 17, no. 4, p. 816–828.

Garnier, J., Leporcq, B., Sanches, N., and Philippon, X., 1999, Biogeochemical mass balances (C, N, P, Si) in three large reservoirs of the Seine Basin (France): Biogeochemistry, v. 47, p. 119–146.

Garnier, J., Sferratore, A., Meybeck, M., Billen, G., and Durr, H., 2006, Modeling silica transfer process in watersheds, in Ittekot, V., Unger, D., Humborg, C., and An, N.T., eds., The silicon cycle—Human perturbations and impacts on aquatic systems: Washington, D.C., Island Press, 139–162 p.

Gassman, P.W., Reyes, M.R., Green, C.H., and Arnold, J.G., 2007, The soil and water assessment tool—Historical development applications, and future research directions: Transactions of the American Society of Agricultural and Biological Engineers, v. 50, no. 4, p. 1211–1240.

Gehl, R.J., Schmidt, J.P., Maddux, L.D., and Gordon, W.B., 2005, Corn yield response to nitrogen rate and timing in sandy irrigated soils: Agronomy Journal, v. 97, p. 1230–1238.

Gentry, L.E., Below, F.E., David, M.B., and Bergerou, J.A., 2001, Source of the soybean N credit in maize production: Plant and Soil, v. 236, p. 175–184.

Gentry, L.E., David, M.B., Royer, T.V., Mitchell, C.A., and Starks, K.M., 2007, Phosphorus transport pathways to streams in tile-drained agricultural watersheds: Journal of Environmental Quality, v. 36, p. 408–415.

Gilroy, E.J., Hirsh, R.M., and Cohn, T.A., 1990, Mean square error of regression-based constituent transport estimates: Water Resources Research, v. 26, p. 2069–2077.

Gitau, M.W., Gburek, W.J., and Jarrett, A.R., 2005, A tool for estimating best management practice effectiveness for phosphorus pollution control: Journal of Soil and Water Conservation, v. 60, no. 1, p. 1–10.

Gollehon, N., Caswell, M., Ribaudo, M., Kellogg, R., Lander, C., and Letson, D., 2001, Confined animal production and manure nutrients: U.S. Department of Agriculture, Economic Research Service, Resource Economics Division, Agriculture Information Bulletin No. 771.

Goni, M.A., Ruttenberg, K.C., and Eglinton, T.I., 1998, A reassessment of the sources and importance of land-derived organic matter in surface sediments from the Gulf of Mexico: Geochimica et Cosmochimica Acta, v. 62, no. 18, p. 3055–3075.

Gooday, A.J., Jorissen, F., Levin, L.A., Middelburg, J.J., Naqvi, S.W.A., Rabalais, N.N., Scranton, and Zhang, M., 2009, Historical records of coastal eutrophication-induced hypoxia: BioScience, v. 6, no. 8, p. 1707–1745.

Goolsby, D.A., Battaglin, W.A., Aulenbach, B.T., and Hooper, R.P., 2001, Nitrogen input to the Gulf of Mexico: Journal of Environmental Quality, v. 30, no. 2, p. 329–336.

Goolsby, D.A., Battaglin, W.A., Lawrence, G.B., Artz, R.S., Aulenbach, B.T., Hooper, R.P., Keeney, D.R., and Stensland, G.S., 1999, Flux and sources of nutrients in the Mississippi–Atchafalaya River basin—Topic 3, Report for the integrated assessment of hypoxia in the Gulf of Mexico: National Oceanic and Atmospheric Administration Coastal Ocean Program Decision Analysis Series No. 17, 128 p.

Gordon, E.S., Goni, M.A., Roberts, Q.N., Kineke, G.C., and Allison, M.A., 2001, Organic matter distribution and accumulation on the inner Louisiana shelf west of the Atchafalaya River: Continental Shelf Research, v. 21, p. 1691–1721.

Gowda, P.H., Dalzell, B.J., and Mulla, D.A., 2007, Model based nitrate TMDLs for two agricultural watersheds of southeastern Minnesota: Journal of the American Water Resources Association, v. 43, no. 1, p. 254–263.

Gowda, P.H., and Mulla, D.J., 2006, Modeling alternative agricultural management practices for High Island Creek watershed in South-Central Minnesota: Journal of Environmental Hydrology, v. 14, no. 13, p. 1–15.

Graco, M., Farías, L., Molina, V., Gutiérrez, D., and Nielsen, L.P., 2001, Massive developments of microbial mats following phytoplankton blooms in a naturally eutrophic bay—Implications for nitrogen cycling: Limnology & Oceanography, v. 46, p. 821–832.

Graff, C.D., Sadeghi, A.M., Lowrance, R.R., and Williams, R.G., 2005, Quantifying the sensitivity of the riparian ecosystem management model (REMM) to changes in climate and buffer characteristics common to conservation practices: Transactions of the American Society of Agricultural Engineers, v. 48, no. 4, p. 1377–1387.

Grandy, A.S., Loecke, T.D., Parr, S., and Robertson, G.P., 2006, Long-term trends in nitrous oxide emissions, soil nitrogen, and crop yields of till and no-till cropping systems: Journal of Environmental Quality, v. 35, p. 1487–1495.

Graham, R.L., Nelson, R., Sheehan, J., Perlack, R.D., and Wright, L.L., 2007, Current and potential U.S. corn stover supplies: Agronomy Journal, v. 99, p. 1–11.

Green, C.H., Tomer, M.D., DiLuzio, M., and Arnold, J.G., 2006a, Hydrologic evaluation of the soil and water assessment tool for a large tile-drained watershed in Iowa: Transactions of the American Society of Agricultural and Biological Engineers, v. 49, no. 2, p. 413–422.

Green, R.E., Bianchi, T.S., Dagg, M.J., Walker, N.D., and Breed, G.A., 2006b, An organic carbon budget for the Mississippi River turbidity plume and plume contributions to air-sea CO_2 fluxes and bottom water hypoxia: Estuaries, v. 29, no. 4, p. 579–597.

Green, R.E., Breed, G.A., Dagg, M.J., and Lohrenz, S.E., 2008, Modeling the response of primary production and sedimentation to variable nitrate loading in the Mississippi River plume, Continental Shelf Research, v. 28, p. 1451–1465.

Green, W.R., and Haggard, B.E., 2001, Phosphorus and nitrogen concentrations and loads at Illinois River south of Siloam Springs, Arkansas, 1997–1999: U.S. Geological Survey Water Resources Investigations Report 01-4217, 12 p., available online at: http://pubs.er.usgs.gov/usgspubs/wri/wri014217.

Greenhalgh, S., and Sauer, A., 2003, Awakening the dead zone: An investment for agriculture, water quality, and climate changes: World Resources Institute, WRI Issue Brief, February, 24 p.

Gregg, M.C., 1999, Uncertainties and limitations in measuring ϵ and χ_T: Journal of Atmospheric and Oceanic Technology, v. 16, p. 1483–1490.

Grizzetti, B., Bouraoui, F., and De Marsily, G., 2005, Modeling nitrogen pressure in river basins—A comparison between a statistical approach and the physically-based SWAT model: Physics and Chemistry of the Earth, Parts A/B/C, v. 30, nos. 8–10, p. 508–517.

Groffman, P., Altabet, M.A., Böhlke, J.K., Butterbach-Bahl, K., David, M.B., Firestone, M.K., Giblin, A.E., Kana, T.M., Nielsen, L.P., and Voytek, M.A., 2006, Methods for measuring denitrification—Diverse approaches to a difficult problem: Ecological Applications, v. 16, p. 2091–2122, available online at: https://darchive.mblwhoilibrary.org/bitstream/1912/1425/1/Groffman%20et%20al.%20%20282006%29%20-%20denitrification%20methods%20paper.pdf.

Hagy, J.D., Boynton, W.R., Keefe, C.W., and Wood, K.V., 2004, Hypoxia in Chesapeake Bay, 1950–2001—Long-term change in relation to nutrient loading and river flow: Estuaries, v. 27, p. 634–658.

Harding, L.W., Jr., 1994, Long-term trends in the distribution of phytoplankton in Chesapeake Bay—Roles of light, nutrients, and streamflow: Marine Ecology Progress Series, v. 104, p. 267–291.

Harding, L.W., Jr., Mallonee, M.E., and Perry, E.S., 2002, Toward a predictive understanding of primary productivity in a temperate, partially stratified estuary: Estuarine Coastal & Shelf Science, v. 55, p. 437–463.

Haufler, J., ed., 2005, Fish and wildlife benefits of Farm Bill Conservation Programs—2000–2005 update: The Wildlife Society, Technical Review 05-2.

Haygarth, P.M., Chapman, P.J., Jarvis, S.C., and Smith, R.V., 1998, Phosphorus budgets for two contrasting grassland farming systems in the UK: Soil Use and Management, v. 14, p. 160–167.

Hazen, E.L., Craig, J.K., Good, C.P., and Crowder, L.B., 2009, Vertical distribution of fish biomass in hypoxic waters on the Gulf of Mexico shelf: Marine Ecology-Progress Series, v. 375, p. 195–207.

Heckrath, G., Brookes, P.C., Paulton, P.R., and Goulding, K.W.T., 1995, Phosphorus leaching from soils containing different phosphorus concentrations in the Broadbalk experiment: Journal of Environmental Quality, v. 24, p. 904–910.

Helmers, M., 2007, Iowa State University agricultural drainage, available online at: http://www3.abe.iastate.edu/agdrainage/.

Helsel, D.R., and Hirsch, R.M., 2002, Statistical methods in water resources: U.S. Geological Survey techniques of water-resources investigations, Book 4, chap. A3, available online at: http://water.usgs.gov/pubs/twri/twri4a3/.

Helton, A.M., 2006, An inter-biome comparison of stream network nitrate dynamics: Athens, GA, University of Georgia, M.S. thesis, 143 p.

Hendon, L.A., Carlson, E.A., Manning, S., and Brouwer, M., 2006, Cross-talk between pyrene and hypoxia signaling pathways in embryonic *Cyprinodon variegates*: Paper presented at Hypoxia Effects on Living Resources in the Gulf of Mexico, September 25–26, 2006: New Orleans, Louisiana, Tulane University, sponsored by National Oceanic and Atmospheric Administration Center for Sponsored Coastal Ocean Research.

Herbert, R.A., 1999, Nitrogen cycling in coastal marine systems: FEMS Microbiology Reviews, v. 23, p. 563–590.

Hernandez, M.E., and Mitsch, W.J., 2006, Influence of hydrologic pulses, flooding frequency, and vegetation on nitrous oxide emissions from created riparian marshes: Wetlands, v. 26, no. 3, p. 862–877.

Hernes, P.J., and Benner, R., 2003, Photochemical and microbial degradation of dissolved lignin phenols—Implications for the fate of terrigenous dissolved organic matter in marine environments: Journal of Geophysical Research—Oceans, v. 108, no. C9, Article Number 3291.

Hetland, R.D., and DiMarco, S.F., 2008, How does the character of oxygen demand control the structure of hypoxia on the Texas-Louisiana continental shelf?: Journal of Marine Systems, doi:10.1016/j.jmarsys.2007.03.002, v. 70, p. 49–62.

Hey, D.L., Kenimer, A.L., and Barrett, K.R., 1994, Water quality improvement by four experimental wetlands: Ecological Engineering, v. 3, p. 381–397.

Hey, D.L., Montgomery, D.L., Urban, L.S., Prato, T., Zarwell, R., Forbes, A., Martell, M., Pollack, J., and Steele, Y., 2004, Flood damage reduction in the Upper Mississippi River Basin—An ecological alternative: The Wetlands Initiative, Chicago, IL, 44 p., available online at: http://www.wetlands-initiative.org/images/UMRBFinalReport.pdf.

Hoffman, B.S., Brouder, S.M., and Turco, R.F., 2004, Tile spacing impacts on *Zea mays* L. yield and drainage water nitrate load: Ecological Engineering, v. 23, p. 251–267.

Holland, E.F., Braswell, B.H., Sulzman, J., and Lamarque, J., 2005, Nitrogen deposition to the United States and Western Europe—Synthesis of observations and models: Ecological Applications, v. 15, p. 38–57.

Holland, E.F., Dentener, B., Braswell, B.H., and Sulzman, J., 1999, Contemporary and pre-industrial global reactive nitrogen budgets: Biogeochemistry, v. 4, p. 7–43.

Holling, C.S., 1978, Adaptive Environmental Assessment and Management, Chichester, UK: John Wiley.

Hong, N., White, J.G., Weisz, R., Crozier, C.R., Gumpertz, M.L., and Cassel, D.K., 2006, Remote sensing-informed variable-rate nitrogen management of wheat and corn—Agronomic and groundwater outcomes: Agronomy Journal, v. 98, p. 327–338.

Howarth, R.W., 1984, The ecological significance of sulfur in the energy dynamics of salt marsh and marine sediments: Biogeochemistry, v. 1, p. 5–27.

Howarth, R.W., 1998, An assessment of human influences on inputs of nitrogen to the estuaries and continental shelves of the North Atlantic Ocean: Nutrient Cycling in Agroecosystems, v. 52, p. 213–223.

Howarth, R.W., 2006, Atmospheric deposition and nitrogen pollution in coastal marine ecosystems, *in* Visgilio, G.R., and Whitelaw, D.M., eds., Acid in the environment—Lessons learned and future prospects: New York, NY, Springer, 97–116 p.

Howarth, R.W., Billen, G., Swaney, D., Townsend, A., Jarworski, N., Lajtha, K., Downing, J.A., Elmgren, R., Caraco, N., Jordan, T., Berendse, F., Freney, J., Kueyarov, V., Murdoch, P., and Zhao-liang, Zhu., 1996, Riverine inputs of nitrogen to the North Atlantic Ocean—Fluxes and human influences: Biogeochemistry, v. 35, p. 75–139.

Howarth, R.W., Boyer, E.W., Pabich, W.J., and Galloway, J.N., 2002, Nitrogen use in the United States from 1961–2000 and potential future trends: Ambio, v. 31, p. 88–96.

Howarth, R.W., Jensen, H.S., Marino, R., and Postma, H., 1995, Transport to and processing of P in near-shore and oceanic waters, *in* Tiessen, H., ed., Phosphorus in the global environment: New York, NY, Wiley, 323–345 p.

Howarth, R.W., and Marino, R., 2006, Nitrogen as the limiting nutrient for eutrophication in coastal marine ecosystems—Evolving views over 3 decades: Limnology & Oceanography, v. 51, p. 364–376.

Howarth, R.W., Marino, R., Swaney, D.P., and Boyer, E.W., 2006a, Wastewater and watershed influences on primary productivity and oxygen dynamics in the lower Hudson River estuary, *in* Levinton, J.S., and Waldman, J.R., eds., The Hudson river estuary: Cambridge, UK, Cambridge University Press, 121–139 p.

Howarth, R.W., Ramakrishna, K., Choi, E., Elmgren, R., Martinelli, L., Mendoza, A., Moomaw, W., Palm, C., Boy, R., Scholes, M., and Zhao-Liang, Zhu, 2005, Nutrient management, responses assessment, *in* Ecosystems and human well-being—Volume 3, Policy responses, The millennium ecosystem assessment: Washington, D.C., Island Press, chap. 9, 295–311 p.

Howarth, R.W., Swaney, D.P., Boyer, E.W., Marino, R., Jaworski, N. and Goodale, C., 2006b, The influence of climate on average nitrogen export from large watersheds in the Northeastern United States: Biogeochemistry, v. 79, p. 163–186.

Hu, X., McIssac, G.F., David, M.B., and Louwers, C.A.L., 2007, Modeling riverine nitrate export from an east-central Illinois watershed using SWAT: Journal of Environmental Quality, v. 36, p. 996–1005.

Humborg, C., Conley, D.J., Rahm, L., Wulff, F., Cociasu, A., and Ittekkot, V., 2000, Silicon retention in river basins—Far-reaching effects on biogeochemistry and aquatic food webs in coastal marine environments: Ambio, v. 29, p. 45–50.

IATP, 2006, Staying home: How ethanol will change U.S. corn exports: Minneapolis, MN, Institute for Agriculture and Trade Policy, 26 p.

Inwood, S.E., Tank, J.L., and Bernot, M.J., 2005, Patterns of denitrification associated with land use in 9 mid-western headwater streams: Journal of the North American Benthological Society, v. 24, p. 227–245.

Isik, M., and Khanna, M., 2002, Variable rate nitrogen application under uncertainty: Implications for profitability and nitrogen use: Journal of Agricultural and Resource Economics, v. 27, no. 1, p. 61–76.

Isik, M., and Khanna, M., 2003, Stochastic technology, risk preferences and adoption of site-specific technologies: American Journal of Agricultural Economics, v. 85, no. 2, p. 305–317.

James, E.E., 2005, Factors influencing the adoption and nonadoption of the conservation reserve enhancement program in the Cannonsville Watershed, New York: University Park, PA, The Pennsylvania State University, Department of Agricultural Economics and Rural Sociology, M.Sc. dissertation, 174 p.

James, E.E., Kleinman, P.J.A., Veith, T.L., Stedman, R., and Sharpley, A.N., 2007, Phosphorus contributions from pastured dairy cattle to streams in the Cannonsville watershed: Journal of Soil and Water Conservation, v. 62, p. 40–47.

Jarosz, E., and Murray, S.P., 2005, Velocity and transport characteristics of the Louisiana-Texas coastal current: Geophysical Monograph, American Geophysical Union, Circulation in the Gulf of Mexico Observations and Models, v. 161, p. 143–156.

Jaynes, D.B., Colvin, T.S., Karlen, D.L., Cambardella, C.A., and Meek, D.W., 2001, Nitrate loss in subsurface drainage as affected by nitrogen fertilizer rate: Journal of Environmental Quality, v. 30, p. 1305–1314.

Jaynes, D.B., Dinnes, D.L., Meek, D.W., Karlen, D.L., Cambardella, C.A., and Colvin, T.S., 2004, Using the late spring nitrate test to reduce nitrate loss within a watershed: Journal of Environmental Quality, v. 33, p. 669–677.

Jaynes, D.B., and James, D.E., 2007, The extent of farm drainage in the United States: Poster presentation at Soil and Water Conservation Society 2007 Annual Conference, July 21–25, 2007, Tampa, FL, available online at: http://www.ars.usda.gov/Research/Researchltrm?modecode=36-25-15-00.

Jaynes, D.B., and Karlen, D.L., 2005, Sustaining soil resources while managing nutrients, in Proceedings of the Upper Mississippi River Sub-Basin Hypoxia Nutrient Committee Workshop, Ames, IA, September 26–28, 2005, p. 141–150, at www.umrshnc.org/files/Hypwebversion.pdf, accessed June 28, 2007.

Jha, M., Arnold, J.G., Gassman, P.W., Giorgi, F., and Gu, R.R., 2006, Climate change sensitivity assessment on Upper Mississippi River Basin streamflows using SWAT: Journal of American Water Resources Association, v. 42, no. 4, p. 997–1016.

Jochem, F.J., McCarthy, M.J., and Gardner, W.S., 2004, Microbial ammonium recycling in the Mississippi River plume during the drought spring of 2000: Journal of Plankton Research, v. 26, p. 1265–1275.

Joye, S.B., and Hollibaugh, J.T., 1995, Influence of sulfide inhibition on nitrification and nitrogen regeneration in sediments: Science, v. 270, p. 623–625.

Justić, D., Bierman, V.J., Jr., Scavia, D., and Hetland, R., 2007, Forecasting Gulf's hypoxia: The next 50 years?, Estuaries and Coasts, v. 30, p. 791–801.

Justić, D., Rabalais, N.N., and Turner, R.E., 1996, Effects of climate change on hypoxia in coastal waters—A doubled CO_2 scenario for the northern Gulf of Mexico: Limnology & Oceanography, v. 41, p. 992–1003.

Justić, D., Rabalais, N.N., and Turner, R.E., 2002, Modeling the impacts of decadal changes in riverine nutrient fluxes on coastal eutrophication near the Mississippi River delta: Ecological Modeling, v. 152, p. 33–46.

Justić, D., Rabalais, N.R., and Turner, R.E., 2003a, Simulated responses of the Gulf of Mexico hypoxia to variations in climate and anthropogenic nutrient loading: Journal of Marine Systems, v. 42, p. 115–126.

Justić, D., Rabalais, N.N., Turner, R.E., and Dortch, Q., 1995, Changes in nutrient structure of river-dominated coastal waters—Stoichiometeric nutrient balance and its consequences: Estuarine, Coastal Shelf Science, v. 40, p. 339–356.

Justić, D., Rabalais, N.N., Turner, R.E., and Wiseman, W.J., Jr., 1993, Seasonal coupling between riverborne nutrients, net productivity and hypoxia: Marine Pollution Bulletin No. 26, p. 184–189.

Justić, D., Turner, R.E., and Rabalais, N.N., 2003b, Climate influences on riverine nitrate flux—Implications for coastal marine eutrophication and hypoxia: Estuaries, v. 26, no. 1, p. 1–11.

Kantha, L.H., 2005, Barotropic tides in the Gulf of Mexico, in Sturges, W., and Lugo-Fernandez, A., eds., Circulation in the Gulf of Mexico—Observations and models: Geophysical monograph: Washington, D.C., American Geophysical Union, Volume 161, 159–163 p.

Kantha, L.H., and Clayson, C.A., 2000, Small scale processes in geophysical fluid flows: San Diego, CA, Academic Press, 668 p.

Kanwar, R.S., Colvin, T.S., and Karlin, D.L., 1997, Ridge, moldboard, chisel and no-till effects on tile water quality beneath two cropping systems: Journal of Production Agriculture, v. 10, p. 227–234.

Karlen, D.L., Dinnes, D.L., Jaynes, D.B., Hurburgh, C.R., Cambardella, C.A., Colvin, T.S., and Rippke, G.R., 2005, Corn response to late-spring nitrogen management in the Walnut Creek watershed: Agronomy Journal, v. 97, p. 1054–1061.

Karlen, D.L., Kramer, L.A., and Logsdon, S.D., 1998a, Field-scale nitrogen balances associated with long-term continuous corn production: Agronomy Journal, v. 90, p. 644–650.

Karlen, D.L., Kumar, A., Kanwar, R.S., Cambardella, C.A., and Colvin, T.S., 1998b, Tillage system effects on 15 years carbon-based and simulated N budgets in a tile-drained Iowa field: Soil & Tillage Research, v. 48, p. 155–165.

Karlson, K., Rosenberg, R., and Bonsdorff, E., 2002, Temporal and spatial large-scale effects of eutrophication and oxygen deficiency on benthic fauna in Scandinavian and Baltic waters—A review: Oceanography & Marine Biology, An Annual Review, v. 40, p. 427–489.

Keene, W.C., Montag, J.A., Maben, J.A., Southwell, M., Leonard, J., Church, T.M., Moody, J.L., and Galloway, J.N., 2002, Organic nitrogen in precipitation over eastern North America: Atmospheric Environment, v. 36, p. 4529–4540.

Kellogg, R.L., 2006, Presentation to SAB hypoxia advisory panel: Agricultural Economist, Resources Inventory and Assessment Division, USDA, Beltsville, MD.

Kellogg, L.E., and Bridgham, S.D., 2003, Phosphorus retention and movement across an ombrotrophic-minerotrophic peatland gradient: Biogeochemistry, v. 63, p. 299–315.

Kellogg, R.L., Lander, C.H., Moffitt, D.C., and Gollehon, N., 2000, Manure nutrients relative to the capacity of cropland and pastureland to assimilate nutrients—Spatial and temporal trends for the United States: U.S. Department of Agriculture, Natural Resources Conservation Service and Economic Research Service, Resource Assessment and Strategic Planning Working Paper 98-1, 140 p., available online at: http://www.nrcs.usda.gov/technical/land/pubs/manntr.pdf.

Kemp, W.M., Sampou, P., Caffrey, J., and Mayer, M., 1990, Ammonium recycling versus denitrification in Chesapeake Bay sediments: Limnology & Oceanography, v. 351, p. 545–563.

Khan, S.A., Mulvaney, R., and Hoeft, R.G., 2001, A simple soil test for detecting sites that are nonresponsive to nitrogen fertilization: Soil Science Society of America Journal, v. 65, p. 1751–1760.

Khanna, M., 2001, Sequential adoption of site-specific technologies and its implications for nitrogen productivity: A double selectivity model: American Journal of Agricultural Economics, v. 83, p. 35–51.

Khanna, M., Isik, M., and Winter-Nelson, A., 2000, Investment in site-specific crop management under uncertainty: Implications for nitrate pollution control and environmental policy: Agricultural Economics, v. 24, no. 1, p. 9–21.

Khanna, M., Yang, W., Farnsworth, R., and Onal, H., 2003, Cost effective targeting of CREP to improve water quality with endogenous sediment deposition coefficients: American Journal of Agricultural Economics, v. 85, p. 538–553.

Khoshmanesh, A., Hart, B.T., Duncan, A., and Beckett, R., 1999, Biotic uptake and release of phosphorus by a wetland sediment: Environmental Technology, v. 29, p. 85–91.

King, D., 2005, Crunch time for water quality trading: Choices, v. 20, p. 71–75.

Kirsch, K., Kirsch, A., and Arnold, J.G., 2002, Predicting sediment and phosphorus loads in the Rock River Basin using SWAT: Transactions of the American Society of Agricultural and Biological Engineers, v. 45, no. 6, p. 1757–1769.

Kitchen, N.R., Sudduth, K.A., and Drummond, S.T., 1999, Soil electrical conductivity as a crop productivity measure for claypan soils: Journal of Production Agriculture, v. 12, p. 607–617.

Kladivko, E.J., Frankenberger, J.R., Jaynes, D.B., Meek, D.W., Jenkinson, B.J., and Fausey, N.R., 2004, Nitrate leaching to subsurface drains as affected by drain spacing and changes in crop production system: Journal of Environmental Quality, v. 33, p. 1803–1813.

Kline, K.L., Dale, V.H. Lee, R. and Leiby, P., 2009, In defense of biofuels, done right: Issues in Science and Technology, v. 25, no. 3, p. 75–84.

Kling, C., Secchi, S., Jha, M., Feng, H., Gassman, P., and Kurkalova, L., 2006, Upper Mississippi River basin modelling system, Part 3—Conservation practice scenario results, *in* Singh, V., and Xu, Y., eds., Coastal hydrology and processes: Canada, Water Resources Publication, 127–134 p.

Koch, B., Khosla, R., Frasier, W.M., Westfall, D.G., and Inman, D., 2004, Economic feasibility of variable-rate nitrogen application utilizing site-specific management zones: Agronomy Journal, v. 96, p. 1572–1580.

Kovacic, D.A., David, M.B., Gentry, L.E., Starks, K.M., and Cooke, R.A., 2000, Effectiveness of constructed wetlands in reducing nitrogen and phosphorus export from agricultural tile drainage: Journal of Environmental Quality, v. 29, p. 1262–1274.

Krom, M.D., and Berner, R.A., 1980, Adsorption of phosphate in anoxic marine sediments: Limnology & Oceanography, v. 25, p. 797–806.

Laboski, C.A.M., Sawyer, J.E., Walters, D.T., Bundy, L.G., Hoeft, R.G., Randall, G.W., and Andraski, T.W., 2006, Evaluation of the Illinois soil nitrogen test in the north central region, *in* Proceedings of the North Central Extension-Industry Soil Fertility Conference, Des Moines, IA, v. 22, p. 86–93.

Laboski, C.A.M., Sawyer, J.E., Walters, D.T., Bundy, L.G., Hoeft, R.G., Randall, G.W., and Andraski, T.W., 2008, Evaluation of the Illinois soil nitrogen test in the North Central Region of the United States: Agronomy Journal, v. 100, p. 1070–1076.

Lambert, D.M., Lowenberg-DeBoer, J., and Malzer, G.L., 2006, Economic analysis of spatial-temporal patterns in corn and soybean response to nitrogen and phosphorus: Agronomy Journal, v. 98, p. 43–54.

Landeck-Miller, R.E., and St. John, J.P., 2006, Modeling primary production in the Lower Hudson River Estuary, *in* Levington, J.S., and Waldman, J.R., eds., The Hudson river estuary, New York, NY, Cambridge University Press, 140–153 p.

Lane, R.R., Day, J.W., Marx, B., Reyes, E., and Kemp, G.P., 2002, Seasonal and spatial water quality changes in the outflow plume of the Atchafalaya River, Louisiana, USA: Estuaries, v. 25, no. 1, p. 30–42.

Lanyon, L.E., 2005, Phosphorus, animal nutrition and feeding: Overview, *in* Sims, J.T., and Sharpley, A.N., eds., Phosphorus: Agriculture and the environment, Madison, WI, American Society of Agronomy Monograph Series No. 46, 561–586 p.

Larson, R.S., 2001, Water quality trends of the Illinois waterway system upstream of Peoria including the Chicago metropolitan area: Champaign, IL, Illinois State Water Survey, Contract Report 2001-03.

Leavitt, P.R., and Hodgson, D.A., 2001, Sedimentary pigments, *in* Smol, J.P., Birks, H.J.B, and Last, W.M., eds., Tracking environmental change using lake sediments, Terrestrial, algal, and siliceous indicators: The Netherlands, Kluwer Academic Publishers, Volume 3, 295–325 p.

Lee, K.N. 1999. Appraising Adaptive Management. Conservation Ecology 3(2): 3. Available online at http://www.consecol.org/vol3/iss2/art3, accessed February 13, 2003.

Lee, K.H., Isenhart, T.M., and Schultz, R.C., 2003, Sediment and nutrient removal in an established multi-species riparian buffer: Journal of Soil and Water Conservation, v. 58, no. 1, p. 1–7.

Lee, K.H., Isenhart, T.M., Schultz, R.C., and Mickelson, S.K., 2000, Multispecies riparian buffers trap sediment and nutrients during rainfall simulations: Journal of Environmental Quality, v. 29, no. 4, p. 1200–1205.

Lemunyon, J.L., and Gilbert, R.G., 1993, The concept and need for a phosphorus assessment tool: Journal of Production Agriculture, v. 6, p. 483–496.

Levin, L.A., Ekau, W., Gooday, A.J., Jorissen, F., Middelburg, J.J., Naqvi, S.W.A., Neira, C., Rabalais, N.N., and Zhang, J., 2009, Effects of natural and human-induced hypoxia on coastal benthos: BioScience, v. 6, no. 1, p. 2063–2098.

Lewandrowski, J., Peters, M., Jones, C., House, R., Sperow, M., Eve, M., and Paustian, K., 2004, Economics of sequestering carbon in the U.S. agricultural sector: U.S. Department of Agriculture, Economic Research Service, Technical Bulletin No. 1909, p. 69.

Lindau, C.W., Delaune, R.D., Scaroni, A.E., and Nyman, J.A., 2008, Denitrification in cypress swamp within the Atchafalaya River Basin, Louisiana: Chemosphere, v. 70, no. 5, p. 886–894.

Liu, H.B., Dagg, J.M. Campbell, L., and Urban-Rich, J., 2004, Picophytoplankton and bacterioplankton in the Mississippi River plume and its adjacent waters: Estuaries, v. 27, no. 1, p. 147–156.

Lohrenz, S.E., Dagg, M.L., and Whitledge, T.E., 1990, Enhanced primary production at the plume/oceanic interface of the Mississippi River: Continental Shelf Research, v. 10, no. 7, p. 639–664.

Lohrenz, S.E., Fahnenstiel, G.L., and Redalje, D.G., 1994, Spatial and temporal variations of photosynthetic parameters in relation to environmental conditions in northern Gulf of Mexico coastal waters: Estuaries, v. 17, p. 779–795.

Lohrenz, S.E., Fahnenstiel, G.L., Redalje, D.G., and Lang, G.A., 1992, Regulation and distribution of primary production in the northern Gulf of Mexico, *in* Program, N.C.O., ed., Nutrient enhanced coastal ocean productivity: College Station, TX, Texas Sea Grant Publications, NECOP Workshop Proceedings, October 1991, p. 95–104.

Lohrenz, S.E., Fahnenstiel, G.L., Redalje, D.G., Lang, G.A., Chen, X.G., and Dagg, M.J., 1997, Variations in primary production of northern Gulf of Mexico continental shelf waters linked to nutrient inputs from the Mississippi River: Marine Ecology-Progress Series, v. 155, p. 45–54.

Lohrenz, S.E., Fahnenstiel, G.L., Redalje, D.G., Lang, G.A., Dagg, M.J., Whitledge, T.E., and Dortch, Q., 1999a, Nutrients, irradiance, and mixing as factors regulating primary production in coastal waters impacted by the Mississippi River plume: Continental Shelf Research, v. 19, p. 1113–1141.

Lohrenz, S.E., Redalje, D.G., Cai, W.J., Acker, J., and Dagg, M., 2008, A retrospective analysis of nutrients and phytoplankton productivity in the Mississippi River plume: Continental Shelf Research, v. 28, no. 12, p. 1466–1475.

Lohrenz, S.E., Wiesenburg, D.A., Arnone, R.A., and Chen, X.G., 1999b, What controls primary production in the Gulf of Mexico?, *in* Kumpf, H., Steidinger, K., and Sherman, K., eds., The Gulf of Mexico large marine ecosystem: Malden, MA, Blackwell Science, Inc., Assessment, Sustainability and Management, 151–170 p.

Lory, J.A., and Scharf, P.C., 2003, Yield goal versus delta yield for predictiong fertilizer nitrogen need in corn: Agronomy Journal, v. 95, p. 994–999.

Loureiro, M.L., McCluskey, J.J., and Mittelhammer, R.C., 2001, Assessing consumers preferences for organic, eco-labeled and regular apples: Journal of Agricultural and Resource Economics, v. 26, no. 2, p. 404–416.

Lubowski, R.N., Bucholtz, S., Claassen, R., Roberts, M.J., Cooper, J.C., Gueorguieva, A., and Johansson, R., 2006, Environmental effects of agricultural land-use change: The role of Economics and policy: U.S. Department of Agriculture, Economic Research Service Report No. ERR-25, 82 p., available online at: http://www.ers.usda.gov/Publications/ERR25/.

LUMCON, 2007, Dead zone size near top end: LUMCON News, available online at: http://www.lumcon.edu/Information/news/default.asp?XMLFilename=200707311648.xml, last accessed August 7, 2007.

MacKinnon, J.A., and Gregg, M.C., 2005, Spring mixing: Turbulence and internal waves during restratification on the New England shelf: Journal of Physical Oceanography, v. 35, p. 2425–2443.

Mallarino, A.P., and Schepers, J.S., 2005, Role of precision agriculture in phosphorus management practices, *in* Sims, J.T., and Sharpley, A.N., eds., Phosphorus: Madison, WI, Agriculture and the Environment, American Society of Agronomy Monograph Series No. 46, Crop Science Society of America and Soil Science Society of America, 881–908 p.

Maloney, M., and Brady, G., 1988, Capital turnover and marketable emission rights: Journal of Law and Economics, v. 31, p. 203–226.

Mamo, M., Malzer, G.L., Mulla, D.J., Huggins, D.R., and Strock, J., 2003, Spatial and temporal variation in economically optimum nitrogen rate for corn: Agronomy Journal, v. 95, p. 958–964.

Mann, L., and Tolbert, V., 2000, Soil sustainability in renewable biomass plantings: Ambio, v. 29, no. 8, p. 492–498.

Marquez, C.O., Cambardella, C.A., Isenhart, T.M., and Schultz, R.C., 1999, Assessing soil quality in a riparian buffer strip system by testing organic matter fractions: Agroforestry Systems, v. 44, p. 133–140.

MART, 2006a, Management Action Review Team Report: U.S. Environmental Protection Agency, Mississippi River/Gulf of Mexico Watershed Nutrient Task Force, Management Action Reassessment Team, 31 p., available online at: http://www.epa.gov/msbasin/taskforce/MART.pdf.

MART, 2006b, Reassessment of point source nutrient mass loadings to the Mississippi River basin: U.S. Environmental Protection Agency, Mississippi River/Gulf of Mexico Watershed Nutrient Task Force, Management Action Reassessment Team, 31 p., available online at: http://www.epa.gov/msbasin/taskforce/Point_Source_Mass_Loading.pdf.

Maryland Department of the Environment, 2005, Chesapeake Bay Restoration Program: Baltimore, MD.

Mayer, P.M., Reynolds, S.K., McCutchen, M.D., and Canfield, T.J., 2006, Riparian buffer width, vegetative cover, and nitrogen removal effectiveness—A review of current science and regulations: Cincinnati, OH, U.S. Environmental Protection Agency, EPA/600/R-05/118.

Mazurek, J., 2002, Government-sponsored voluntary programs for firms—An initial survey, *in* Dietz, T., and Stern, P.C., eds., New tools for environmental protection: Education, information, and voluntary measures: Washington, D.C., The National Academies Press, National Academy of Sciences, p. 219–234.

McDowell, L.L., and McGregor, K.C., 1984, Plant nutrient losses in runoff from conservation tillage corn: Soil Tillage Research, v. 4, p. 79–91.

McDowell, R.W., and Sharpley, A.N., 2001, A comparison of fluvial sediment phosphorus (P) chemistry in relation to location and potential to influence stream P concentrations: Aquatic Geochemistry, v. 7, p. 255–265.

McDowell, R.W., and Sharpley, A.N., 2003, Uptake and release of phosphorus from overland flow in a stream environment: Journal of Environmental Quality, v. 32, p. 937–948.

McDowell, R.W., Sharpley, A.N., and Kleinman, P.J.A., 2002, Integrating phosphorus and nitrogen decision management at watershed scales: Journal of American Water Resources Association, v. 38, no. 2, p. 479–491.

McGraw, T., and Hemb, R., 1995, Fertility variability in the Minnesota River Valley watershed in 1993 as determined from grid soil testing results on 52,000 acres on commercial fields, *in* Robert, P.C., Rust, R.H., and Larson, W.E., eds., Site-specific management for agricultural systems: Minneapolis, MN, American Society of Agronomy, Crop Science Society of America and Soil Science Society of America, 2nd International Conference.

McIsaac, G.F., 2006, Net anthropogenic nitrogen inputs (NANI) to the Mississippi River basin: Presented at Science Symposium—Sources, transport, and fate of nutrients in the Mississippi and Atchafalaya River basins, November 7–9, 2006, Minneapolis, MN.

McIsaac, G.F., David, M.B., Gertner, G.Z., and Goolsby, D.A., 2001, Nitrate flux in the Mississippi River: Nature, v. 414, p. 166–167.

McIsaac, G.F., David, M.B., Gertner, G.Z., and Goolsby, D.A., 2002, Relating net nitrogen input in the Mississippi River basin to nitrate flux in the lower Mississippi River—A comparison of approaches: Journal of Environmental Quality, v. 31, p. 1610–1622.

McIsaac, G.F., and Hu, X., 2004, Net N input and riverine N export from Illinois agricultural watersheds with and without extensive tile drainage: Biogeochemistry, v. 70, p. 251–271.

McKee, B.A., Aller, R.C., Allison, M.A., Bianchi, T.S., and Kineke, G.C., 2004, Transport and transformation of dissolved and particulate materials on continental margins influenced by major rivers—Benthic boundary layer and seabed processes: Continental Shelf Research, v. 24, p. 899–926.

McLaughlin, S.B., and Kszos, L.A., 2005, Biomass and bioenergy development of switchgrass (*Panicum virgatum*) as a bioenergy feedstock in the United States: Biomass & Bioenergy, v. 28, p. 515–535.

McLaughlin, S., and Walsh, M., 1998, Evaluating environmental consequences of producing herbaceous crops for bioenergy: Biomass & Bioenergy, v. 14, no. 4, p. 317–324.

McMahon, G., Alexander, R.B., and Qian, S., 2003, Support of total maximum daily load programs using spatially referenced regression models: Journal of Water Resources Planning and Management, v. 129, p. 315–329.

Mee, L.D., 2006, Reviving dead zones: Scientific American, v. 295, p. 78–85.

Meisinger, J.J., and Delgado, J.A., 2002, Priciples for managing nitrogen leaching: Journal of Soil and Water Conservation, v. 57, no. 6, p. 485–499.

Melillo, J.M., and Cowling, E.B., 2002, Reactive nitrogen and public policies for environmental protection: AMBIO, A Journal of the Human Environment, v. 31, no. 2, p. 141–149.

Mellor, G.L., and Yamada, T., 1982, Development of a turbulence closure model for geophysical fluid problems: Reviews of Geophysics and Space Physics, v. 20, p. 851–875.

Meretsky, V.J., Wegner, D.L., and Stevens, L.E., 2000, Balancing endangered species and ecosystems—A case study of adaptive management in Grand Canyon: Environmental Management, v. 25, no. 6, p. 579–586.

Metropolitan Council, 2004, Regional progress in water quality—Analysis of water quality data from 1976 to 2002 for the major rivers in the Twin Cities: St. Paul, MN, Metropolitan Council, Publication 32-04-045.

Middelburg, J.J., and Levin, L.A., 2009, Coastal hypoxia and sediment biogeochemistry: BioScience, v. 6, no. 7, p. 1273–1293.

Mitsch, W.J., Day, J.W., Jr., Gilliam, W., Groffman, P.M., Hey, D.L., Randall, G.W., and Wang, N., 2001, Reducing nitrogen loading to the Gulf of Mexico from the Mississippi River basin—Strategies to counter a persistent ecological problem: Bioscience, v. 51, p. 373–388.

Mitsch, W.J., Day, J.W., Jr., Gilliam, J.W., Groffman, P.M., Hey, D.L., Randall, G.W., and Wang, N., 1999, Reducing nutrient loads, especially nitrate-nitrogen, to surface water, ground water, and the Gulf of Mexico: Topic 5 report for the integrated assessment on hypoxia in the Gulf of Mexico: Silver Spring, MD, National Oceanic and Atmospheric Administration Coastal Ocean Program Decision Analysis Series No. 19, 111 p., available online at: http://oceanservice.noaa.gov/products/hypox_t5final.pdf.

Mitsch, W.J., Day, J.W., Jr., Zhang, L., and Lane, R.R., 2005a, Nitrate-nitrogen retention in wetlands in the Mississippi River Basin: Ecological Engineering, v. 24, p. 267–278.

Mitsch, W.J., Zhang, L., Anderson, C.J., Altor, A.E., and Hernandez, M.E., 2005b, Creating riverine wetlands—Ecological succession, nutrient retention, and pulsing effects: Ecological Engineering, v. 25, p. 510–527.

Montagna, P., Hodges, B., Maidment, D., and Minsker, B., 2006, Long-term studies of hypoxia in Corpus Christi Bay: The cybercollaboratory testbed: Paper presented at Hypoxia Effects on

Living Resources in the Gulf of Mexico, September 25–26, 2006: New Orleans, Louisiana, Tulane University, sponsored by National Oceanic and Atmospheric Administration Center for Sponsored Coastal Ocean Research.

Moomaw, W.R., 2002, Energy, industry and nitrogen—Strategies for decreasing reactive nitrogen emissions: AMBIO, A Journal of the Human Environment, v. 31, no. 2, p. 184–189.

Morey, S.L., Martin, P.J., O'Brien, J.J., Wallcraft, A.A., Zavala-Hidalgo, J., 2003a, Export pathways for river discharged fresh water in the northern Gulf of Mexico: Journal of Geophysical Research, v. 108, no. C10, p. 3303, doi:1029/2002JC001674.

Morey, S.L., Schroeder, W.W., O'Brien, J.J., and Zavala-Hidalgo, J., 2003b, The annual cycle of riverine influence in the eastern Gulf of Mexico basin: Geophysical Research Letters, v. 30, p. 1867, doi:10.1029/2003GL017348.

Morgenstern, R., and Pizer, W., eds., 2007, Reality check—The nature and performance of voluntary environmental programs in the United States, Europe, and Japan: Washington, D.C., Resources for the Future Press, 200 p.

Morse, J.W., and Eldridge, P.M., 2007, A non-steady state diagenetic model for changes in sediment biogeochemistry in response to seasonally hypoxic/anoxic conditions in the "dead zone" of the Louisiana shelf: Marine Chemistry, v. 106, p. 239–255.

Morse, J.W., and Rowe, G.T., 1999, Benthic biogeochemistry beneath the Mississippi River plume: Estuaries, v. 22, p. 206–214.

Mortimer, C.H.J., 1941, The exchange of dissolved substances between mud and water in lakes—I and II: Journal of Ecology, v. 29, p. 280–329.

Mosier, A., Kroeze, C., Nevison, C., Oenema, O., Seitzinger, S., and van Cleemput, O., 1998, Closing the global N2O budget: Nitrous oxide emissions through the agricultural nitrogen cycle: Nutrient Cycling in Agroecosystems, v. 52, nos. 2–3, 225–248 p., available online at: http://www.springerlink.com/content/lu04703848m261w5/.

MR/GMWNTF, 2001, Action plan for reducing, mitigating, and controlling hypoxia in the northern Gulf of Mexico: Washington, D.C., Mississippi river/Gulf of Mexico Watershed Nutrient Task Force, 36 p., available online at: http://www.epa.gov/msbasin/taskforce/pdf/actionplan.pdf.

Mueller, D.H., Wendt, R.C., and Daniel, T.C., 1984, Phosphorus losses as affected by tillage and manure application: Soil Science Society of America Journal, v. 48, p. 901–905.

Mulholland, P.J., Steinman, A.D., Marzolf, E.R., Hart, D.R., and DeAngelis, D.L., 1994, Effect of periphyton biomass on hydraulic characteristics and nutrient cycling in streams: Oecologia, v. 98, p. 40–47.

Mulvaney, R.L., Khan, S.A., Hoeft, R.G., and Brown, H.M., 2001, A soil organic nitrogen fraction that reduces the need for nitrogen fertilization: Soil Science Society of America Journal, v. 65, p. 1164–1172.

Murdock, L.W., Howe, P.L., and Schwab, G.J., 2002, Variable rate nitrogen fertilizer for corn grown in Kentucky, *in* Proceedings of the North-Central Extension-Industry Soil Fertility Conference, Brookings, SD, Potash and Phosphate Institute, v. 18, p. 81–85.

Nangia, V., Gowda, R.H., Mulla, D.J., and Sands, G.R., 2008, Water quality modeling of fertilizer management impacts on nitrate losses in tile drains at the field scale: Journal of Environmental Quality, v. 37, no. 2, p. 296–307.

NASA-SeaWiFS, 2007, Sea-viewing Wide Field-of-view Sensor (SeaWiFS) Project: National Aeronautics and Space Administration, available online at: http://oceancolor.gsfc.nasa.gov/SeaWiFS/.

Nassauer, J.I., Santelmann, M.V., and Scavia, D., eds., 2007, From the corn belt to the Gulf societal and environmental implications of alternative agricultural futures: Resources for the Future Press.

National Research Council, 2002, New tools for environmental protection: Education, information and voluntary measures. Committee on the Human Dimensions of Global Change. *in* Dietz, T., and Stern, P.C., eds., Division of Behavioral and Social Sciences and Education, Washington, D.C., National Academy Press.

National Academy of Sciences, 2007, Mississippi River water quality and the Clean Water Act: Progress, challenges, and opportunities: National Research Council, National Academy of Sciences, Committee on the Mississippi River and the Clean Water Act, 215 p., available online at: http://www.nap.edu/catalog.php?record_id=12051.

National Research Council, 2000, Clean coastal waters—Understanding and reducing the problems from nutrient pollution: Washington, D.C., National Academy of Sciences Press, 405 p.

National Research Council, 2004, Adaptive management for water resource planning: Washington, D.C., National Academy of Science Press, 123 p.

Neal, C., 2001, The potential for phosphorus pollution mediation by calcite precipitation in UK freshwaters: Hydrology and Earth System Sciences, v. 5, p. 119–131.

Neitsch, S.L., Arnold, J.G., Kiniry, J.R., Srinivasan, R., and Williams, J.R., 2004, Soil and water assessment tool input/output file documentation, version 2005: Temple, TX, U.S. Department of Agriculture, Agricultural Research Service, Grassland, Soil and Water Research Laboratory, available online at: ftp://ftp.brc.tamus.edu.pub/outgoing/sammons/swat2005, accessed November 28, 2006.

Nelson, R., Tietenberg, T., and Donihue, M.R., 1993, Differential environmental regulation: Effects on electric utility capital turnover and emissions: Review of Economics and Statistics, v. 75, p. 368–373.

Nielsen, K., Nielsen, L.P., and Rasmussen, P., 1995, Estuarine nitrogen retention independently estimated by the denitrification rate and mass balance methods—A study of Norsminde Fjord, Denmark: Marine Ecology Progress Series, v. 119, p. 273–283.

Nixon, S.W., 1992, Quantifying the relationship between nitrogen input and the productivity of marine ecosystems: Tokyo, Japan, Advanced Marine Technology Conference, no. 5.

Nixon, S.W., Ammerman, J., Atkinson, L., Berounsky, V., Billen, G., Boicourt, W., Boynton, W., Church, T., DiToro, D., Elmgren, R., Garber, J., Giblin, A., Jahnke, R., Owens, N., Pilson, M.E.Q., and Seitzinger, S., 1996, The fate of nitrogen and phosphorus at the land-sea margin of the North Atlantic Ocean: Biogeochemistry, v. 35, p. 141–180.

Nixon, S.W., Kelly, J.R., Furnas, B.N., Oviatt, C.A., and Hale, S.S., 1980, Phosphorus regeneration and the metabolism of coastal marine bottom communities, *in* Tenore, K.R., and Coull, B.C., eds., Marine Benthic dynamics: Columbia, SC, University of South Carolina Press, p. 219–242.

NOAA, 2007, Nutrient Enhanced Coastal Ocean Productivity (NECOP) Program: National Oceanic and Atmospheric Administration, Washington, D.C., available online at: http://www.aoml.noaa.gov/ocd/necop/, accessed April, 2007.

Northcott, W.J., Cooke, R.A., Walker, S.E., Mitchell, J.K., and Hirschi, M.C., 2001, Application of DRAINMOD-N to fields with irregular drainage systems: Transactions of the American Society of Agricultural Engineers, v. 44, no. 2, p. 241–249.

Novak, J.M., Stone, K.C., Watts, D.W., and Johnson, M.H., 2003, Dissolved phosphorus transport during storm and base flow conditions from an agriculturally intensive Southeastern Coastal Plain watershed: Transactions of the American Society of Agricultural Engineers, v. 46, no. 5, p. 1355–1363.

Nowlin, W.D., Jochems, A.E., DiMarco, S.F., Reid, R.O., and Howard, M.K., 2005, Low-frequency circulation over the Texas-Louisiana continental shelf, *in* Sturges, W., and Lugo-Fernandez, A., eds., Circulation in the Gulf of Mexico—Observations and models: Geophysical monograph: Washington, D.C., American Geophysical Union, Volume 161, 219–240 p.

Oberle, S.L., and Keeney, D.R., 1990, Soil type, precipitation, and fertilizer N effects on corn yields: Journal of Production Agriculture, v. 3, p. 522–527.

O'Connor, T., and Whitall, D., 2007, Linking hypoxia to shrimp catch in the northern Gulf of Mexico: Marine Pollution Bulletin, v. 54, no. 4, p. 460–463.

O'Donnell, T.K., and Galat, D.L., 2008, Evaluating success criteria and project monitoring in river enhancement within an adaptive management framework: Environmental Management, v. 41, p. 90–105.

Oey, L.-Y., Ezer, T., Forristall, G., Cooper, C., DiMarco, S., and Fan, S., 2005a, An exercise in forecasting loop current and eddy frontal positions in the Gulf of Mexico: Geophysical Research Letters, v. 32, p. L12611, doi:10.1029/2005GL023253.

Oey, L.-Y., Ezer, T., and Lee, H.-C., 2005b, Loop current rings and related circulation in the Gulf of Mexico—A review of numerical models and future challenges, *in* Sturges, W., and Fernandez, A., eds., Circulation in the Gulf of Mexico—Observations and models: Washington, D.C., American Geophysical Union, 31–56 p.

Ogden, J.C., Davis, S.M., Jacobs, K.J., Barnes, T., and Fling, H.E., 2005, The use of conceptual ecological models to guide ecosystem restoration in south Florida: Wetlands, v. 25, no. 4, p. 795–809.

Oguz, T., and Gilbert, D., 2007, Abrupt transitions of the top-down controlled Black Sea pelagic ecosystem during 1960–2000—Evidence for regime-shifts under strong fishery exploitation and nutrient enrichment modulated by climate-induced variations: Deep-Sea Research, v. 54, p. 220–242.

Ohlmann, J.C., and Niiler, P.P., 2005, Circulation over the continental shelf in the northern Gulf of Mexico: Progress in Oceanography, v. 64, p. 45–81.

Olness, A.E., Smith, S.J., Rhoades, E.D., and Menzel, R.G., 1975, Nutrient and sediment discharge from agricultural watersheds in Oklahoma: Journal of Environmental Quality, v. 4, p. 331–336.

Olson, R.J., Hensler, R.F., Attoe, O.J., Witzel, S.A., and Peterson, L.A., 1970, Fertilizer nitrogen and crop rotation in relation to movement of nitrate nitrogen through soil profiles: Soil Science Society of America Proceedings, v. 34, p. 448–452.

Omay, A.B., Rice, C.W., Maddux, L.D., and Gordon, W.B., 1997, Changes in soil microbial and chemical properties under long-term crop rotation and fertilization: Soil Science Society of America Journal, v. 61, p. 1672–1678.

Opdyke, M.R., David, M.B., and Rhoads, B.L., 2006, The influence of geomorphological variability in channel characteristics on sediment denitrification in agricultural streams: Journal of Environmental Quality, v. 35, p. 2103–2112.

Osborne, S.L., Schepers, J.S., Francis, D.D., and Schlemmer, M.R., 2002, Detection of phosphorus and nitrogen deficiencies in corn using spectral radiance measurements: Agronomy Journal, v. 94, p. 1215–1221.

Osterman, L.E., Poore, R.Z., Swarzenski, P.W., and Turner, R.E., 2005, Reconstructing a 180-yr record of natural and anthropogenic induced hypoxia from the sediments of the Louisiana continental shelf: Geology, v. 33, no. 4, p. 329–332.

Overholtz, W.J., 2002, The Gulf of Maine–Georges Bank Atlantic herring (*Clupea harengus*): spatial pattern analysis of the collapse and recovery of a large marine fish complex: Fisheries Research, v. 57, p. 237–254.

Owens, L.B., Edwards, W.M., and Van Keuren, R.W., 1997, Runoff and sediment losses resulting from winter feeding on pastures: Journal of Soil and Water Conservation, v. 52, p. 194–197.

Owens, L.B., Malone, R.W., Shiptalo, M.J., Edwards, W.M., and Bonta, J.V., 2000, Lysimeter study of nitrate leaching from a corn-soybean rotation: Journal of Environmental Quality, v. 29, p. 467–474.

Paerl, H.W., and Fulton, R.S., III, 2006, Ecology of harmful cyanobacteria, *in* Graneli, E., and Turner, J., eds., Ecology of harmful marine algae: Berlin, Springer-Verlag, 69–78 p.

Paerl, H.W., Valdes, L.M., Piehler, M.F., and Stow, C.A., 2006b, Assessing the effects of nutrient management in an estuary experiencing climatic change—The Neuse River Estuary, NC, USA: Environmental Management, v. 37, p. 422–436.

Pakulski, J.D., Benner, R., Whitledge, T., Amon, R., Eadie, B., Cifuentes, L., Ammerman, J., and Stockwell, D., 2000, Microbial metabolism and nutrient cycling in the Mississippi and Atchafalaya River plumes: Estuarine and Coastal Shelf Science, v. 50, no. 2, p. 173–184.

Paludan, C., and Blicher-Mathiesen, G., 1996, Losses of inorganic carbon and nitrous oxide from a temperate freshwater wetland in relation to nitrate loading: Biogeochemistry, v. 35, no. 2, p. 305–326.

Parkin, T.B., and Kaspar, T.C., 2006, Nitrous oxide emissions from corn-soybean systems in the Midwest: Journal of Environmental Quality, v. 35, p. 1496–1506.

Parma, A.M., and the NCEAS Working Group on Population Management, 1998, What can adaptive management do for our fish, forests, food, and biodiversity?: Integrative Biology, v. 1, p. 16–26.

Patrick, W.H., Jr., Gotoh, S., and Williams, B.G., 1973, Strengite dissolution in flooded soils and sediments: Science, v. 179, p. 564–565.

Patterson, B.D., 1987, The principle of nested subsets and its implications for biological conservation: Conservation Biology, v. 1, p. 323–334.

Pavelis, G.A., ed., 1987, Farm drainage in the United States—History, status, and prospects: U.S. Department of Agriculture, Economic Research Service, Misc. Pub. 1455, 170 p.

Perez, A.N., Oehlers, L., and Walter, R.B., 2006, Detection of hypoxia-related proteins in Medaka (*Oryzias latipes*) by difference gel electrophoresis and identification by sequencing of peptides using MALDI-TOF mass spectrometry: Paper presented at Hypoxia Effects on Living Resources in the Gulf of Mexico, September 25–26, 2006, Tulane University, New Orleans, Louisiana, sponsored by National Oceanic and Atmospheric Administration Center for Sponsored Coastal Ocean Research.

Perlack, R.D., and Turhollow, A.F., 2003, Feedstock cost analysis of corn stover residues for further processing: Energy, v. 28, no. 14, p. 1395–1403.

Perlack, R.D., Wright, L.L., Turhollow, A.F., Graham, R.L., Stokes, B.J., Erbach, D.C., 2005, Biomass as feedstock for a bioenergy and bioproducts industry: The technical feasibility of a billion-ton annual supply: Oak ridge National Laboratory, 78 p., available online at: http://feedstockreview.ornl.gov/pdf/billion_ton_vision.pdf.

Peters, R.H., 1986, The role of prediction in limnology: Limnology and Oceanography, v. 31, p. 1143–1159.

Phipps, R.G., 1997, Nitrate removal capacity of constructed wetlands: Ames, IA, Iowa State University, Ph.D. dissertation, 68 p.

Phipps, R.G., and Crumpton, W.G., 1994, Factors affecting nitrogen loss in experimental wetlands with different hydrologic loads: Ecological Engineering, v. 3, p. 399–408.

Pierson, S.T., Cabrere, M.L., Evanylo, G.K., Kuykendall, H.A., Hoveland, C.S., McCann, M.A., and West, L.T., 2001, Phosphorus and ammonium concentrations in surface runoff from grasslands fertilized with broiler litter: Journal of Environmental Quality, v. 30, p. 1784–1789.

Platon, E., and Sen Gupta, B.K., 2001, Benthic foraminiferal communities in oxygen depleted environments of the Louisiana Continental Shelf, *in* Rabalais, N.N., and Turner, R.E., eds., Coastal hypoxia—Consequences for living resources and ecosystems: Washington, D.C., American Geophysical Union, Coastal and Estuarine Study Series, Volume 58, 147–163 p.

Platon, E., Sen Gupta, B.K., Rabalais, N.N., and Turner, R.E., 2005, Effect of seasonal hypoxia on the benthic foraminiferal community of the Louisiana inner continental shelf—The 20th century record: Marine Micropaleontology, v. 54, p. 263–283. Popova, Y.A., Keyworth, V.G., Haggard, B.E., Storm, D.E., Lynch, R.A., and Payton, M.E., 2006, Stream nutrient limitation and sediment interactions in the Eucha-Spavinaw basin, USA: Journal of Soil and Water Conservation, v. 61, p. 105–115.

Popova, Y.A., Keyworth, V.G., Haggard, B.E., Storm, D.E., Lynch, R.A., and Payton, M.E., 2006, Stream nutrient limitation and sediment interactions in the Eucha-Spavinaw basin, USA: Journal of Soil and Water Conservation, v. 61, p. 105–115.

Pote, D.H., Daniel, T.C., Nichols, D.J., Sharpley, A.N., Moore, P.A., Jr., Miller, D.M., and Edwards, D.R., 1999, Relationship between phosphorus levels in three Ultisols and phosphorus concentrations in runoff: Journal of Environmental Quality, v. 28, p. 170–175.

Pote, D.H., Daniel, T.C., Sharpley, A.N., Moore, P.A., Jr., Edwards, D.R., and Nichols, D.J., 1996, Relating extractable soil phosphorus to phosphorus losses in runoff: Soil Science Society of America Journal, v. 60, p. 855–859.

Power, J.F., Wiese, R., and Flowerday, D., 2000, Managing nitrogen for water quality—Lessons from management systems evaluation area: Journal of Environmental Quality, v. 29, p. 355–366.

PPI/PPIC/FAR, 2005, Soil test levels in North America—Summary update: Norcross, GA, Potash & Phosphate Institute, PPI/PPIC/FAR Technical Bulletin No. 2005-1.

PPI/PPIC/FAR, 2002, Plant nutrient use in North American agriculture—Producing food and fiber, preserving the environment and integrating organic and inorganic sources: Norcross, GA, Potash & Phosphate Institute, PPI/PPIC/FAR Technical Bulletin No. 2002-1.

Preston, S.D., and Brakebill, J.W., 1999, Application of spatially referenced regression modeling for the evaluation of total nitrogen loading in the Chesapeake Bay watershed: U.S. Geological Survey Water-Resources Investigations Report No. 99-4054, 12 p., available online at http://md.water.usgs.gov/publications/wrir-99-4054/html/index.htm.

Prospero, J.M., Barrett, K., Church, T., Dentener, F., Duce, R.A., Galloway, J.N., Levy, H., Moody, J., and Quinn, P., 1996, Atmospheric deposition of nutrients to the North Atlantic Basin: Biogeochemistry, v. 35, p. 27–73.

Rabalais, N.N., 2006, Benthic communities and the effects of hypoxia in Louisiana coastal waters: Paper presented at Hypoxia Effects on Living Resources in the Gulf of Mexico, September 25–26, 2006, Tulane University, New Orleans, Louisiana, sponsored by National Oceanic and Atmospheric Administration Center for Sponsored Coastal Ocean Research.

Rabalais, N.N., Atilla, N., Normandeau, C., and Turner, R.E., 2004, Ecosystem history of Mississippi River-influenced continental shelf revealed through preserved phytoplankton pigments: Marine Pollution Bulletin, v. 49, p. 537–547.

Rabalais, N.N., Lohrenz, S.E., Redalje, D.G., Dortch, Q., Justić, D., Turner, R.E., Qureshi, N.A., Dagg, M.J., Eadie, B.J., and Fahnensteil, G.L., 1999b, Nutrient-enhanced coastal productivity and ecosystem responses, *in* Wiseman, W.J., Jr., Rabalais, N.N., Dagg, M.J., and Whitledge, T.E., eds., Nutrient enhanced coastal ocean productivity in the northern Gulf of Mexico: Silver Spring, MD, National Oceanic and Atmospheric Administration Coastal Ocean Program, Decision Analysis Series No. 14, U.S. Department of Commerce, National Ocean Service, Center for Sponsored Coastal Research, chap. 4, 51–78 p.

Rabalais, N.N., Smith, L.E., Harper Jr., D.E., and Justic, D., 2001, Effects of seasonal hypoxia on continental shelf benthos, *in* Rabalais, N.N., and Turner, R.E., eds., Coastal hypoxia – Consequences for living resources and ecosystems: Washington, D.C., American Geophysical Union, Coastal and Estuarine Studies, v. 58, p. 211–240.

Rabalais, N.N., and Turner, R.E., 2001, Hypoxia in the northern Gulf of Mexico—Description, causes, and change, *in* Rabalais, N.N., and Turner, R.E., eds., Coastal hypoxia—Consequences for living resources and ecosystems: Washington, D.C., American Geophysical Union, Coastal and Estuarine Studies, Volume 58, 454 p.

Rabalais, N.N., and Turner, R.E., 2006, Oxygen depletion in the Gulf of Mexico adjacent to the Mississippi River, *in* Neretin, L.N., ed., Past and present marine water column anoxia: NATO Science Series, IV-Earth and Environmental Sciences: The Netherlands, Kluwer, 225–245 p.

Rabalais, N.N., Turner, R.E., Dortch, Q., Justić, D., Bierman, V.J., and Wiseman, W.J., 2002, Nutrient-enhanced productivity in the northern Gulf of Mexico—Past, present and future: Hydrobiologia, v. 475, p. 39–63.

Rabalais, N.N., Turner, R.E., Justić, D., Dortch, Q., and Wiseman, W.J., 1999a, Characterization of hypoxia: Topic 1 report for the integrated assessment of hypoxia in the Gulf of Mexico: Silver Spring, MD, National Oceanic and Atmospheric Administration Coastal Ocean Program Decision Analysis Series No. 15, 203 p., available online at: http://oceanservice.noaa.gov/products/hypox_tfinal.pdf.

Rabalais, N.N., Turner, R.E., Sen Gupta, B.K., Platon, E., and Parsons, M.L., 2007a, Sediments tell the history of eutrophication and hypoxia in the northern Gulf of Mexico: Ecological Applications, Supplement, v. 17, no. 5, p. 129–143.

Rabalais, N.N., Turner, R.E., Sen Gupta, B.K., Boesch, D.F., Chapman, P., and Murrell M.C., 2007b, Hypoxia in the northern Gulf of Mexico: Does the science support the plan to reduce, mitigate, and control hypoxia?: Estuaries and Coasts, v. 30, no. 5, p. 753–772.

Ragauskas, A.J., Williams, C.K., Davison, B.H., Britovsek, G., Cairney, J., Eckert, C.A., Fredrick, W.J., Jr., Hallet, J.P., Leak, D.J., Liotta, C.L., Mielenz, J.R., Murphy, R., Templer, R., and

Tschaplinski, T., 2006, The path forward for biofuels and biomaterials: Science, v. 311, p. 484–489.

Ragueneau, O., Conley, D.J., Leynaert, A., Longphuirt, S.N., and Slomp, C.P., 2006a, Responses of coastal ecosystems to anthropogenic perturbations of silicon cycling, in Ittekot, V., Unger, D., and Humborg, C., eds., The silicon cycle: Human perturbations and impacts on aquatic systems: Washington, D.C., Island Press, 197–213 p.

Ragueneau, O., Conley, D.J., Leynaert, A., Longphuirt, S.N., and Slomp, C.P., 2006b, Role of diatoms in silica cycling and coastal marine food webs, in Ittekot, V., Unger, D., and Humborg, C., eds., The silicon cycle—Human perturbations and impacts on aquatic systems: Washington, D.C., Island Press, 163–195 p.

Raloff, J., 2004, Dead waters—Massive oxygen-starved zones are developing along the world's coasts: Science News Online, June 5, 2004, v. 165, no. 23, p. 8, available online at: http://sciencenews.org/articles/20040605/bob9.asp, accessed June 28, 2007.

Randall, G.W., Huggins, D.R., Russelle, M.P., Fuchs, D.J., Nelson, W.W., and Anderson, J.L., 1997, Nitrate losses through subsurface tile drainage in conservation reserve program, alfalfa, and row crop systems: Journal of Environmental Quality, v. 26, p. 1240–1247.

Randall, G.W., and Sawyer, J., 2005, Nitrogen application timing, forms and additives, in Gulf Hypoxia and Local Water Quality Concerns Workshop, Ames, IA, September 27–28, 2005, p. 73–84.

Randall, G.W., and Vetsch, J.A., 2005, Nitrate losses in subsurface drainage from a corn-soybean rotation as affected by fall vs. spring application of nitrogen and nitrapyrin: Journal of Environmental Quality, v. 34, no. 2, p. 590–597.

Randall, G.W., Vetsch, J.A., and Huffman, J.R., 2003, Nitrate losses in subsurface drainage from a corn-soybean rotation as affected by time of nitrogen application and use of nitrapyrin: Journal of Environmental Quality, v. 32, p. 1764–1772.

Raun, W.R., Solie, J.B., Stone, M.L., Martin, K.L., Freeman, K.W., Mullen, R.W., Zhang, H., Schepers, J.S., and Johnson, G.V., 2005, Optical sensor based algorithm for crop nitrogen fertilization: Communications in Soil Science and Plant Analysis, v. 36, nos. 19–20, p. 2759–2781.

Redalje, D.G., Lohrenz, S.E., and Fahnenstiel, G.L., 1992, Phytoplankton dynamics and the vertical flux of organic carbon in the Mississippi River Plume and inner Gulf of Mexico shelf region, in Falkowski, P.G., and Woodhead, A.D., eds., Primary productivity and biogeochemical cycles in the sea: New York, NY, Plenum Press, 526 p.

Redalje, D.G., Lohrenz, S.E., and Fahnenstiel, G.L., 1994b, The relationship between primary production and the vertical export of particulate organic matter in a river impacted coastal ecosystem: Estuaries, v. 17, p. 829–838.

Redalje, D.G., Lohrenz, S.E., and Fahnenstiel, G.L., 1994a, The vertical export of particulate and dissolved organic carbon from the surface waters of the northern Gulf of Mexico shelf: Journal of Mississippi Academy of Sciences, v. 39, p. 63.

Reddy, K.R., Kadlec, R.H., Flag, E., and Gale, P.M., 1999, Phosphorous retention in streams and wetlands—A review: Environmental Science and Technology, Critical Review, v. 29, p. 83–146.

Reddy, K.R., Wetzel, R.G., and Kadlec, R.H., 2005, Biogeochemistry of phosphorous in wetlands, in Phosphorous—Agriculture and the environment: Madison, WI, Agronomy Monograph No. 46, American Society of Agronomy, Crop Science Society of America, Soil Science Society of America, p. 263–316.

Redfield, A.C., 1958, The biological control of chemical factors in the environment: American Scientist, v. 46, p. 205–222.

Reetz, H.F., Jr., Murrell, T.S., and Murrell, L.J., 2001, Site-specific nutrient management—Production examples: Norcross, GA, Potash & Phosphate Institute, Better Crops, v. 85, no. 1, p. 12–13, 17.

Regan, H.M., Colyvan, and Burgman, M.A., 2002, A taxonomy and treatment of uncertainty for ecology and conservation biology: Ecological Applications, v. 12, p. 618–628.

Reichelderfer, K., 1985, Do USDA program participants contribute to soil erosion: U.S. Department of Agriculture, Economic Research Service, Agricultural Economic Report No. 532, 83 p.

Renaud, M., 1986, Hypoxia in Louisiana coastal waters during 1983: implications for fisheries: Fishery Bulletin, v. 84, p. 19–26.

Rhoads, D.C., Boesch, D.F., Tang, Z., Xu, F., Huang, L., and Nilsen, K.J., 1985, Macrobenthos and sedimentary facies on the Changjiang delta platform and adjacent continental shelf, East China Sea: Continental Shelf Research, v. 4, p. 189–213.

Ribaudo, M., 1989, Targeting the Conservation Reserve Program to maximize water quality benefits: Land Economics, v. 65, p. 320–332.

Ribaudo, M., Heimlich, R., Claassen, R., and Peters, M., 2001, Least-cost management of non-point source pollution—Source reduction versus interception strategies for controlling nitrogen loss in the Mississippi basin: Ecological Economics, v. 37, p. 183–197.

Ribaudo, M., Heimlich, R., and Peters, M., 2005, Nitrogen sources and Gulf hypoxia—Potential for environmental credit trading: Ecological Economics, v. 52, p. 159–168.

Richards, R.P., and Baker, D.B., 2002, Trends in water quality in LEASEQ rivers and streams (northwestern Ohio), 1975–1995: Journal of Environmental Quality, v. 31, p. 90–96.

Richards, R.P., Baker, D.B., Kramer, J.W., Ewing, D.E., Merryfield, B.J., and Miller, N.L., 2001, Storm discharge, loads, and average concentrations in northwest Ohio rivers, 1975–1995: Journal of the American Water Resources Association, v. 37, p. 423–438.

Richardson, C.J., 1999, The role of wetlands in storage, release, and cycling of phosphorus on the landscape—A 25-year retrospective, *in* Reddy, K.R., ed., Phosphorus biogeochemistry in sub-tropical ecosystems: Boca Raton, FL, CRS Press/Lewis Publishers, 47–68 p.

Richardson, W.B., Strauss, E.A., Bartsch, L.A., Monroe, E.M., Cavanaugh, J.C., Vingum, L., and Soballe, D.M., 2004, Denitrification in the upper Mississippi River—Rates, controls, and contribution to nitrate flux: Canadian Journal of Fisheries and Aquatic Science, v. 61, p. 1102–1112.

Risser, P.G., 1985, Toward a holistic management perspective: Bioscience, v. 35, p. 414–418.

Rizzo, W.M., Lackey, G.L., and Christian, R.R., 1992, Significance of euphotic, subtidal sediments to oxygen and nutrient cycling in a temperate estuary: Marine Ecology Progress Series, v. 86, p. 51–61.

Robertson, W.D., Blowes, D.W., Placek, C.J., and Cherry, J.A., 2000, Long-term performance of in situ reactive barriers for nitrate remediation: Ground Water, v. 38, p. 689–695.

Robertson, D.M., Schwarz, G.E., Saad, D.A., and Alexander, R.B., 2009, Incorporating Uncertainty into the ranking of sparrow model nutrient yields from Mississippi/Athchafalaya River Basin watersheds: Journal of the American Water Resources Association, v. 45, no. 2, p. 534–549.

Robinson, C.A., Cruse, R.M., and Ghaffarzadeh, M., 1996, Cropping system and nitrogen effects on mollisol organic carbon: Soil Science Society of America Journal, v. 60, p. 264–269.

Rose, G.A., deYoung, B., Kulka, D.W., Goddard, S.V., and Fletcher, G.L., 2000, Distribution shifts and overfishing the northern cod: a view from the ocean: Canadian Journal of Fisheries and Aquatic Sciences, v. 57, p. 644–664.

Rose, K.A., Adamack, A.T., Murphy, C.A., Sable, S.E., Kolesar, S.E., Craig, J.K., Breitburg, D.L., Thomas, P. Brouwer, M.H., Cerco, C.F., and Diamond, S., 2009, Does hypoxia have population-level effects on coastal fish? Musings from the virtual world: Journal of Experimental Marine Biology and Ecology, v. 381, p. S188–S203.

Rowe, G.T., and Chapman, P., 2002, Continental shelf hypoxia—Some nagging questions: Gulf of Mexico Science, v. 20, p. 155–160.

Rowe, G.T., Cruz-Kaegi, M.L., Morse, J.W., Boland, G.S., and Escobar Briones, E.G., 2002, Sediment community metabolism associated with continental shelf hypoxia, northern Gulf of Mexico: Estuaries, v. 25, no. 6, p. 1097–1106.

Royer, T.V., David, M.B., and Gentry, L.E., 2006, Timing of riverine export of nitrate and phosphorus from agricultural watersheds in Illinois—Implications for reducing nutrient loading to the Mississippi River: Environmental Science and Technology, v. 40, p. 4126–4131.

Royer, T.V., Tank, J.L., and David, M.B., 2004, The transport and fate of nitrate in headwater, agricultural streams in Illinois: Journal of Environmental Quality, v. 33, p. 1296–1304.

Runkel, R.L., Crawford, C.G., and Cohn, T.A., 2004, Load estimator (LOADEST)—A FORTRAN program for estimating constituent loads in streams and rivers: U.S. Geological Survey Techniques and Methods, Book 4, chap. A5, 69 p., available online at: http://pubs.usgs.gov/tm/2005/tm4A5/.

Russell, A.E., Laird, D.A., Parkin, T.B., and Mallarino, A.P., 2005, Impact of nitrogen fertilization and cropping system on carbon sequestration in Midwestern mollisols: Soil Science Society of America Journal, v. 69, p. 413–422.

Russelle, M.P., Lamb, J.F.S., Montgomery, B.R., Elsenheimer, D.W., Miller, B.S., and Vance, C.P., 2001, Alfalfa rapidly remediates excess inorganic nitrogen at a fertilizer spill site: Journal of Environmental Quality, v. 30, p. 30–36.

Russo, R.C., Rashleigh, B., and Ambrose, R.B., 2008, Watershed management in the United States. Chapter 11 *in* Gonenc, I.E. Vadineanu, A., Wolflin, J.P., and Russo R.C., Eds., Sustainable use and development of watersheds. Springer, New York, NY, pgs. 173–198.

Sands, G.R., Song, I., Busman, L.M., and Hansen, B.J., 2008, The effects of subsurface drainage depth and intensity on nitrate loads in the northern cornbelt: Transactions of American Society of Agricultural and Biological Engineers, v. 51, no. 3, p. 937–946.

Santhi, C., Arnold, J.G., Williams, J.R., Dugas, W.A., Srinivasan, R., and Hauck, L.M., 2001, Validation of the SWAT model on a large river basin with point and nonpoint sources: Journal of the American Water Resources Association, v. 37, no. 5, p. 1169–1188.

Saunders, D.C., and Kalff, J., 2001, Nitrogen retention in wetlands, lakes and rivers: Hydrobiologia, v. 443, p. 205–212.

Sawyer, J.E., and Nafziger, E.D., 2005, Regional approach to making nitrogen fertilizer rate decisions for corn, *in* Proceedings of North Central Extension-Industry Soil Fertility Conference, v. 21, p. 16–24.

Sawyer, J.E., and Randall, G.W., 2008, Nitrogen rates, *in* Final Report: Gulf Hypoxia and Local Water Quality Concerns Workshop. Saint Joseph, MI: American Society of Agricultural and Biological Engineers.

Scavia, D., and Donnelly, K.A., 2007, Reassessing hypoxia forecasts for the Gulf of Mexico: Environmental Science and Technology, v. 41, p. 8111–8117.

Scavia, D., Justić, D., and Bierman, V.J., Jr., 2004, Reducing hypoxia in the Gulf of Mexico—Advice from three models: Estuaries, v. 27, no. 3, p. 419–425.

Scavia, D., Kelly, E.L.A., and Hagy, J.D., III, 2006, A simple model for forecasting the effects of nitrogen loads on Chesapeake Bay hypoxia: Estuaries and Coasts, v. 29, no. 4, p. 674–684.

Scavia, D., Rabalais, N.N., Turner, R.E., Justić, D., and Wiseman, W.J., Jr., 2003, Predicting the response of Gulf of Mexico hypoxia to variations in Mississippi River nitrogen load: Limnology and Oceanography, v. 48, no. 3, p. 951–956.

Schaller, J.L, Royer, T.V., David, M.B., and Tank, J.L., 2004, Denitrification associated with plants and sediments in an agricultural stream: Journal of the North American Benthological Society, v. 23, no. 4, p. 667–676.

Scharf, P.C., Brouder, S.M., and Hoeft, R.G., 2006a, Chlorophyll meter readings can predict nitrogen need and yield response of corn in the North-Central USA: Agronomy Journal, v. 98, p. 655–665.

Scharf, P.C., Kitchen, N.R., Sudduth, K.A., and Davis, J.G., 2006b, Spatially variable corn yield is a weak predictor of optimal nitrogen rate: Soil Science Society of America Journal, v. 70, p. 2154–2160.

Scharf, P.C., Kitchen, N.R., Sudduth, K.A.,Davis, J.G., Hubbard, V.C., and Lory, J.A., 2005, Field-scale variability in optimal nitrogen fertilizer rate for corn: Agronomy Journal, v. 97, p. 452–461.

Scheffer, M., Carpenter, S., Folley, J.A., Folke, C., and Walker, B., 2001, Catastrophic shifts in ecosystems: Nature, v. 413, p. 591–596.

Schipper, L.A., Barkle, G.F., and Vojvodic-Vukovic, M., 2005, Maximum rates of nitrate removal in a denitrification wall: Journal of Environmental Quality, v. 34, p. 1270–1276.

Schipper, L.A., Barkle, G.F., Hadfield, J.C., Vojvodic-Vukovic, M., and Burgess, C.P., 2004, Hydraulic constraints on the performance of a groundwater denitrification wall for nitrate removal from shallow groundwater: Journal of Contaminant Hydrology, v. 69, p. 263–279.

Schipper, L., and Vojvodic-Vukovic, M., 1998, Nitrate removal from groundwater using a denitrification wall amended with sawdust—Field trial: Journal of Environmental Quality, v. 27, p. 664–668.

Schipper, L.A., and Vojvodic-Vukovic, M., 2001, Five years of nitrate removal, denitrification and carbon dynamics in a denitrification wall: Water Research, v. 35, p. 3473–3477.

Schlegel, A.J., Grant, C.A., and Havlin, J.L., 2005, Challenging approaches to nitrogen fertilizer recommendations in continuous cropping systems in the Great Plains: Agronomy Journal, v. 97, p. 391–398.

Schlesinger, W.H., and Melack, J.M., 1981, Transport of organic carbon in the world's rivers: Tellus, v. 33, p. 171–187.

Schmidt, J.P., Schmitt, M.A., Randall, G.W., Lamb, J.A., Orf, J.H., and Gollany, H., 2000, Swine manure application to nodulating and non-nodulating soybean: Agronomy Journal, v. 92, p. 987–992.

Schmitt, M.A., Schmidt, D.R., and Jacobson, L.D., 1996, A manure management survey of Minnesota swine producers—Effect of farm size on manure application: Applied Engineering in Agriculture, v. 12, no. 5, p. 595–599.

Schultz, R.C., Isenhart, T.M., Simpkins, W.W., and Colletti, J.P., 2004, Riparian forest buffers in agroecosystems—Lessons learned from the Bear Creek Watershed, central Iowa, USA: Agroforestry Systems, v. 61, p. 35–50.

Scott, C.A., Walter, M.F., Brooks, E.S., Boll, J., Hes, M.B., and Merrill, M.D., 1998, Impacts of historical changes in land use and dairy herds in water quality in the Catskills Mountains: Journal of Environmental Quality, v. 27, p. 1410–1417.

Segerson, K., and Miceli, T., 1998, Voluntary environmental agreements: Good or bad news for environmental protection?: Journal of Environmental Economics and Management, v. 36, p. 109–130.

Seitzinger, S.P., and Giblin, A.E., 1996, Estimating denitrification in North Atlantic continental shelf sediments: Biogeochemistry, v. 35, p. 235–260.

Seitzinger, S.P., Styles, R.V., Boyer, E.W., Alexander, R.B., Billen, G., Howarth, R.W., Mayer, B., and van Breemen, N., 2002, Nitrogen retention in rivers—Model development and application to watersheds in the northeastern U.S.A.: Biogeochemistry, v. 57–58, p. 199–237.

Sferratore, A., Garnier, J., Billen, G., Conley, D.J., and Pinault, S., 2006, Diffuse and point sources of silica in the Seine River watershed: Environmental Science & Technology, v. 40, p. 6630–6635.

Shapiro, C.A., Ferguson, R.B., Hergert, G.W., Dobermann, A.R., and Wortmann, C.S., 2003, Fertilizer suggestions for corn: University of Nebraska-Lincoln Extension, G174, available online at: http://elkhorn.unl.edu/epublic/live/g174/build/#target2.

Sharpley, A.N., 1997, Rainfall frequency and nitrogen and phosphorus in runoff from soil amended with poultry litter: Journal of Environmental Quality, v. 26, p. 1127–1132.

Sharpley, A.N., Daniel, T., Gibson, G., Bundy, L., Cabrera, M., Sims, T., Stevens, R., Lemunyon, J., Kleinman, P.J., and Parry, R., 2006a, Best management practices to minimize agricultural phosphorus impacts on water quality: U.S. Department of Agriculture, Agricultural Research Service ARS-163, 52 p.

Sharpley, A.N., Lleinman, P.J.A., and McDowell, R.W., 2001, Innovative management of agricultural phosphorus to protect soil and water resources: Communications in Soil Science and Plant Analysis, v. 32, p. 1071–1100.

Sharpley, A.N., Meisinger, J.J., Breeuwsma, A., Sims, J.T., Daniel, T.C., and Schepers, J.S., 1998, Impacts of animal manure management on ground and surface water quality, *in* Hatfield, J.L., and Stewart, B.A., eds., Animal waste utilization—Effective use of manure as a soil resource: Boca Raton, FL, Ann Arbor Press, 173–242 p.

Sharpley, A.N., Schmidt, J.P., and Hergert, L., 2006b, Nutrient management practices, *in* Schnepf, M., and Cox, C., eds., Environmental benefits of conservation on cropland—The status of our knowledge: Ankeny, IA, Society of Soil and Water Conservation, 149–193 p.

Sharpley, A.N., and Smith, S.J., 1994, Wheat tillage and water quality in the Southern Plains: Soil Tillage Research, v. 30, p. 33–38.

Sharpley, A.N., Smith, S.J., and Bain, R., 1993, Effect of poultry litter application on the nitrogen and phosphorus content of Oklahoma soils: Soil Science Society of America Journal, v. 57, p. 1131–1137.

Sharpley, A.N., Smith, S.J., Zollweg, J.A., and Coleman, G.A., 1996, Gully treatment and water quality in the Southern Plains: Journal of Soil and Water Conservation, v. 51, p. 512–517.

Sharpley, A.N., Weld, J.L., Beegle, D.B., Kleinman, P.J.A., Gburek, W.J., Moore, P.A., Jr., and Mullins, G., 2003, Development of phosphorus indices for nutrient management planning strategies in the United States: Journal of Soil and Water Conservation, v. 58, no. 3, p. 137–152.

Sharpley, A.N., Withers, P.J.A., Abdalla, C., and Dodd, A., 2005, Strategies for the sustainable management of phosphorus, *in* Sims, J.T., and Sharpley, A.N., eds., Phosphorus—Agriculture and the environment: American Society of Agronomy Monograph, p. 1069–1101.

Shehandeh, H., Wright, A.L., Hons, F.M., and Lascano, R.J., 2005, Spatial and temporal variation of soil nitrogen parameters related to soil texture and corn yield: Agronomy Journal, v. 97, p. 772–782.

Shepard, R., 2005, Nutrient management planning—Is it the answer to better management?: Journal of Soil and Water Conservation, v. 60, p. 171–176.

Shibles, R.M., 1998, Soybean nitrogen acquisition and utilization, *in* Proceedings of the 28th North Central Extension-Industry Soil Fertility Conference, St. Louis, MO, Nov. 11–12, 1998: Brookings, SD, Potash & Phosphate Institute, p. 5–11.

Shirmohammadi, A., Chaubey, I., Bosch, D.D., Muñoz-Carpena, R., Dharmasri, C., Arabi, M., Wolfe, M.L,. Frankenberger, J., Graff, C., Sohrabi, T.M., Shirmohammadi, A., 2006, Uncertainty in TMDL models: Transactions of the American Society of Agricultural and Biological Engineers, v. 49, no. 4, p. 1033–1049.

Simpson, J.H., and Hunter, J.R., 1974, Fronts in the Irish Sea: Nature, v. 250, p. 404–406.

Sims, J.T., 1997, Agricultural and environmental issues in the management of poultry wastes—Recent innovations and long-term challenges, *in* Rechcigl, J.E., and MacKinnon, H.C., eds., Uses of by-products and wastes in agriculture: Washington, D.C., American Chemical Society, 72–90 p.

Sims, J.T., Joern, B.C., and Simard, R.R., 1998, Phosphorus losses in agricultural drainage—Historical perspective and current research: Journal of Environmental Quality, v. 27, p. 277–293.

Sims, J.T., and Kleinman, P.J.A., 2005, Managing agricultural phosphorus for environmental protection, *in* Sims, J.T., and Sharpley, A.N., eds., Phosphorus—Agriculture and the environment: Madison, WI, American Society of Agronomy Monograph, American Society of Agronomy, 1021–1068 p.

Singh, R., and Helmers, M.J., 2006, Subsurface drainage and its management in the upper Midwest tile landscape, *in* Proceedings of the 2006 EWRI Congress, American Society of Civil Engineers, Omaha, NE, May 21–25, 2006.

Skaggs, R.W., Breve, M.A., and Gilliam, J.W., 1995, Predicting effects of water table management on loss of nitrogen from poorly drained soils: European Journal of Agronomy, v. 4, no. 4, p. 441–451.

Skaggs, R.W., Youssef, M.A., and Chescheir, G.M., 2003, Effect of subsurface drain depth on nitrogen losses from drained lands: Transactions of the American Society of Agricultural Engineering, v. 46, p. 237–244.

Skaggs, R.W., Youssef, M.A., Chescheir, G.M., Gilliam, J.W., 2005, Effect of drainage intensity on nitrogen losses from drained lands: Transactions of the American Society of Agricultural Engineering, v. 48, p. 2169–2177.

Sklar, F.H., Chimney, M.J., Newman, S., McCormick, P., Gawlick, D., Miao, S.L., McVoy, C., Said, W., Newman, J., Coronado, C., Crozier, G., Korvela, M., and Rutchey, K., 2005, The ecological-societal underpinnings of Everglades restoration: Frontiers in Ecology and the Environment, v. 3, no. 3, p. 161–169.

Sloth, N.P., Blackburn, H., Hansen, L.S., Risgaard-Petersen, N., and Lomstein, B.A., 1995, Nitrogen cycling in sediments with different organic loading: Marine Ecology Progress Series, v. 116, p. 163–170.

Smith, R.A., and Alexander, R.B., 2000, Sources of nutrients in the nation's watersheds, *in* Managing nutrients and pathogens from animal agriculture, *in* Proceedings from the Natural Resource, Agriculture, and Engineering Service Conference for Nutrient Management Consultants, Extension Educators, and Producer Advisors, Camp Hill, PA, March 28–30, 2000.

Smith, R.A., Schwarz, G.E., and Alexander, R.B., 1997, Regional interpretation of water quality monitoring data: Water Resources Research, v. 33, p. 2781–2798.

Smith, S.J., Sharpley, A.N., Berg, W.A., Naney, J.W., and Coleman, G.A., 1992, Water quality characteristics associated with Southern Plains grasslands: Journal of Environmental Quality, v. 21, p. 595–601.

Smith, S.V., and Hollibaugh, J.T., 1989, Carbon-controlled nitrogen cycling in a marine 'macrocosm'—An ecosystem-scale model for managing cultural eutrophication: Marine Ecology Progress Series, v. 52, p. 103–109.

Snyder, C.S., 2006, Phosphorus and potassium budgets and soil test levels in the Mississippi-Atchafalaya River Basin: Better Crops, v. 90, no. 1, p. 19–21, available online at: http://www.ipni.net/ppiweb/bcrops.nsf/$webindex/D62FD5F4335D283E85257110001556B9/$file/06-1p19.pdf.

Snyder, C.S., Bruulsema, T.W., Sharpley, A.N., and Beegle, D.B., 1999, Site-specific use of the environmental phosphorus index tool: Norcross, GA, Potash & Phosphate Institute, SSMG-1, 4 p., available online at: http://www.ppi-ppic.org/ppiweb/ppibase.nsf/b369c6dbe705dd13852568 e3000de93d/1a2c31b028f949238525695300581e03/$FILE/SSMG1.pdf.

Snyder, C.S., and Leep, R.H., 2007, Fertilization, *in* Barnes, R.F., Nelson, C.J., Moore, K.J., and Collins, M., eds., Forages—Volume II, The science of grassland agriculture, 6th edition: Ames, IA, Blackwell Publishing, chap. 24, 355–378 p.

Snyder, C.S., Randall, G.W., Almond, R.E., and Hoeft, R.G., 2001, Fall nitrogen management for agronomic response and environmental protection: Fall fertilization facts—Opportunities and considerations: Norcross, GA, Potash & Phosphate Institute, available online at: http://www.ppi-ppic.org/ppiweb/ppibase.nsf/$webindex/BFA77B79E6C8028C8525694E002D096D!opendocument

Sogbedji, M.J., and McIsaac, G.F., 2006, Evaluation of the ADAPT model for simulating nitrogen dynamics in a tile drained agricultural watershed in central Illinois: Journal of Environmental Quality, v. 35, p. 1914–1923.

Sørensen, J., Rasmussen, L.K., and Koike, I., 1987, Micromolar sulfide concentrations alleviate acetylene blockage of nitrous oxide reduction by denitrifying *Pseudomonas fluorescens*: Canadian Journal of Microbiology, v. 33, p. 1001–1005.

Spieles, D.J., and Mitsch, W.J., 2000, The effects of season and hydrologic and chemical loading on nitrate retention in constructed wetlands—A comparison of low and high nutrient riverine systems: Ecological Engineering, v. 14, p. 77–91.

Sprague, L.A., Clark, M.L., Rus, D.L., Zelt, R.B., Flynn, J.L., and Davis, J.W., 2006, Nutrient and suspended-sediment trends in the Missouri River Basin, 1993–2003: U.S. Geological Survey Scientific Investigations Report No. 2006-5231, 80 p., available online at: http://pubs.usgs.gov/sir/2006/5231/.

St. John, J.P., Fitzpatrick, J.J., and Landeck Miller, R.E., in press, TMDL modeling for Long Island Sound: New York Water Environment Association, Clearwaters, v. 37, no. 3, p. 14–24.

Stadmark, J., and Leonardson, L., 2005, Emissions of greenhouse gases from ponds constructed for nitrogen removal: Ecological Engineering, v. 25, p. 542–551.

Stevens, R.J., and Laughlin, R.J., 1998, Measurement of nitrous oxide and di-nitrogen emissions from agricultural soils: Nutrient Cycling in Agroecosystems, v. 52, p. 131–139, available online at: http://www.springerlink.com/content/g61700k53q415214/.

Stone, M.C., Hotchkiss, R.C., Hubbard, C.M., Fontaine, T.A., Mearnes, L.O., and Arnold, J.G., 2001, Impacts of climate change on Missouri River basin water yield: Journal of the American Water Resources Association, v. 37, no. 5, p. 1119–1130.

Stow, C.A., Qian, S.S., and Craig, J.K., 2005, Declining threshold for hypoxia in the Gulf of Mexico: Environmental Science and Technology, v. 39, p. 716–723.

Sugg, Z., 2007, Assessing U.S. farm drainage: Can GIS lead to better estimates of subsurface drainage extent?: Water Resources Institute, Washington, D.C., 8 p., available online at: http://www.wri.org/biodiv/pubs_description.cfm?pid=4324#pdf_files.

Sutula, M., Bianchi, T.S., and McKee, B.A., 2004, Effect of seasonal sediment storage in the lower Mississippi River on the flux of reactive particulate phosphorus to the Gulf of Mexico: Limnology & Oceanography, v. 49, p. 2223–2235.

Swaney, D.P., Shermana, D.M., and Howarth, R.W., 1996, Modeling water, sediment, and organic carbon discharges in the Hudson/Mohawk basin—Coupling two terrestrial sources: Estuaries, v. 19, no. 4, p. 833–847.

Swarzenski, P.W., Campbell, P.L., Osterman, L.E., and Poore, R.Z., 2008, A 1000-year sediment record of recurring hypoxia off the Mississippi River: The potential role of terrestrially-derived organic matter inputs: Marine Chemistry, v. 109, no. 1–2, p. 130–142.

Switzer, T.S., Chesney, E., and Baltz, D.M., 2006, Habitat selection by flatfishes along gradients of environmental variability—Implications for susceptibility to hypoxia in the northern Gulf of Mexico: Paper presented at Hypoxia Effects on Living Resources in the Gulf of Mexico, September 25–26, 2006, Tulane University, New Orleans, Louisiana, sponsored by National Oceanic and Atmospheric Administration Center for Sponsored Coastal Ocean Research.

Sylvan, J.B., Dortch, Q., Nelson, D.M., Maier Brown, A.F., Morrison, W., and Ammerman, J.W., 2006, Phosphorus limits phytoplankton growth on the Louisiana shelf during the period of hypoxia formation: Environmental Science and Technology, v. 40, no. 24, p. 7548–7553, available online at: http://pubs3.acs.org/acs/journals/supporting_information.page?in_manuscript=es061417t.

Taheripour, F., Khanna, M., and Nelson, C., 2007, Welfare impacts of alternative public policies for agricultural pollution control in an open economy: A general equilibrium framework: Working paper, August, 27, 2007

Taylor, J.C., and Miller, J.M., 2001, Physiological performance of juvenile southern flounder, *Paralichthys lethostigma* (Jordan and Gilbert, 1884), in chronic and episodic hypoxia: Journal of Experimental Marine Biology and Ecology, v. 258, p. 195–214.

Teisl, M., Roe, B., and Hicks, R., 2002, Can eco-labels tune a market?—Evidence from dolphin-safe labeling: Journal of Environmental Economics and Management, v. 43, p. 339–359.

Terry, D., 2006, Fertilizer tonnage reporting in the U.S.—Basis and current need: Better Crops, v. 90, no. 4, p. 15–17, available online at: http://www.ipni.net/ppiweb/bcrops.nsf/$webindex/F0BE4489F424A3FE8525721400271DCB/$file/06-4p15.pdf.

Tetra Tech, Inc., 1998, Documentation of phase I and phase II activities in support of point source nutrient loading analysis in the Mississippi River system: Prepared for USEPA Nonpoint-source Control Branch, contract no. 68-C7-0014, Washington, D.C.

The Encyclopedia of Earth, http://www.eoearth.org/; viewed December 9, 2009.

Thogersen, J., 2002, Promoting 'green' consumer behavior with eco-labels, *in* Dietz, T., and Stern, P.C., eds., New tools for environmental protection: Education, information, and voluntary measures: Washington, D.C., The National Academies Press, National Academy of Sciences, 83–104 p.

Thronson, A., and Quigg, A., 2008, Fifty-five years of fish kills in Coastal Texas: Estuaries and Coasts, v. 31, no. 4, p. 802–813.

Thom, W.O., and Sabbe, W.E., 1994, Soil sampling procedures for the southern region of the United States: Lexington, KY, Kentucky Agricultural Experiment Station, Southern Cooperative Series Bulletin No. 377.

Thorpe, S.A., 2004, Langmuir circulation: Annual Review of Fluid Mechanics, v. 36, p. 55–79.
Tiwari, A.K., Risse, L.M., and Nearing, M.A., 2000, Evaluation of WEEP and its comparison with USLE and RUSLE: Transactions of the American Society of Agricultural Engineers, v. 43, no. 5, p. 1129–1135.
Tolbert, V., 1998, Guest editorial: Biomass & Bioenergy, v. 14, no. 4, p. 301–306.
Tufekcioglu, A., Raich, J.W., Isenhart, T.M., and Schultz, R.C., 2003, Biomass, carbon and nitrogen dynamics of multi-species riparian buffers within an agricultural watershed in Iowa, USA: Agroforestry Systems, v. 57, p. 187–198.
Turner, R.E., 1999, A comparative mass balance budget (C, N, P and suspended solids) for a natural swamp and overland flow systems, *in* Vymazal, J., ed., Nutrient cycling and retention in natural and constructed wetlands: Leiden, The Netherlands, Backhuys Publishing, 61–71 p.
Turner, R.E., 2005, Nitrogen and phosphorus concentration and retention in water flowing over freshwater wetlands, *in* Fredrickson, L., King, S.L., and Kaminski, R.M., eds., Ecology and management of bottomland hardwood systems: Columbia, MO, University of Missouri Press, The State of Our Understanding, 57 66 p.
Turner, R.E., Milan, C.S., and Rabalais, N.N., 2004, A retrospective analysis of trace metals, C, N and diatom remnants in the Mississippi River delta shelf: Marine Pollution Bulletin, v. 49, p. 548–556.
Turner, R.E., Qureshi, N., Rabalais, N.N., Dortch, Q., Justic, D., Shaw, R.F., and Cope, J., 1998, Fluctuating silicate—Nitrate ratios and coastal plankton food webs, *in* Proceedings of the National Academy of Sciences, v. 95, p. 13048–13051.
Turner, R.E., and Rabalais, N.N., 1991, Changes in Mississippi River water quality this century— Implications for coastal food webs: Bioscience, v. 41,p. 140–148.
Turner, R.E., and Rabalais, N.N., 1994, Coastal eutrophication near the Mississippi River delta: Nature, v. 368, p. 619–621.
Turner, R.E., Rabalais, N.N., Alexander, R.B., McIsaac, G., and Howarth, R.W., 2007, Characterization of nutrient, organic carbon, and sediment loads and concentrations from the Mississippi River into the Northern Gulf of Mexico: Estuaries and Coasts, v. 30, p. 773–790.
Turner, R.E., Rabalais, N.N., and Justić, D., 1999, Long-term watershed and water quality changes in the Mississippi River system, *in* Wiseman, W.J., Jr., Rabalais, N.N., Dagg, M.J., and Whitledge, T.E., eds., Nutrient enhanced coastal ocean productivity in the northern Gulf of Mexico: Silver Spring, Maryland, National Oceanic and Atmospheric Administration Coastal Ocean Program, Decision Analysis Series No. 14, U.S. Department of Commerce, National Ocean Service, Center for Sponsored Coastal Research, chap. 3, 37–50 p.
Turner, R.E., Rabalais, N.N., and Justić, D., 2006, Predicting summer hypoxia in the northern Gulf of Mexico—Riverine N, P, and Si loading: Marine Pollution Bulletin, v. 52, p. 139–148.
Turner, R.E., Rabalais, N.N., and Justic, D., 2008, Gulf of Mexico hypoxia: Alternate states and a legacy: Environmental Science & Technology, v. 42, no. 7, p. 2323–2327.
Twilley, R.R., Cowan, J., Miller-Way, T., Montagna, P.A., and Mortazavi, B., 1999, Benthic nutrient fluxes in selected estuaries in the Gulf of Mexico, *in* Bianchi, T.S., Pennock, J.R., and Twilley, R.R., eds., Biogeochemistry of Gulf of Mexico Estuaries: New York, NY, Wiley, 163–209 p.
Udawatta, R.P., Motavalli, P.P., and Garrett, H.E., 2004, Phosphorus loss and runoff characteristics in three adjacent agricultural watersheds with claypan soils: Journal of Environmental Quality, v. 33, p. 1709–1719.
UMRSHNC, 2006, Gulf hypoxia and local water quality concerns workshop—A workshop assessing tools to reduce agricultural nutrient losses to water resources in the corn belt, *in* Workshop Proceedings, Iowa State University, Ames, IA, September 26–28, 2005, p. 205, available online at: http://www.umrshnc.org/index.php?option=com_content&task=view&id=19&Itemid=34.
U.S. Department of Agriculture, 2003, Cost associated with development and implementation of Comprehensive Nutrient Management Plans—Part 1 – Nutrient management, land treatment, manure and wastewater handling and storage, and recordkeeping: U.S.

Department of Agriculture, Natural Resource Conservation Service, 220 p., available online at: http://www.nrcs.usda.gov/technical/land/pubs/cnmp1.html.

U.S. Department of Energy, 2006, Clean coal and natural gas power systems: Available online at: http://www.fe.doe.gov/programs/powersystems/index.html.

U.S. Environmental Protection Agency, 2000a, State compendium—Programs and regulatory activities related to animal feeding operations: Washington, D.C., USEPA Office of Water, Office of Waste Management.

U.S. Environmental Protection Agency, 2000b, National air pollution trends, 1900–1998: Available online at: http://www.epa.gov/ttn/chief/trends/trends98/trends98.pdf.

U.S. Environmental Protection Agency, 2000c, Ambient water quality criteria recommendations: Information supporting the development of state and tribal nutrient criteria: Rivers and streams in Ecoregion VI: Washington, D.C., USEPA, Office of Water, EPA 822-B-00-017, 91 p., available online at: http://www.epa.gov/waterscience/criteria/nutrient/ecoregions/rivers/rivers_6.pdf.

U.S. Environmental Protection Agency, 2000d, Nutrient criteria technical guidance manual: Rivers and Streams: Chapter 1 – Introduction: Washington, D.C., USEPA Office of Water, Office of Science and Technology, EPA-822-B-00-002, 16 p., available online at: http://www.epa.gov/waterscience/criteria/nutrient/guidance/rivers/chapter_1.pdf.

U.S. Environmental Protection Agency, 2002, National water quality inventory—2000 Report to Congress: Washington, D.C., USEPA, Office of Water, EPA-841-R-02-001, available online at: http://www.epa.gov/305b/2000report/.

U.S. Environmental Protection Agency, 2003b, Economic analyses of nutrients and sediment reduction actions to restore Chesapeake Bay water quality: Annapolis, MD, EPA Chesapeake Bay Program Office, 162 p., available online at: http://www.chesapeakebay.net/tribtools.htm.

U.S. Environmental Protection Agency, 2003a, Nutrient reduction technology cost estimations for point sources in the Chesapeake Bay watershed: Annapolis, MD, EPA Chesapeake Bay Program Office, 132 p., available online at: http://www.chesapeakebay.net/tribtools.htm.

U.S. Environmental Protection Agency, 2004, Managing manure guidance for concentration animal feeding operations (CAFOs): Washington, D.C., USEPA, Office of Water, EPA-821-B-04-009, U.S. Government Printing Office, available online at: http://cfpub.epa.gov/npdes/afo/info.cfm#manure.

U.S. Environmental Protection Agency, 2004a, Water quality trading assessment handbook—Can water quality trading advance your watershed's goals?: U.S. Environmental Protection Agency, Office of Water, Office of Wetlands, Oceans, and Watersheds, EPA 841-B-04-001, 120 p., available online at: http://www.epa.gov/owow/watershed/trading/handbook/docs/NationalWQTHandbook_FINAL.pdf.

U.S. Environmental Protection Agency, 2004b, What is the status of point source nitrogen reduction in the Chesapeake Bay watershed?: Annapolis, MD, EPA Chesapeake Bay Program Office Fact Sheet, 2 p.

U.S. Environmental Protection Agency, 2005, National management measures to control nonpoint source pollution from, urban areas: Washington, D.C., U.S. Environmental Protection Agency, Office of Water, EPA-841-B-05-004.

U.S. Environmental Protection Agency, 2006b, Ecoregional nutrient criteria documents for rivers & streams: Available online at: http://www.epa.gov/waterscience/criteria/nutrient/ecoregions/rivers/.

U.S. Environmental Protection Agency, 2006a, National Emissions Inventory (NEI) Air Pollutant Emissions Trends Data: Available online at: http://www.epa.gov/ttn/chief/trends/index.html.

U.S. Environmental Protection Agency, 2007, The Long Island Sound study at EPA New England: Available online at: http://www.epa.gov/boston/eco/lis/epane.html.

U.S. Government Accounting Office, 1997, Global warming—Information on the results of four of EPA's voluntary climate change programs: Washington, D.C., U.S. Government Printing Office, GAO:RCED-97-163, 32 p., available online at: http://www.gao.gov/archive/1997/rc97163.pdf.

Vache, K.B., Eilers, J.M., and Santelman, M.V., 2002, Water quality modeling of alternative agricultural scenarios in the U.S. corn belt: Journal of the American Water Resources Association, v. 38, no. 2, p. 773–787.

Vahtera, E., Conley, D., Gustufsson, B.G., Kuosa, H., Pitkanen, H., Savchuk, O.P., Tamminen, T., Vitasalo, M., Voss, M., Wasmund, N., and Wulff, F., 2007, Internal ecosystem feedbacks enhance nitrogen-fixing cyanobacteria blooms and complicate management in the Baltic Sea: Ambio, v. 36, no. 12, p. 186–194.

Van Driel, P.W., Merkley, L.C., and Robertson, W.D., 2006, Denitrification of agricultural drainage using wood-based reactors: Transactions of the ASABE, v. 49,p. 565–573.

Van Liew, M.W., Veith, T.L., Bosch, D.D., and Arnold, J.G., 2006, Suitability of SWAT for the Conservation Effects Assessment Project—A comparison on USDA ARS watersheds: Journal of Hydrologic Engineering, v. 12, no. 2, p. 173–183.

Vaquer-Sunyer, R., and Duarte, C.M., 2008, Thresholds of hypoxia for marine biodiversity: Proceedings of the National Academy of Sciences of the United States of America, v. 105, no. 40, p. 15452–15457.

Varvel, G.E., 2006, Soil organic carbon changes in diversified rotations of the western cornbelt: Soil Science Society of America Journal, v. 70, p. 426–433.

Varvel, G.E., Schepers, J.S., and Francis, D.D., 1997, Ability for in-season correction of nitrogen deficiency in corn using chlorophyll meters: Soil Science Society of America Journal, v. 61, p. 1233–1239.

Vollenweider, R.A., 1976, Advances in defining critical loading levels of phosphorus in lake eutrophication: Memorie dell' Istituto Italiano di Idrobiologia, v. 33, p. 53–83.

Von Holle, C.K., 2005, Agricultural nitrogen use and producer attitudes in tile-drained watersheds of east-central Illinois: University of Illinois, Urbana, IL, Urbana-Champaign M.S. thesis, 72 p.

Walker, N.D., and Rabalais, N.N., 2006, Relationships among satellite chlorophyll a, river inputs, and hypoxia on the Louisiana continental shelf, Gulf of Mexico: Estuaries and Coasts, v. 29, no. 6B, p. 1081–1093.

Wang, X., and Melesse, A.M., 2005, Evaluation of the SWAT model's snowmelt hydrology in a northwestern Minnesota watershed: Transactions of the American Society of Agricultural and Biological Engineers, v. 48, no. 4, p. 1359–1376.

Wannamaker, C.M., and Rice, J.A., 2000, Effects of hypoxia on movements and behavior of selected estuarine organisms from the southeastern United States: Journal of Experimental Marine Biology and Ecology, v. 249, p. 45–163.

Watershed Agriculture Council, 2004, History of the Watershed Agriculture Council: Available online at: http://www.nycwatershed.org/index_wachistory.html.

Wawrik, B., Paul, J.H., Bronk, D.A., and Gray, J.D., 2004, High rates of ammonium recycling drive phytoplankton productivity in the offshore Mississippi River plume: Aquatic Microbial Ecology, v. 35, p. 175–184.

Wedwick, S., Lakhani, B., Stone, J., Waller, P., and Artiola, J., 2001, Development and sensitivity analysis of the GLEAMS-IR model: Transactions of the American Society of Agricultural Engineers, v. 44, no. 5, p. 1095–1104.

Weed, D.A.J., and Kanwar, R.S., 1996, Nitrate and water present in and flowing from root zone soil: Journal of Environmental Quality, v. 25, p. 709–719.

Wells, M.C., Ju, Z., Heater, S.J., and Walter, R.B., 2006, Microarray gene expression analyses in Medaka (*Oryzias latipes*) exposed to hypoxia: Paper presented at Hypoxia Effects on Living Resources in the Gulf of Mexico, September 25–26, 2006 Tulane University, New Orleans, Louisiana, sponsored by National Oceanic and Atmospheric Administration Center for Sponsored Coastal Ocean Research.

Westerman, P.W., Donnely, T.L., and Overcash, M.R., 1983, Erosion of soil and poultry manure— A laboratory study: Transactions of the American Society of Agricultural Engineers, v. 26, p. 1070–1078, 1084.

Wilhelm, W.W., Johnson, J.M.F., Hatfield, J.L., Voorhees, W.B., and Linden, D.R., 2004, Crop and soil productivity response to corn residue removal—A literature review: Agronomy Journal, v. 96, no. 1, p. 1–17.

Wiseman, W.J., Rabalais, N.N., Turner, R.E., Dinnel, S.P., and MacNaughton, A., 1997, Seasonal and interannual variability within the Louisiana coastal current: stratification and hypoxia: Journal of Marine Systems, v. 12, p. 237–248.

Wiseman, Jr., W.J., Rabalais, N.N., Turner, R.E., and Justic, D., 2004, Hypoxia and the physics of the Louisiana coastal current, in Nihoul, J.C.J., Zavialov, P.O., and Micklin, P.P., eds., Dying and dead seas—Climate versus anthropogenic causes: The Netherlands, Kluwer Academic Publishers, 359–372 p.

Wisner, R., 2007, Iowa Farm Outlook: Presentation at Tel Aviv University, May 15, 2007, available online at: http://www.econ.iastate.edu/facultywisner/documents/telavivethanolpresentation-wisner07.pdf.

Wittry, D.J., and Mallarino, A.P., 2002, Use of variable-rate technology for agronomic and environmental phosphorus-based liquid swine manure management, in Robert, P.C., Rust, R.H., and Larson, W.E., eds., 6th International Conference on Site-Specific Management for Agricultural Systems Proceedings, American Society of Agronomy, Minneapolis, MN, July 14–17, 2002.

Wittry, D.J., and Mallarino, A.P., 2004, Comparison of uniform- and variable-rate phosphorus fertilization for corn-soybean rotations: Agronomics Journal, v. 96, p. 26–33.

Wollheim, W.M., Vorosmarty, C.J., Peterson, B.J., Seitzinger, S.P., and Hopkinson, C.S., 2006, Relationship between river size and nutrient removal: Geophysical Research Letters, v. 33, p. L06410, doi:10.1029/2006GL025845.

Wortmann, C., Helmers, M., Mallarino, A.P., Barden, C., Devlin, D., Pierzynski, G., Lory, J., Massey, R., Holz, J., Shapiro, C., and Kovar, J., 2005, Agricultural phosphorus management and water quality protection in the Midwest: Lincoln, NE, University of Nebraska-Lincoln Extension and CSREES-USDA, Heartland Regional Water Coordination Initiative, Regional Publication 187, 24 p., available online at: http://www.ianrpubs.unl.edu/epublic/live/rp187/build/rp187.pdf.

Wu, RSS, 2002, Hypoxia: from molecular responses to ecosystem responses: Marine Pollution Bulletin, v. 45, p. 35–45.

Wu, J., Adams, R., Kling, C., and Tanaka, K., 2004, Assessing the costs and environmental consequences of agricultural land use changes—A site-specific, policy-scale modeling approach: American Journal of Agricultural Economics, v. 86, p. 26–41.

Wu, J., and Babcock, B., 1999, Metamodeling potential nitrate water pollution in the central United States: Journal of Environmental Economics and Management, v. 28, p. 1916–1928.

Wu, J., and Tanaka, K., 2005, Reducing nitrogen runoff from the Upper Mississippi River basin to control hypoxia in the Gulf of Mexico—Easements or taxes?: Marine Resource Economics, v. 20, p. 121–144.

Wysocki, L.A, Bianchi, T.S., Powell, R., and Reuss, N., 2006, Spatial variability in the coupling of organic carbon, nutrients, and phytoplankton pigments in surface waters and sediments of the Mississippi River plume: Estuarine, Coastal and Shelf Science, v. 69, p. 47–63.

Xu, Y., 2006, Total nitrogen inflow and outflow from a large river swamp basin to the Gulf of Mexico: Hydrological Sciences – Journal – des Sciences Hydrologiques, v. 51, no. 3, p. 531–542.

Xue, Y., David, M.B., Gentry, L.E., and Kovacic, D.A., 1998, Kinetics and modeling of dissolved phosphorus export from a tile-drained agricultural watershed: Journal of Environmental Quality, v. 27, p. 917–922.

Yang, W., Khanna, M., and Farnsworth, R., 2005, Effectiveness of conservation programs in Illinois and gains from targeting: American Journal of Agricultural Economics, v. 5, p. 1248–1255.

Yang, W., Khanna, M., Farnsworth, R., and Onal, H., 2004, Is geographical targeting cost-effective: The case of the Conservation Reserve Enhancement Program in Illinois: Review of Agricultural Economics, v. 27, p. 70–88.

Yang, W., Khanna, M., Farnsworth, R., and Onal, H., 2003, Integrating economics, environmental and GIS modeling to target cost effective land retirement in multiple watersheds: Ecological Economics, v. 46, p. 249–267.

Yuan, Y., Bingner, R.L., and Rebich, R.A., 2001, Evaluation of AnnAGNPS on Mississippi Delta MSEA watersheds: Transactions of the ASAE, v. 44, no. 5, p. 1183–1190.

Yuan, T., Bingner, R.L., Theurer, F.D., 2006, Subsurface flow component for AnnAGNPS: Applied Engineering in Agriculture, v. 22, no. 2, p. 231–241.

Yuan, Y., Bingner, F.D., Theurer, F.D., Ribich, R.A., and Moore, P.A., 2005, Phosphorus component in AnnAGNPS: Transactions of the American Society of Agricultural Engineers, v. 48, no. 6, p. 2145–2154.

Zavala-Hidalgo, J., Morey, S.L., and O'Brien, J.J., 2003, Seasonal circulation on the western shelf of the Gulf of Mexico using a high-resolution numerical model: Journal of Geophysical Research, v. 108, no. C12, p. 3389, doi:10.1029/2003JC001879.

Zhang, L., and Mitsch, W.J., 2000, Hydrologic budgets of the two Olentangy River experimental wetlands, 1994–99, *in* Mitsch, W.J., and Zhang, L., eds., Olentangy River Wetland Research Park at the Ohio State University, Annual Report 1999, p. 41–46.

Zhang, L., and Mitsch, W.J., 2001, Water budgets of the two Olentangy River experimental wetlands in 2000, *in* Mitsch, W.J., and Zhang, L., eds., Olentangy River Wetland Research Park at the Ohio State University, Annual Report 2000, p. 17–28.

Zhang, L., and Mitsch, W.J., 2002, Water budgets of the two Olentangy River experimental wetlands in 2001, *in* Mitsch, W.J., and Zhang, L., eds., Olentangy River Wetland Research Park at the Ohio State University, Annual Report 2001, p. 23–34.

Zhang, L., and Mitsch, W.J., 2004, Water budgets of the two Olentangy River experimental wetlands in 2003, *in* Mitsch, W.J., Zhang, L., and Tuttle, C., eds., Olentangy River Wetland Research Park at the Ohio State University, Annual Report 2003, p. 39–52.

Zimmerman, A.R., and Nance, J.M. 2001, Effects of hypoxia on the shrimp industry of Louisiana and Texas, *in* Rabalais, N.N., and Turner, R.E., eds., Coastal hypoxia—Consequences for living resources coastal and estuarine studies: Washington, D.C., American Geophysical Union, chap. 15, Volume 58, 293–310 p.

Zou, E., 2006, Impacts of hypoxia on physiology and toxicology of the brown shrimp *Penaeus aztecus*: Paper presented at Hypoxia Effects on Living Resources in the Gulf of Mexico, September 25–26, 2006, Tulane University, new Orleans, Louisiana, sponsored by National Oceanic and Atmospheric Administration Center for Sponsored Coastal Ocean Research.

Subject Index

Note: Locators followed by 'f' and 't' refer to figure and table respectively.
Appendix page numbers are given with an 'A' after the page number.

A

Acoustic Doppler Current Profilers (ADCP), 20
Action Plan, 3–4, 9, 44–45
  primary hypoxia management goals, 4
Adaptive management, 7, 111
  elements directly relevant to goal setting and research needs, 111
  integrated modeling and monitoring, 113
  small watershed scale, conceptual models, 113
  successful implementation of, 112
Adjusted maximum likelihood estimate (AMLE), 55–56, 65
Agricultural conservation programs, 135–137
Agricultural drainage, 143
  alternative drainage system design and management, 143–145
  bioreactors, 145
  in corn belt, 53
  current extent and patterns of, 51–52
  estimated extent of, 52f
Agricultural management options, anticipated benefits associated with different, 200t
Anaerobic ammonia oxidation (ANNAMOX) process, 40
Animal feeding operations (AFO), 159, 227A
Animal production systems, 158–163, 227A
  alternative manure management technologies, 162–163
  crop selected to receive manure application, 230A
  estimates of manure production and N and P loss to water and air, 162t
  farming system and nutrient budget, 228A
  intensification of animal feeding operations, 227A
  intensity and duration of grazing, 230–231A
  managing manures, 229A
  manure as component of N and P mass balances, 159–161
  nutrient budgets, 227A
  nutrient surpluses, 228–229A
  rate and frequency of application, 230A
  recoverable manure N, 158f
  recoverable manure P, 159f
  remedial strategies, 161–162
  stream-bank fencing, 231A
  system development and nutrient flows, 158–159
  targeting remedial strategies within MARB, 229A
Atchafalaya and Mississippi Rivers, modeled surface salinity showing freshwater plumes from, 17f
Atchafalaya flow
  change in relative importance of, 15f
Atchafalaya River plume vs. Mississippi River plume, 15–16
Atmospheric deposition, 183–185, 222–227A

B

"Back-of-the-envelope" calculation, Atchafalaya River flow, 16
Bacterial pigments (Louisiana shelf), 10
Basin-wide co-benefits, research assessing, 128–129
Basin-wide integrated economic-biophysical models, 123t
Benthic foraminiferal community, 10–11
Benthic organisms, and hypoxia, 42
Bierman model, 47
Biochemical oxygen demand (BOD), 4
Biogeochemical processes, 40
Biophysical models for agricultural nonpoint sources, large-scale integrated economic and, 125–127

277

"Bird's foot" delta, 18
Bottom layer hypoxia, 15
   development of, 40
   Gulf of Mexico NECOP data, 28
Buoyancy fluxes, 14

C

"Cap and trade," 138
CASTnet program, 226A
Cellulosic ethanol, 191
Characterization of hypoxia, 205–207
   current state of forecasting, 46–49
   denitrification, P burial, and nutrient recycling, 38–41
   goals and management options
      scientific basis for, 209–211
   limiting factors and role of Si, 29–31
   physical context
      changes in Mississippi river hydrology and their effects on vertical mixing, 15–18
      oxygen budget, 12–13
      shelf circulation: local *versus* regional, 20–21
      vertical mixing as function of stratification and vertical shear, 13–15
      zones of hypoxia controls, 18–20
   possible regime shift in Gulf of Mexico, 41–44
   processes in formation of hypoxia in Gulf of Mexico, 9
      historical patterns and evidence for hypoxia on shelf, 9–11
   role of N and P in controlling primary production
      N and P fluxes to NGOM background, 23–24
      N and P limitation, 24–28
   single *vs.* dual nutrient removal strategies, 44–45
   sources of organic matter to hypoxic zone, 31–33
      advances in organic matter understanding, 34–36
      sources of organic matter to NGOM, 33
      synthesis efforts regarding organic matter sources, 37–38
Clean Water Act (CWA), 118, 233A
Coastal goal, Task Force, 4
Coastal marine ecosystems
   hypoxia in, 42
   N and P in, 39
   P cycle of, sulfate reduction, 39
Colored dissolved organic matter (CDOM), 31–32
Complex physical models, 49
Complex water quality models, 48
Confined Animal Feeding Operations (CAFO) regulations, 159
   status of implementation of permits under 2003, 160t
Conservation buffers, 151–153
   installed in six subbasins of MARB, 154t
Conservation effects assessment program (CEAP), 4, 135
Conservation reserve enhancement program (CREP), 93, 135
Conservation reserve program (CRP), 93, 135
Conservation security program (CSP), 135
Conservation technical assistance (CTA), 93
Controlled- and slow-release N fertilizers (CRN), 172
Cropping systems, 155–157
Cross-shore dispersion processes, effects of, 13
Current fluctuations seaward, 21

D

Denitrification
   and hypoxia, 40
   and N removal, 89
   P burial, and nutrient recycling, 38–41
Diatoms, 30–31
Dissolved inorganic nitrogen (DIN), 41
Dissolved inorganic phosphorus (DIP), 41
Dissolved organic carbon (POC), 31–32
Dissolved reactive P (DRP) concentrations, 91
Drainage, agricultural, 143–146
Drainage, tile, *see* Tile drainage

E

Econometric model, Wu and Tanaka, 126
Econometric modeling approach (Kling), 126
Emissions and water quality trading programs, 137–138
Emission trading, 137
Empirical models, 46
Environmental Quality Incentive Program (EQIP), 93, 136
EPIC model, 121, 126
ESTIMATOR, 55
Eutrophication, Chesapeake Bay, 27
Eutrophication-induced hypoxia, 41
Eutrophication/stratification as drivers of hypoxia, 15

Subject Index 279

F

Fertilizer
  sources, 164–165
  use and application technology, 165–172
Fisheries, impact of hypoxia in, 43
Flow-weighted average nitrate concentrations, 54f
  and reduced N *versus* percent cropland, 54f
Fluvial sediments, 91
Flux estimation method, change in, 54–56
Food and Agricultural Policy Institute (FAPRI), 190
Freshwater wetlands, 146
  nitrogen, 146–149
  observed *vs.* predicted NO_3 mass removal, 148f
  percent mass nitrate removal in wetlands, 147f
  phosphorus, 149–150

G

Global material cycles, 222A
GOM hypoxia analysis, 94–95
Gradient Richardson number, 13
Grain *vs.* cellulosic ethanol and water quality, 193–195
Grassland Reserve Program (GRP), 136
Green surface mixed layer model, 48
Gulf of Mexico
  attributes of models used for delivering nutrients to, 98–100t
  economic value, 43
  NECOP data, 28
  regime shift in, 41–44
Gulf of Mexico Coastal Ocean Observing System (GCOOS), 20
Gulf of Mexico hypoxia, models addressing, 47

H

Harmful Algal Bloom and Hypoxia Research and Control Act (1998), 3
Hydrodynamic models, three-dimensional, 13
Hypoxia, 1
  abstracts on, 215–222A
  characterization, *see* characterization of hypoxia
  historical patterns, 9–11
  on living resources, effects of, 215–222A
  in NGOM, conceptual framework for, 112f
  occurrence of, 1
  organic matter from major rivers, 32
Hysteresis effect, 44

I

IBIS/THMB Model, 104–105
  comparison with SPARROW and SWAT, 106
  targeting, 106–107
  uncertainty, 107–108
In-field nutrient management
  changes in consumption of principal fertilizer N, 165f
  controlled-release fertilizers, 172
  corn yields in six leading corn-producing states, 171f
  effect of variable-rate *vs.* uniform rate application of fertilizer P, 180f
  effect of variable-rate *vs.* uniform rate application of liquid swine manure, 179f
  effects of N management on soil resource sustainability, 173–176
  fertilizer N consumption as anhydrous ammonia, 165f
  fertilizer sources, 164–165
  fertilizer use and application technology, 165–172
  fraction of annual fertilizer N tonnage in Illinois sold, 168f
  nutrient management planning strategies, 181–182
  percentage of N fertilized corn acreage, 166f
  precision agriculture management tools for nitrogen, 176–177
  precision agriculture management tools for phosphorus, 178–181
  USDA ARMS data for three states with highest fall N application, 167f
  variability in soil test P levels in typical farmer fields in Minnesota, 178f
  watershed-scale fertilizer management, 172
Institute for Agricultural and Trade Policy (IATP), 190
Integrated Assessment, 3–5, 9–10, 32, 51, 93, 94, 96–97
  assessment and review of cost estimates, 121–125
  predictions from modeling system, 121–122
papers on sources of OM to Gulf of Mexico, 34–35t
point sources, 84
post 2000, 33

Integrated economic-biophysical models, policies and findings from, 124–125t

J
Justić model, 47

L
Landsat data, Land cover based on, 53f
Landscape design, principles of, 129–131
  avoiding effects of land use on ecological processes, 131
  avoid land uses that deplete natural resources over broad area, 130–131
  compatible land-use and -management practices, 131
  examine impacts of local decisions in regional context, 129–130
  plan for long-term change and unexpected events, 130
  preserve rare landscape elements/habitats/species, 130
Late Spring Nitrate Test (LSNT), 172
Linear and multiple regression models, 47
Load Estimator (LOADEST), 55–56
Loop Current/Loop Current Eddy System, 20
Louisiana–Texas continental shelf
  Atchafalaya River discharge, 100% of, 16
  Mississippi River discharge, half of, 16
  physical, biological, and chemical reasoning behind delineation of, 20
  strength of stratification on, 15
  surface salinity, temperature, and stratification on, 17

M
Management Action Review Team Report (MART), 4
Marine phytoplankton, 33
Mesoplankton fecal pellets, 36
"Microbial loop," 31
Minimum variance unbiased estimator (MVUE), 55
Mississippi–Atchafalaya River basin (MARB), 1
  annual nutrient fluxes
    for five large subbasins, 67t
    for 10 subbasins in, 68t
  annual patterns, 56–61
    comparison (percent and absolute basis) of MARB nitrate-N fluxes, 57f
    flow, available phosphorus, and silicate, 60f
    flow and available nitrogen, 59f
    nitrogen, 56–58
    nutrient ratios, 58–61
    phosphorus and silicate, 58
    ratio of N:P and dissolved silicate to dissolved inorganic N, 61f
  annual yield estimates, 67–72
  area of major crops planted in, 76
  average annual nutrient yields
    for five large subbasins, 68t
    for nine subbasins, 69t
  cumulative fluxes of five major subbasins, 65
  ethanol and water quality in, 190–195
  extent of, 2f
  land-use activities influencing water quality, 2–3
  location of nine large subbasins comprising, 66f
  low-salinity surface water, 9
  N, P and silicate, nutrient fluxes, source areas, 66
  nitrate-N fluxes (1995–2005), 55f
  schematic showing locations of monitoring sites, 58
  seasonal patterns, 61–64
    flow, phosphorus, and silicate flux for, 63f
    flow and nitrogen flux during spring, 62f
    nitrogen, 61
    phosphorus and silicate, 62–64
    ratio of N:P and silicate to dissolved inorganic N, 65
    spring fluxes as percentage of annual fluxes, 64f
  targeting remedial strategies within, 229A
Mississippi effluents, proposed diversions of, 18f
Mississippi river
  hydrologic regime of, 14
  POC and DOC in, 31–32
  spring fluxes and nitrate-N flux for, 73
  tributaries and distributaries, 14
  wetlands created, restored, enhanced in subbasins of, 93t
Mississippi River/Gulf of Mexico Watershed Nutrient Task Force, *see* Task Force
Mississippi River plume *vs.* Atchafalaya River plume, 15–16
Mobile mud, 36

N

National Oceanic and Atmospheric Administration (NOAA), 3
National Pollutant Discharge Elimination System (NPDES), 159
Natural Resources Inventory (NRI), 51
Northeastern Gulf of Mexico Chemical Oceanography Program (NECOP), 14
Nitrogen cycle, 223fA
Nitrogen over-enrichment persists, 29
Nitrogen/phosphorus
 availability of, 38
 in controlling primary production, role of nitrogen and phosphorus limitation, 24–28
 fluxes to NGOM, 23–24
 point source inputs of, 231–233A
  comparison of MART estimated sewage treatment plant annual effluent loads of, 232A
 reduction, setting targets for, 115–119
 baseline for reductions, 116–119
Northern Gulf of Mexico (NGOM), 1
 complexity of, nutrient limitation and, 24
 conceptual framework for hypoxia in, 112f
 different zones during period when hypoxia can occur, 19f
 frequency of hypoxia in, 1985–2005, 2f, 19
 hypoxia and, overview, 1–3
 increased N loading, hypoxia, 32
 N and P in driving primary production, 45
 NASA-SeaWiFS image of, 26f
 physics of, complex, 49
 reducing hypoxia in, effects of different practices have following characteristics, 113
 three zones of hypoxia control, 19
"N-rich" calibration approach, 177
Nutrient(s)
 flow diagrams and mass balance
  atmospheric deposition, 222–227A
  global material cycles, 222A
  recycling, denitrification and P burial, 38–41
  removal strategies, single *versus* dual, 44–46
Nutrient application to corn, impacts on, 192–193
Nutrient concentrations and biogeochemical processes, 40

Nutrient Enhanced Coastal Ocean Productivity (NECOP), 27
Nutrient fate, transport, and sources, 5–6, 51
 ability to route and predict nutrient delivery to Gulf, 96–108
 characterization of, 207–209
 mass balance of nutrients, 76–86
 nutrient transport processes, 87–95
 temporal characteristics of stream flow and nutrient flux, 51–56
  MARB annual and seasonal fluxes, 56–64
  subbasin annual and seasonal flux, 65–74
Nutrient loadings
 differential, 30
 excessive, 23
 trend and N:P ratios, 45
Nutrients, managing/co-benefits/consequences
 agricultural drainage, 143–145
 animal production systems, 158–163
 conservation buffers, 151–154
 cropping systems, 155–157
 effective actions for other nonpoint sources, 183–186
 ethanol and water quality in MARB, 190–195
 freshwater wetlands, 146–150
 in-field nutrient management, 164–182
 integrating conservation options, 195–202
 most effective actions for industrial and municipal sources, 186–189
Nutrients, mass balance of, 76
 area of major crops planted in MARB, 76
 cropping patterns, 76
 nonpoint sources, 77–84
  net N inputs for the four major regions of MARB, 79f
  net P inputs for the four major subbasins of MARB, 83f
  nitrogen, 77–80
  nitrogen mass balance components and net N inputs for MARB, 78f
  phosphorus, 80–84
  phosphorus mass balance components and net P inputs for MARB, 82f
 point sources, 84–86
Nutrient transport processes, 87–95
 aquatic processes, 87–92
  nitrogen, 88–90
  N removed in aquatic ecosystems, 90f
  phosphorus, 91–92
  silicate, 92

Nutrient transport (*cont.*)
  freshwater wetlands, 93
  nutrient sources and sinks in coastal wetlands, 94–95

O

Ohio River basin, 69
  net N inputs and annual nitrate-N fluxes and yields for, 70f
  P and particulate/organic P fluxes for, 72f
Organic carbon and biogenic silica, accumulation of, 10
Organic matter (OM), 31
  findings relevant to sources, 35–36
  marine phytoplankton, 33
  Mississippi River derived, 32–33
  quantitative assessments for portions of hypoxic zone, 37
  synthesis efforts regarding sources, 37–38
Organic matter to hypoxic zone, sources of, 31–33
  advances in organic matter understanding: characterization and processes, 34–36
  sources of organic matter to NGOM: post 2000 integrated assessment, 33
  synthesis efforts regarding organic matter sources, 37–38
Oxygen balance equation, 12
Oxygen concentration decrease, 12
Oxygen depletion, *see* hypoxia

P

Paleoecological studies, limiting factor in, 10
Particulate organic carbon (POC), 32, 35
PEB index, 10–11
  plots of, 11f
Phosphorus cycle, 224fA
Phosphorus point source fluxes, 86f
Phytoplankton assemblages, response of, 25f
Phytoplankton community composition, altered, 30
Planktonic primary production, 23
Precision agriculture management tools
  for nitrogen, 176–177
  for phosphorus, 178–181

Q

Quality of life goal, Task force, 4
Quinqueloculina, 11

R

Reducing hypoxia, science and management goals for, 3–4
Regime shifts, 41–44
  due to hypoxia, 43
  in Gulf of Mexico, 43
Regression model, Booth and Campbell, 125–126
Residential and urban sources, 185–186
Respiration, 13
Riparian buffers, 151

S

SAB nitrogen and phosphorus recommendations *vs.* USEPA nitrogen and phosphorus, 234–237A
"Salt wedge" estuaries, 14
Scavia model, 47
Science Advisory Board (SAB), 4–5
Scientific basis for goals and management options
  adaptive management, 111–114
  cost-effective approaches for nonpoint source control, 133
    agricultural subsidies and conservation compliance provisions, 138–140
    eco-labeling and consumer driven demand, 141–142
    emissions and water quality trading programs, 137–138
    existing agricultural conservation programs, 135–137
    taxes, 140–141
    voluntary programs – without economic incentives, 134–135
  options for managing nutrients, co-benefits, and consequences
    agricultural drainage, 143–145
    animal production systems, 158–163
    conservation buffers, 151–154
    cropping systems, 155–157
    effective actions for other nonpoint sources, 183–186
    ethanol and water quality in MARB, 190–195
    freshwater wetlands, 146–150
    in-field nutrient management, 164–182
    integrating conservation options, 195–202
    most effective actions for industrial and municipal sources, 186–189
    protecting water quality and social welfare in Basin, 120–131

Subject Index 283

setting targets for nitrogen and phosphorus reduction, 115–119
Sediment-based denitrification, 27
Sediment nutrient flux submodel, 48
Sediment oxygen demand (SOD), 48
Shelf hydrodynamics, altered, 18
Silicon cycle, 225fA
Simpson–Hunter criterion of tidal mixing, 14
Soil organic carbon (SOC), 173
SPARROW model, 88, 97–102
   advantage, 108
   comparison with SWAT and IBIS/THMB, 106
   targeting, 106–107
   uncertainty, 107–108
   wet-deposition data, 101
Spatial pattern of bottom hypoxia, 21
Spring nitrate-N flux, 74
Statistical and simulation models, 49
Strong density stratification, 14
Strong vertical density gradients, 13
Study Group (hypoxia), 4–7
   to advance science characterizing hypoxia and its causes, 212
   approach, 7–8
   characterization of hypoxia, 5
   characterization of nutrient fate, transport, and sources, 5–6
   to evaluate issues, SAB's request to, 4–7
      characterization of hypoxia, 5
      characterization of nutrient fate, transport, and sources, 5–6
      scientific basis for goals and management options, 6–7
   to reduce size of hypoxic zone, 213
   scientific basis for goals and management options, 6–7
Subbasin nitrate-N yield compared to net N inputs, 69–71
   soil mineralization, 70
   soybean production, 69–70
SWAT Model, 103–104
   comparison with SPARROW and IBIS/THMB, 106
   targeting, 106–107
   uncertainty, 107–108

T
Task force, 3
Three-dimensional numerical circulation models, 18

Tile drainage
   control of, 144, 170, 196–199
   Integrated Assessment, in the, 51
   intensity of, 70, 74
   models of, 97, 104, 192
   N leakage and, 75, 148, 151, 156, 161, 169, 172, 202, 207
   occurrence, xlv
   P export and, 92, 180, 202
Total Maximum Daily Loads (TMDL), 119, 138

U
Upper Mississippi River basin
   nitrogen mass balance components and net N inputs for, 81f
   phosphorus mass balance components and net N inputs for, 85f
Upper Mississippi River subbasin, 69
   Missouri and, 72
   net N inputs and annual nitrate-N fluxes and yields for, 71f
"Upwelling," modified form of, 23
US Department of Agriculture's (USDA), 4
US Environmental Protection Agency (USEPA), 3, 118
   guidance on nutrient criteria, 233–234A
US Mathematical Programming (USMP) model, 121, 126–127

V
Vertical eddy diffusivity, 12
Vertical flux, 35
Vertical mixing, 12
   changes in Mississippi river hydrology and their effects on, 15–18
   as function of stratification and vertical shear, 13–15
   Louisiana-Texas shelf, 13
Vertical transport by upwelling/downwelling, 12

W
Water column eutrophication sub-model, 48
Water quality implications of projected grain-based ethanol production levels, 191–192
   estimated changes in N losses from cropping changes, 192t
Water quality trading, 137

Water remineralisation, high bottom, 15
Watershed-scale fertilizer management, 172
"Weather band," 21
Wetland Reserve Program (WRP), 93, 136

White House National Science and Technology Council (NSTC), 3
Wildlife Habitat Incentive Program (WHIP), 136
Within-basin goal, Task force, 4